Lecture Notes in Computer Science 6381

Peter Csaba Ölveczky (Ed.)

Rewriting Logic and Its Applications

8th International Workshop, WRLA 2010
Held as a Satellite Event of ETAPS 2010
Paphos, Cyprus, March 20-21, 2010
Revised Selected Papers

Springer

Volume Editor

Peter Csaba Ölveczky
University of Oslo
Department of Informatics
Gaustadalléen 23
0373 Oslo
Norway
E-mail: peterol@ifi.uio.no

Library of Congress Control Number: 2010935729

CR Subject Classification (1998): F.3, D.2, D.3, D.1.3, F.1, F.4.1

LNCS Sublibrary: SL 1 – Theoretical Computer Science and General Issues

ISSN 0302-9743
ISBN-10 3-642-16309-2 Springer Berlin Heidelberg New York
ISBN-13 978-3-642-16309-8 Springer Berlin Heidelberg New York

springer.com

Printed in Germany

Typesetting: Camera-ready by author, data conversion by Scientific Publishing Services, Chennai, India
Printed on acid-free paper 06/3180

Preface

This volume contains the proceedings of the Eighth International Workshop on Rewriting Logic and its Applications (WRLA 2010) that was held in Paphos, Cyprus, March 20–21, 2010, as a satellite workshop of the European Joint Conferences on Theory and Practice of Software (ETAPS 2010).

Rewriting logic is a natural semantic framework for representing concurrency, parallelism, communication and interaction, as well as being an expressive (meta)logical framework for representing logics. It can then be used for specifying a wide range of systems and programming languages in various application fields. In recent years, several executable specification languages based on rewriting logic (ASF+SDF, CafeOBJ, ELAN, Maude) have been designed and implemented. The aim of the WRLA workshop series is to bring together researchers with a common interest in rewriting logic and its applications, and to give them the opportunity to present their recent works, discuss future research directions, and exchange ideas.

Previous WRLA workshops were held in Asilomar (1996), Pont-à-Mousson (1998), Kanazawa (2000), Pisa (2002), Barcelona (2004), Vienna (2006), and Budapest (2008), and their proceedings have been published in *Electronic Notes in Theoretical Computer Science.* In addition, selected papers from WRLA 1996 have been published in a special issue of *Theoretical Computer Science*, and selected papers from WRLA 2004 appeared in a special issue of *Higher-Order and Symbolic Computation.*

The year 2010 marks the 20th anniversary of the first papers on rewriting logic. We were very happy to have José Meseguer as an invited speaker reflecting on the past (and future) twenty years of achievements in this area. We were also grateful that Natarajan Shankar could contribute to this special occasion by giving an invited talk. To further celebrate the 20-year-old, selected papers from WRLA 2010 will appear in a special issue of the *Journal of Logic and Algebraic Programming.*

The 13 regular papers presented at WRLA 2010 and included in this volume were selected by the Program Committee out of 29 submissions. Each submission was reviewed by four reviewers. As mentioned, the program also included invited talks by Natarajan Shankar and José Meseguer, as well as tool demonstrations, the now classic rewrite engine competition, and a report on an advanced school on Maude and CafeOBJ.

Many colleagues and friends contributed to the success of WRLA 2010. First, I would like to thank the authors who submitted their work to WRLA 2010 and who, through their contributions, made this workshop a high-quality event. I would also like to thank the program committee members and the external reviewers for their timely and insightful reviews as well as for their involvement in the post-reviewing discussions. The rewrite engine competition at WRLA

2010 was organized by Francisco Durán; many thanks to him for taking the lead on organizing the competition and to all rewrite engine developers who participated in the competition. I am also grateful to my friends who provided me with all kinds of help and useful advice, to the invited speakers, to the WRLA steering committee for their work in getting the proceedings published in Springer's *LNCS* series, to Jan Bergstra for accepting to devote a special issue of *JLAP* to selected papers from WRLA 2010, and to Anna Philippou for taking care of the local arrangements and accommodating our special requests. I also thank Andrei Voronkov for the excellent EasyChair conference system and his prompt feedback to our request. Finally, I thank the Department of Informatics at the University of Oslo for financially supporting the workshop.

I hope that WRLA 2010 provided all the participants with a broad overview of rewriting logic and its research directions, and inspired them to continue to contribute to the success of rewriting logic for the next 20 years, and beyond.

August 2010 Peter Csaba Ölveczky

Workshop Organization

Program Chair

Peter C. Ölveczky	University of Oslo

Steering Committee

Kokichi Futatsugi	JAIST, Tatsunokuchi
Claude Kirchner	INRIA Bordeaux – Sud-Ouest
Narciso Martí-Oliet	Universidad Complutense de Madrid
José Meseguer	University of Illinois at Urbana-Champaign
Ugo Montanari	Università di Pisa
Grigore Roşu	University of Illinois at Urbana-Champaign
Carolyn Talcott	SRI International, Menlo Park
Martin Wirsing	Ludwig-Maximilians-Universität München

Program Committee

Artur Boronat	University of Leicester
Mark van den Brand	Eindhoven University of Technology
Roberto Bruni	Università di Pisa
Manuel Clavel	Universidad Complutense de Madrid
Francisco Durán	Universidad de Málaga
Steven Eker	SRI International, Menlo Park
Santiago Escobar	Universidad Politécnica de Valencia
Kokichi Futatsugi	JAIST, Tatsunokuchi
Claude Kirchner	INRIA Bordeaux – Sud-Ouest
Alexander Knapp	Universität Augsburg
Dorel Lucanu	Alexandru Ioan Cuza University, Iaşi
Salvador Lucas	Universidad Politécnica de Valencia
Narciso Martí-Oliet	Universidad Complutense de Madrid
Ugo Montanari	Università di Pisa
Pierre-Etienne Moreau	École des Mines de Nancy & INRIA Nancy
Thomas Noll	RWTH Aachen
Peter C. Ölveczky	University of Oslo
Miguel Palomino	Universidad Complutense de Madrid
Grigore Roşu	University of Illinois at Urbana-Champaign
Mark-Oliver Stehr	SRI International, Menlo Park
Carolyn Talcott	SRI International, Menlo Park
Eelco Visser	Delft University of Technology

External Reviewers

Alarcon, Beatriz
Alba-Castro, Mauricio
Alpuente, Maria
Andrei, Oana
Bourdier, Tony
Balland, Emilie
Bellia, Marco
Boreale, Michele
Braga, Christiano
Burel, Guillaume
Chiba, Yuki
Engelen, Luc
Franssen, Michael
Gadducci, Fabio
Gaina, Daniel
Goriac, Eugen-Ioan
Gutierrez, Raul
Hausmann, Daniel
Hills, Mark
Lluch Lafuente, Alberto
Meer, Arjan van den
Middeldorp, Aart
Nakamura, Masaki
Ogata, Kazuhiro
Oliver, Javier
Riesco, Adrian
Rocha, Camilo
Romero, Daniel
Santana de Oliveira, Anderson
Serbanuta, Traian
Serebrenik, Alexander
Villanueva, Alicia

Table of Contents

Invited Talks

Termination and Narrowing

Tools

The K Framework

Applications and Semantics

Maude Model Checking and Debugging

Rewrite Engines

Rewriting, Inference, and Proof*

Natarajan Shankar[1]

Computer Science Laboratory
SRI International
Menlo Park CA 94025 USA
shankar@csl.sri.com
http://www.csl.sri.com/~shankar/

Abstract. Rewriting is a form of inference, and one that interacts in several ways with other forms of inference such as decision procedures and proof search. We discuss a range of issues at the intersection of rewriting and inference. How can other inference procedures be combined with rewriting? Can rewriting be used to describe inference procedures? What are some of the theoretical challenges and practical applications of combining rewriting and inference? How can rewriters, decision procedures, and their combination be certified? We discuss these problems in the context of our ongoing effort to use PVS as a metatheoretic framework to construct a proof kernel for justifying the claims of theorem provers, rewriters, model checkers, and satisfiability solvers.

Rewriting is a versatile framework that can be used as a programming notation, a modeling formalism, and as an inference method. It is a crucial component of any effective interactive theorem prover. Rewriting can also be used as framework for prototyping and reasoning about inference procedures. A rewriter is itself a very powerful inference procedure and certifying the claims made by rewriters can be quite challenging. We explore several themes centered around rewriting, inference, and proof. We describe the combination of rewriting and decision procedures employed by SRI's Prototype Verification System (PVS) [ORSvH95] in its simplifier. This inference rule is built into PVS and hence its soundness cannot be taken for granted. We present an architecture for justifying the soundness of such complex inference procedures based on the use of verified reference checkers. We review some of the progress in developing this architecture. We also show how rewriting can be used to define such reference checkers.

There is a long history of work in rewriting in the context of theorem proving. Woody Bledsoe [Ble77] advocated it as a human-oriented method for automated proof search. The Boyer-Moore family of theorem provers [BM79, BM88, KMM00] are well known for induction, but rewriting is one of its big strengths. The Rewrite Rule Laboratory [KZ88] supports both explicit and implicit induction within a rewriting framework. The OBJ family of systems [GW88, GKM$^+$87] employed rewriting as an algebraic specification language. Maude [CDE$^+$99] and ELAN [BKK$^+$96] are descendants of OBJ that support extremely fast rewriting within an expressive rewriting logic framework.

* This research was supported NSF Grants CSR-EHCS(CPS)-0834810 and CNS-0917375. Sam Owre commented on earlier drafts of the paper, and the participants at the 2010 Workshop on Rewriting Logic and Applications, particularly José Meseguer and Peter Ölveczky, offered valuable feedback and advice.

P.C. Ölveczky (Ed.): WRLA 2010, LNCS 6381, pp. 1–14, 2010.

Rewriting also plays a crucial role in the interactive proof assistant of PVS. It is used within the PVS simplifier in conjunction with decision procedures and simplification rules. The simplifier is around 4000 lines of Common Lisp code, and it relies on decision procedures that also run to nearly 4000 lines of code. PVS uses external decision procedures including a BDD package and the Yices SMT solver, but the simplifier is easily the most complicated of the built-in inference procedures. We describe this procedure and examine the challenge of certifying results that are claimed by the simplifier.

Our approach to certifying the results of untrusted inference procedures is developed within the *Kernel of Truth* (KoT) project at SRI. The approach is captured by Figure 1. At the bottom, there is a small, trusted proof kernel, and at the top, we have the untrusted inference procedures. We rely on *verified* reference checkers to check the claims made by untrusted inference procedures [Sha08] relative to the trusted proof kernel. In the KoT approach, we do not verify the verifier, but we instead check verification claims with a verified checker. The approach is driven by the idea that checking is always easier than solving, and checkers are usually easier to verify than solvers. Furthermore, the certificates that are checked by the checkers need not be formal proofs and can be customized to specific classes of problems. The verification of the checker demonstrates the existence of a formal proof corresponding to a valid certificate, but the actual proof need not be explicitly constructed. By using verified checkers, the untrusted procedures can be optimized for speed while avoiding the overhead of proof instrumentation and generation. The untrusted procedures can provide hints or certificates to the verified checkers. The verified checkers are expected to be simple and might therefore perform slower than the untrusted procedures, but this is acceptable since the validation of untrusted results and certificates is done offline. The hints provided by the untrusted procedures should also make it more efficient to check the resulting claims. The verification of the checkers can be performed by the untrusted inference tools, since for these specific claims, we can break the circular dependency by generating and checking the kernel-level formal proofs.

Our KoT approach should be contrasted with existing techniques for building trusted inference tools. The LCF approach used by systems such as Coq [The09], HOL [GM93], Isabelle [Pau94], and Nuprl [CAB+86], relies on proof generation as a way of validating claims. Though proofs are constructed using tactics that generate subgoals from goals, the application of these tactics are valid only when there is a proof of the goal from the subgoals. Proof generation imposes an engineering and performance overhead. Some SAT solvers have been instrumented to generate resolution proofs, and though these proofs can get quite large, the overhead of generating and checking them is quite modest at about 2 to 12% for generation with a checking time that is significantly smaller than the solving time [ZM03]. However, it should be noted that these resolution proofs are actually certificates and not formal proofs, and not all inference procedures are similarly amenable to proof generation.

Some systems use reflection to verify and apply decision procedures. In computational reflection [BM81], the logic is used to formalize a syntactic fragment such as arithmetic or propositional logic and to verify an inference procedure on this representation. A meaning function is used to connect this syntax to formulas in the logic.

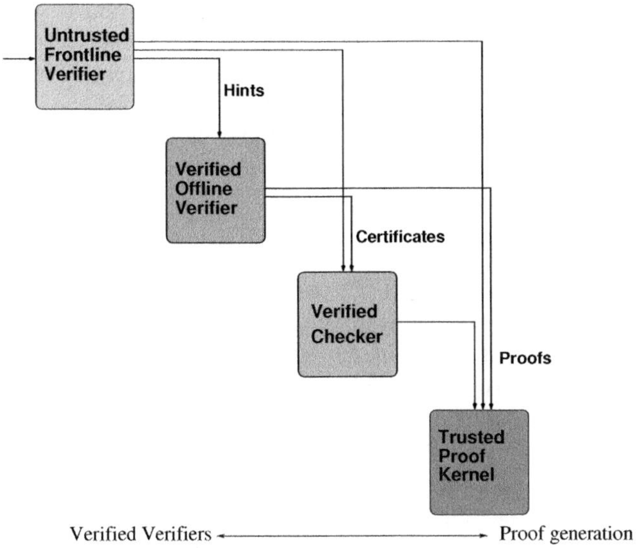

Fig. 1. The Kernel of Truth Architecture

Computational reflection does not require any external devices, but it does require that the inference procedure be executable. It is also possible to verify inference procedures non-reflectively using a different trusted or untrusted inference tool.

Recently, Davis [Dav09] has completed a dissertation where he has verified a fairly sophisticated theorem prover, a simplified version of ACL2, in a reflective manner, by defining 11 layers of proof checkers of increasing sophistication. The most sophisticated of these, level 11, includes mechanisms for induction, rewriting, and simplification. This level 11 proof checker is used to define proof checkers at levels 1 to 11, and to show that proofs at level $i + 1$ can be justified by proofs at level i, for $1 \leq i \leq 10$. These correctness proofs at level 11 can then be translated to level 1 proofs.

Our KoT approach has some similarities to the approach used by Davis, but we focus on building verified checkers for certificates and not on verifying the inference procedures themselves. The main reason for this is that high-end inference tools evolve rapidly. Even if verification were feasible, it would be hard to keep up with the changes to these tools. In contrast, the certificates generated by these tools can be fairly stable. Checking these certificates can be much more efficient than solving the original inference problem. Many different verification tools can share the same certificate format so that the investment in verifying the checkers can be amortized over these multiple uses. Our KoT approach smoothly accomodates the entire spectrum from proof generation to verification so that some inference procedures can be justified by generating proofs corresponding to their claims, and others by verification, but in most cases we would use the middle option of generating certificates that are checked by verified checkers.

In Section 1, we present a brief overview of PVS. Section 2 describes the PVS simplifier which combines rewriting, simplification, and decision procedures. In Section 3, we describe the concept of inference systems as an abstract framework for proving the soundness and completeness of inference procedures. The Kernel of Truth framework is outlined in Section 4 where we describe a kernel proof checker that is being developed within PVS to serve as a reference proof system to justify the correctness of various specialized checkers. Checkers for rewriting and resolution are presented in Section 5, and the current status of the project and future work are covered in Section 6.

1 Brief Background on PVS

The Prototype Verification System (PVS) is a comprehensive framework for interactive and automated verification based on higher-order logic [ORSvH95] where variables can range not only over individuals, as in first-order logic, but also over functions, functions of functions, and so on. Higher-order logic uses types to avoid paradoxes due to self-application. Types are built from base types such as the Booleans `bool` and the real numbers `real`. The type [A→B] represents the type of functions with domain type A and range type B. For example, $[A{\rightarrow}\texttt{bool}]$ represents the type of predicates over the type A, and we abbreviate this as `PRED[`A`]` or as `set[`A`]`. The type $[A_1, \ldots, A_n]$ represents the type of n-tuples where the i'th element has type A_i for $1 \leq i \leq n$. In addition, PVS has predicate subtypes which are of the form $\{x : T|e\}$ which contains the elements x of T satisfying e. With this, we can define subtypes for subranges, rational numbers, integers, even numbers, prime numbers, ordering relations, and order-preserving maps. For example, the subtype of even numbers can be defined as `{i : int | EXISTS (j : int) 2*j = i}`. With predicate subtypes, typechecking and theorem proving become interdependent since the demonstration that an expression like $6+4$ is an even number now requires a proof. The PVS typechecker generates proof obligations corresponding to predicate subtypes called type correctness conditions (TCCs). It also makes use of typing judgements to incorporate forward chaining rules such as the assertion that the sum of even numbers is an even number. PVS also has dependent types such as $[x : A{\rightarrow}B]$, where the range type B can depend on the domain element x. For example, if B is the type $\mathit{multiples}(x)$ containing the integer multiples of x, then the dependent type $[x : \mathit{int}{\rightarrow}\mathit{multiples}(x)]$ contains those functions on the integers that map each integer x to some multiple of x. Arrays are just functions. Update expressions can be applied to update the value of a function, record, or tuple at a specific index, field, or position, respectively. The PVS language has other features like parametric theories and recursive and corecursive datatypes. The PVS language and its type system can be used to embed other methodologies that require the generation of proof obligations, for example, Hoare logic [Hoa69, HJ00] or the B method [Abr96, Muñ99].

Almost all of the PVS language is executable. The only non-executable parts are equality on infinite higher-order types, which includes quantification on infinite domains. The PVS code generator detects when it is safe to evaluate updates destructively and is able to generate efficient code in these cases.

PVS also has an interactive proof checker that builds on various automated procedures for decision procedures, binary decision diagrams, satisfiability modulo theories, and rewriting. The proof checker uses a sequent representation for proof goals such that each proof step either completes the proof of a subgoal or generates new subgoals. Proof strategies are built from primitive inference steps using a strategy language. Strategies can be defined to execute complex patterns of proof steps, like induction followed by simpification and rewriting. The PVS simplifier which combines rewriting, simplification, and the use of decision procedures is described in the next section.

2 Combining Rewriting, Simplification, and Decision Procedures

A simplifier is a crucial part of an interactive proof assistant. It must ensure that the formula as presented to the user should have the expected simplifications applied. Many simplifications such as eliminating multiplication by 0, cancelling common factors in a fraction, and applying distributive laws, are quite natural. Others, such as beta reduction, are usually, but not always, a good idea. Decision procedures need to be employed during simplification. They can be used to propagate known information so that an expression of the form $i = j \Rightarrow A[i := v](j) \neq v$ can be simplified to FALSE. With decision procedures, simplification now becomes contextual. For example, the context $i = j$ is used in simplifying the expression $A[i := v](j) \neq v$. It also makes sense to integrate rewriting into the simplifier. Many simplifications can be expressed using rewriting. Conversely, rewriting also exploits simplification since we can assume that the expression being rewritten is in simplified form. Also, conditions in the application of conditional rewrite rules can often be discharged by simplification.

The PVS simplifier employs an inside-out strategy where sub-expressions are simplified before the expression is analyzed. The simplifier also carries a context for the decision procedure that is incrementally extended with new assertions. For example, when simplifying the branches of a conditional expression, the condition is asserted positively in the `THEN` branch, and negatively in the `ELSE` branch. The ground decision procedures can be used to decide if a given formula (that is, a boolean expression) is true or false (or not known to be either) with respect to the current context and relative to theories such as those of equality over uninterpreted function symbols and linear arithmetic. In a sequent of the form $a_1 \ldots a_m \vdash b_1 \ldots b_n$, the a_i are simplified and recorded as being true, and the b_i are simplified and recorded as being false. The simplifications are described below. The recording process can yield a refutation in which case the sequent has been proved. The ground decision procedure is a Shostak combination [Sho84, RS01, SR02] of the theory of equality with uninterpreted function symbols, quantifier-free integer and real linear arithmetic equalities and inequalities, array and function updates, and tuples and records.

The simplifier performs a range of simplifications. One set of simplifications applies to redexes as in the following examples (drawn from the PVS Prover Guide [SORSC99]).

1. Lambda redex: `(lambda x : x * x)(2)` $\longrightarrow 2 * 2$
2. Record redex : `b((# a:= 1, b:= 2, c:= 3 #))` $\longrightarrow$ `2`
3. Tuple redex : `proj_2((1, 2, 3))` $\longrightarrow$ `2`
4. Function update redex: For function `f`,

`(f WITH [(i) := 3])(i)` $\Longrightarrow$ `3`

`(f WITH [(0) := 3])(1)` $\Longrightarrow$ `f(1)`

5. Record update redex: For record `r`,

```
a(r WITH [(a) := 3]) ⟹ 3
a(r WITH [(b) := 2]) ⟹ a(r)
```

6. Cotuple redex:

```
in_2(out_2(x)) ⟹ x
out_2(in_2(x)) ⟹ x
in_2?(in_2(x)) ⟹ TRUE
in_1?(in_2(x)) ⟹ FALSE
```

7. Datatype redex: `car(cons(1, null))` ⟹ `1`
8. Recognizer redex:

```
cons?(null) ⟹ FALSE
cons?(cons(1, null)) ⟹ TRUE
```

9. Subtype redex: `even?(i)` ⟹ `TRUE`, if `even?` is one of the subtype predicates in the type of `i`.

Several of the above simplifications can be expressed as rewrite rules, but some like record and tuple reduction are generic across records and tuple types and require a family of rewrite rules depending on the actual type of the record or tuple. The second of the function update reductions requires a disequality to be established in the general case, and is therefore not directly representable as a rewrite rule. The remaining simplifications are summarized in brief. The PVS prover guide [SORSC99] contains more details.

A second set of simplifications addresses arithmetic expressions which are placed into an ordered sum-of-products form while grouping similar monomials and eliminating multiplication by 0 or 1. These simplifications yield a normal form for polynomials.

A third set of simplifications applies to Boolean expressions involving conjunction, disjunction, implication, equivalence, and negation.

A fourth set of simplifications applies to conditional and case expressions to prune infeasible branches and merge equivalent branches. For conditional expressions, the test is added to the context when simplifying the `THEN` branch, and its negation is added when simplifying the `ELSE` branch.

A final set of simplifications applies to quantified expressions. Note that when the body of a lambda-expression or a quantified expression is simplified, the type constraints on the bound variables are assumed, i.e., added to the context. Examples of simplifications applied to quantified expressions are

```
      (EXISTS x: x = 5) ⟹ TRUE
(EXISTS x, y, z: x = y + z AND f(x, y, z))
                          ⟹ (EXISTS y, z: f(y + z, y, z))
 (EXISTS (x: T): TRUE) ⟹ TRUE
(FORALL (x: T): FALSE) ⟹ FALSE
```

The last two simplifications only happen when the type `T` is known to be nonempty.

Rewriting. The PVS simplifier uses conditional rewriting to simplify expressions with respect to definitions and lemmas that have been installed by the user. A conditional rewrite rule triggers only when the conditions *simplify* to `TRUE`. Some rewrite rules such as recursive definition might loop if applied unconditionally. Therefore, if the right-hand side expression of a recursive definition is a conditional expression, then the top-level condition must simplify to `TRUE` or `FALSE` in order for the rewrite to occur. This mix of decision procedures and rewriting means that context comes into play in simplifying these conditions. Since the instantiable variables in a rewrite rule can have type constraints, proof obligations are generated corresponding to these type constraints on the actual instantiation for these variables. These proof obligations must also be discharged by the simplifier. The type constraints on the instantiable variables are therefore treated as conditions. Matching on the left-hand side of the rewrite rules also uses decision procedures to, for example, match a term a with $i + 3$, where i is a natural number, when it is possible to demonstrate that a is at least 3 in the given context. Decision procedures can also be used to check if a pattern of the form $f(x, x)$ matches an instance of the form $f(a, b)$ where $a = b$ is known in the context. Pattern matching is also lifted to the higher-order level through the use of Miller's higher-order patterns [Mil90] which turns out to be very useful for rewrite rules involving higher-order operations such as `map` and `reduce`.

3 Rewriting and Inference Systems

We now present *inference systems* as an abstract framework for presenting and reasoning about inference procedures [SR02, Sha05, dMDS07, Sha09]. Inference systems can be represented using rewriting logic. An inference system is a triple $\langle \Psi, \Lambda, \vdash \rangle$ consisting of a set Ψ of inference states, a mapping Λ from an inference state to a formula, and a binary inference relation $\vdash$ between inference states. For each formula ϕ, there must be at least one state ψ such that $\Lambda(\psi) = \phi$. There is a special unsatisfiable inference state $\bot$. Given an input formula, the inference system is used to construct a sequence of logical states $\psi_0 \vdash \ldots \vdash \bot$ where the input formula is represented by the first state. The inference relation $\vdash$ must be

1. **Conservative:** If $\psi \vdash \psi'$, then $\Lambda(\psi)$ and $\Lambda(\psi')$ must be equisatisfiable.
2. **Progressive:** For any subset S of Ψ, there is a state $\psi \in S$ such that there is no $\psi' \in S$ where $\psi \vdash \psi'$.
3. **Canonizing:** If $\psi \in \Psi$ is irreducible, that is, there is no ψ' such that $\psi \vdash \psi'$, then either $\psi \equiv \bot$ or $\Lambda(\psi)$ is satisfiable.

Inference systems are presented in the form of inference rules. For example, the inference system for ordered resolution on a set of ordered clauses (deleting tautologies, i.e., clauses containing a literal and its negation) is given in Figure 2.

In a number of cases, inference rules can be given as rewrite rules. In the case of resolution, the rewrite rules operate on an inference state that is a set of clauses. One rewrite rule adds a new clause obtained by resolving two clauses, and the other rewrite rule detects a contradiction.

Inference systems can be given for a variety of decision procedures including SAT and SMT solvers. These inference systems can also be used to construct satisfying

Res	$\dfrac{K, k \vee \kappa_1, \overline{k} \vee \kappa_2}{K, k \vee \kappa_1, \overline{k} \vee \kappa_2, \kappa_1 \vee \kappa_2}$	$\kappa_1 \vee \kappa_2 \notin K$ $\kappa_1 \vee \kappa_2$ is not tautological
Contrad	$\dfrac{K}{\bot}$ if $p, \neg p \in K$ for some p	

Fig. 2. Inference System for Ordered Resolution

assignments when the input formulas are satisfiable, and proofs when the input formulas are unsatisfiable. Theory solvers for theories such as equality, arithmetic, and arrays can also be expressed as inference systems. While some of these require specialized data structures like hash tables and linked pointer structures, it is possible to prototype such solvers within a rewriting system like Maude.

4 A Kernel of Truth

Some applications, particularly safety-critical and security-critical ones, need the claims made using complex inference tools to be certified. For inference, the expected standard for certification is that of a proof. We have already noted that it is possible to construct inference procedures that are proof generating. In many cases, the overhead of proof generation is not significant, although the representation of these proofs can become large. State-of-the-art inference tools are constantly being modified and improved, and proof generation is an added burden. For example, in the case of the PVS simplifier described in Section 2, we would have to combine proofs from many different sources.

We outline a lighter approach to certifying inference claims. In our approach, we use a kernel proof checker as the reference standard. In our case, we use PVS to define a proof checker for first-order logic with the axioms of ZFC. Such a proof checker can be defined in about 500 lines of PVS. Though this proof checker is executable, we mostly use it to demonstrate the existence of proofs that are, in the usual case, not explicitly constructed. The point of the reference proof checker is to demonstrate the correctness of other checkers. These checkers can range from reference implementations of inference procedures to those that check certificates for specific classes of problems. We illustrate this with checkers for resolution proofs and certificates from rewriting.

The reference proof checker has some notable features. One, it uses one-sided sequents. This is mainly to reduce the size and complexity of the proof calculus. The kernel itself could be used to justify the correctness of a two-sided sequent calculus. The system we use is quite similar to the one given in Shoenfield [Sho67]. Second, it uses two kinds of function and predicate symbols. *Interpreted* function and predicate symbols are used for defined operations such as those for equality, set membership, and arithmetic. Uninterpreted function and predicate symbols are used as schematic operations. These can be substituted by lambda-expressions of the appropriate arity. Such lambda-expressions do not have a first-class status in the logic, but are merely used to instantiate the schematic operators. Schematic operators have several uses. They can be used to introduce eigenvariables corresponding to the sequent rule for universal quantification as schematic constants. Uninterpreted predicates of arity 0 can serve as

Ax	$\dfrac{}{\vdash A, \neg A, \Delta}$
$\neg\neg$	$\dfrac{\vdash A, \Delta}{\vdash \neg\neg A, \Delta}$
$\vee$	$\dfrac{\vdash A, B, \Delta}{\vdash A \vee B, \Delta}$
$\neg\vee$	$\dfrac{\vdash \neg A, \Delta \quad \vdash \neg B, \Delta}{\vdash \neg(A \vee B), \Delta}$
Cut	$\dfrac{\vdash A, \Delta \quad \vdash \neg A, \Delta}{\vdash \Delta}$

Fig. 3. A Sequent Calculus for Propositional Logic

$\exists$	$\dfrac{\vdash A[t/x], \Delta}{\vdash \exists x.A, \Delta}$
$\neg\exists$	$\dfrac{\vdash \neg A[c/x], \Delta}{\neg\exists x.A, \Delta}$

Fig. 4. Sequent proof rules for quantification. The schematic constant c must not occur in Δ.

propositional atoms. Uninterpreted functions and predicate can also be used to capture schematic axioms and theorems. For example, the comprehension axiom scheme of set theory can be written as

$$\forall y.\exists z.\forall x.(x \in z \iff x \in y \wedge p(x)),$$

where p is a schematic predicate. Here, p can be replaced by a lambda-expression of the form $\lambda w.A$, that contains no free variables (but may contain schematic function and predicate symbols) to yield $\forall y.\exists z.\forall x.x \in z \iff x \in y \wedge A[x/w]$. Similarly, the replacement axiom scheme can be written as

$$\forall w.(\forall x \in w.\exists! y.q(x, y, w)) \Rightarrow \exists z.\forall y.(y \in z \iff \exists x \in w.q(x, y, w)),$$

where q is a schematic predicate.

With this proof checker, we can also define the concept of an LCF-style *tactic*. A *theorem* is a sequent that has a proof. A tactic then is an operation that maps a *conclusion* sequent $\vdash \Delta$ to a list of premise sequents $\vdash \Delta_1, \ldots, \vdash \Delta_n$ such that the conclusion is a theorem if the premise sequents are theorems. This concept of a tactic can be given as a type in PVS.

The basic judgement is given by a one-sided sequent of the form $\vdash A_1, \ldots, A_n$. We have a contraction rule that allows $\vdash \Gamma$ to be derived from $\vdash \Delta$ when $\Delta \subseteq \Gamma$. The basic propositional connectives are negation and disjunction, and these are used to define the other connectives. The propositional proof rules are shown in Figure 3 and the quantifier rules are shown in Figure 4.

As noted earlier, the language admits schematic function and predicate symbols that can be instantiated. For n-ary uninterpreted function symbol f, let $\Delta[\lambda\overline{x}.s/f]$, where $\overline{x}$ represents the sequence $x_1, \ldots, x_n$, be the result of replacing each subterm $f(t_1, \ldots, t_n)$ in Δ with $s[t_1/x_1, \ldots, t_n/x_n]$. Similarly, for n-ary uninterpreted predicate symbol p, let $\Delta[p \leftarrow \lambda\overline{x}.A]$ be the result of replacing each subformula $p(t_1, \ldots, t_n)$ by $A[t_1/x_1, \ldots, t_n/x_n]$ while renaming bound variables in A as needed to avoid variable capture. We then have a function instantiation rule that allows $\vdash \Delta[\lambda\overline{x}.s/f]$ to be derived from $\vdash \Delta$, and the predicate instantiation rule allows $\vdash \Delta[\lambda\overline{x}.s/f]$ to be derived from $\vdash \Delta$.

For equality, we have inference rules corresponding to reflexivity and congruence. The rules for transitivity and symmetry can be derived from reflexivity and predicate instantiation.

5 A Verified Checker for Rewriting and Other Inference Procedures

We now describe some ongoing work on building a certified checker for rewriting. Here, we rely only on the first-order logic part of the kernel checker. A proof system for certifying rewriting is given by Rosu, Eker, Lincoln, and Meseguer [RELM03]. We describe a checker for rewriting that we are currently verifying. A term is either a variable or an n-ary function symbol, with $0 \leq n$, applied to a sequence of n terms. The free variables $vars(s)$ of a term s is the set of all the variables that occur in the term. A path π is a a finite sequence of natural numbers. Given a term s and a path π, the subterm $s|_\pi$ of s at π is defined as s itself, when π is empty, and as $s_i|_{\pi'}$, where $\pi \equiv i, \pi'$ and $s \equiv f(s_1, \ldots, s_n)$. The result of replacing the subterm $s|_\pi$ by a term t is represented by $s|_{\pi \leftarrow t}$. We restrict our attention to unconditional rewrite rules of the form $\forall\overline{x}.l = r$, where $\overline{x}$ is a sequence of distinct variables that contains all and only the variables in $vars(l)$ and $vars(r) \subseteq vars(l)$. The rewriter takes a set of rewrite rules and applies them to rewrite an expression e to e'. We represent the certificate as a sequence of triples $\langle \tau_1, \ldots, \tau_m \rangle$, where each triple τ_i of the form $\langle R_i, \pi_i, \sigma_i \rangle$ consists of a rewrite rule R_i, a path π_i, and a substitution σ_i. Such a sequence of triples is a valid certificate for the claim $e = e'$ if there is a sequence of terms $e_0, \ldots, e_m$ such that $e \equiv e_0$ and $e' \equiv e_m$, and $e_{i+1} \equiv e_i|_{\pi_{i+1} \leftarrow \sigma_{i+1}(r_{i+1})}$, where l_{i+1} and r_{i+1} are the left-hand and right-hand sides of the rewrite rule R_{i+1} and $e_i|_{\pi_{i+1}} \equiv \sigma_{i+1}(l_{i+1})$ for $0 \leq i < m$.

The checker for such a certificate has to check the validity conditions above. We define $rwcheck(R, \langle \tau_1, \ldots, \tau_m \rangle, e, e')$ to check that $\langle \tau_1, \ldots, \tau_m \rangle$ yields a sequence of replacements from e to e' using the rewrite rules in R. We then have to prove the metatheorem that whenever $rwcheck(R, \langle \tau_1, \ldots, \tau_m \rangle, e, e')$ holds, there is a formal proof of $\vdash \neg R, e = e'$, where $\neg R$ is the formula-wise negation of each element of R. Once this metatheorem is proved, there is no need to actually generate this proof. For a rewrite step from e_i to e_{i+1}, the formal justification goes as follows. First, we establish the instantiation rule where we can derive $\vdash \sigma(l = r)$ from $\vdash \forall\overline{x}.l = r$, where σ maps each variable in $\overline{x}$ is mapped to a ground term, i.e., a term with no free variables. The proof of $e = e'$ then follows by applying congruence to $\vdash \sigma_{i+1}(l_{i+1} = r_{i+1})$ to derive $\vdash e_i = e_{i+1}$, and then transitivity to establish $\vdash e_0 = e_m$ from the sequents $\vdash e_i = e_{i+1}$, for $0 \leq i < m$.

In the more general situation of proof development in first-order and higher-order logic, rewriting can occur within the body of a quantification. The validity checker for certificates requires a more powerful metatheorem that uses the *equality theorem* [Sho67] that justifies the replacement of one term by a provably equal one within a formula.

Other specialized proof calculi can be similarly justified using certificate formats that are validated by verified checkers. For example, the proof system for resolution can be justified as follows. We have to show that when a set of clauses K yields a refutation, then the negation of the formula corresponding to K is provable. For some applications, we have to represent the derivation of clauses and not just refutational proofs. Each resolution step where a clause κ is derived from the clauses κ_1 and κ_2 is represented by the proof of the sequent $\vdash \neg\kappa_1, \neg\kappa_2, \kappa$. This sequent is easily proved from Ax, $\vee$, and $\neg\vee$ steps. A complete resolution refutation from a set of clauses K is represented by the proof of the sequent $\vdash \neg K$. Such a proof is constructed from those of the individual resolution inferences using the Cut rule. The resolution calculus is used as a kernel for justifying a SAT solver as a verified checker.

In addition to certificates and proof formats, the KoT kernel can also be used as a foundation for other logics. For example, a proof calculus for a modal logic or a higher-order logic can be justified in terms of its set-theoretic semantics. Once this is done, we can use the proof checker for the new logic as a kernel for other checkers. We plan to use this to justify proof calculi for various higher-order logics (including PVS) as well as modal, temporal, and program logics. We also plan to develop certificate checkers for SAT and SMT solvers, model checkers, and program analyzers.

6 Conclusions

Modern implementations of inference procedures like rewriting and propositional and theory satisfiability are extremely sophisticated and not easily amenable to formal verification. This makes it difficult to certify the results obtained by these procedures. It is even more difficult to certify results that are obtained by a combination of these tools. We address the challenge of certifying results from untrusted tools by relying on verified checkers that can be used to validate certificates generated by these tools. Our *Kernel of Truth* approach takes a middle ground between proof generation, where the untrusted tools are required to generate formal proofs, and verification, where only verified inference tools are used. Both these extreme cases are feasible within the KoT framework, but we also allow the more practical alternative of verifying checkers that use other logics or representations of certificates. Our approach is also similar to *translation validation* [PSS98] where each source-to-binary translation by a compiler is verified. These individual translations are often easily verified, whereas the verification of an entire compiler is a monumental task [SC98]. The KoT approach similarly avoids directly verifying inference procedures in favor of the checking the individual claims made by them.

The idea of verifying the inference steps of PVS within PVS itself might seem circular and vacuous. However, this is the one instance where we exploit the availability of explicit formal proofs to break the circularity. This is done by using the KoT framework itself to generate the complete formal proof for the verification of the correctness of any checkers. This proof can be checked by the kernel proof checker so that we need not trust the inference procedures used by PVS.

As mentioned at the beginning, the work we have described is still at a very early stage. We have defined the kernel proof checker for first-order logic described in Section 4 and have used it to prove some basic metatheorems. In earlier work from 2007 with Marc Vaucher, we also verified a sophisticated SAT solver using PVS. With Andrei Dan and Antoine Toubhans, we have developed a verified certificate checker for the proof traces generated by PicoSAT [Bie08]. A similar verified checker has been developed by Darbari, Fischer, and Marques-Silva [DFMS10]. We are working toward verified certificate checkers for rewriting, satisfiability modulo theories (SMT), and simplifiers. We hope to eventually develop a range of certificate checkers so that inference tools like PVS, Yices [DdM06], and SAL [dMOR+04] can be instrumented to generate checkable certificates.

The approach of checking the results of a computation with a verified checker is not restricted to inference tools. Result checking can be applied to a wide range of computations where there is a specific correctness claim associated with the output. Such an approach has been already been advocated by Mehlhorn [Meh03] in what he calls the *Reliable Algorithmic Software Challenge*. Generating efficiently checkable certificates both for inference and non-inference procedures is itself an interesting challenge. At the Workshop on Rewriting Logic and Applications, José Meseguer posed the challenge of certifying that a unifier is the most general unifier. It is easy to certify that a substitution is in fact a unifier, but it is often important to know that it is in fact the most general one. There are many ways to approach such challenges. One approach is to verify a unification algorithm[Pau84]. Another approach is to demonstrate that no generalization of the unifier is a valid unifier. A third approach is for the unification procedure to generate a trace that demonstrates how any solution to the unification problem can be transformed into an instance of the unifier, but generating and checking such traces is still a problem. A similar challenge arises with graph algorithms where we must demonstrate that a path is indeed the shortest path or that a target vertex is unreachable. In earlier work [Sha10], we showed how fixpoints can be used to construct efficient certificates for such graph search algorithms. The KoT project is a response to Mehlhorn's challenge for the specific case of inference procedures but we are also interested in extending it to a larger class of computations. Compared to the goal of verifying software, the Kernel of Truth framework has the more limited ambition of checking computations in a verifiable manner.

References

[Abr96] Abrial, J.-R.: The B-Book: Assigning Programs to Meanings. Cambridge University Press, Cambridge (1996)

[Bie08] Biere, A.: PicoSAT essentials. JSAT 4(2-4), 75–97 (2008)

[BKK+96] Borovansky, P., Kirchner, C., Kirchner, H., Moreau, P.E., Vittek, M.: ELAN: A logical framework based on computational systems. In: Proc. of the First Int. Workshop on Rewriting Logic, vol. 4. Elsevier, Amsterdam (1996)

[Ble77] Bledsoe, W.W.: Non-resolution theorem proving. Artificial Intelligence 9, 1–36 (1977)

[BM79] Boyer, R.S., Moore, J.S.: A Computational Logic. Academic Press, New York (1979)

[BM81] Boyer, R.S., Moore, J.S.: Metafunctions: Proving them correct and using them efficiently as new proof procedures. In: Boyer, R.S., Moore, J.S. (eds.) The Correctness Problem in Computer Science. Academic Press, London (1981)

[BM88] Boyer, R.S., Moore, J.S.: A Computational Logic Handbook. Academic Press, New York (1988)

[CAB+86] Harper, R.W., Howe, D.J., Knoblock, T.B., Mendler, N.P., Panangaden, P., Sasaki, J.T., Smith, S.F.: Implementing Mathematics with the Nuprl Proof Development System. Prentice Hall, Englewood Cliffs (1986), Nuprl home page, `http://www.cs.cornell.edu/Info/Projects/NuPRL/`

[CDE+99] Clavel, M., Durán, F., Eker, S., Lincoln, P., Martí-Oliet, N., Meseguer, J., Quesada, J.F.: The Maude system. In: Narendran, P., Rusinowitch, M. (eds.) RTA 1999. LNCS, vol. 1631, pp. 240–243. Springer, Heidelberg (1999)

[Dav09] Davis, J.C.: A Self-Verifying Theorem Prover. PhD thesis, Computer Science Department, The University of Texas at Austin (December 2009)

[DdM06] Dutertre, B., de Moura, L.: The Yices SMT solver (2006), `http://yices.csl.sri.com/`

[DFMS10] Darbari, A., Fischer, B., Marques-Silva, J.: Industrial-strength certified sat solving through verified sat proof checking. In: Int. Colloq. on Theoretical Aspects of Computing (ICTAC) (2010) (to appear)

[dMDS07] de Moura, L., Dutertre, B., Shankar, N.: A tutorial on satisfiability modulo theories. In: Damm, W., Hermanns, H. (eds.) CAV 2007. LNCS, vol. 4590, pp. 20–36. Springer, Heidelberg (2007)

[dMOR+04] de Moura, L., Owre, S., Rueß, H., Rushby, J., Shankar, N., Sorea, M., Tiwari, A.: SAL 2. In: Alur, R., Peled, D.A. (eds.) CAV 2004. LNCS, vol. 3114, pp. 496–500. Springer, Heidelberg (2004), SAL home page, `http://sal.csl.sri.com/`

[GKM+87] Goguen, J., Kirchner, C., Megrelis, A., Meseguer, J., Winkler, T.: An introduction to OBJ3. In: Kaplan, S., Jouannaud, J.-P. (eds.) CTRS 1987. LNCS, vol. 308, pp. 258–263. Springer, Heidelberg (1988)

[GM93] Gordon, M.J.C., Melham, T.F. (eds.): Introduction to HOL: A Theorem Proving Environment for Higher-Order Logic. Cambridge University Press, Cambridge (1993), HOL home page, `http://www.cl.cam.ac.uk/Research/HVG/HOL/`

[GW88] Goguen, J.A., Winkler, T.: Introducing OBJ. Technical Report SRI-CSL-88-9, Computer Science Laboratory, SRI International, Menlo Park, CA (August 1988)

[HJ00] Huisman, M., Jacobs, B.: Java program verfication via a hoare logic with abrupt termination. In: Maibaum, T. (ed.) FASE 2000. LNCS, vol. 1783, pp. 284–303. Springer, Heidelberg (2000)

[Hoa69] Hoare, C.A.R.: An axiomatic basis for computer programming. ACM Comm. 12(10), 576–583 (1969)

[KMM00] Kaufmann, M., Manolios, P., Strother Moore, J.: Computer-Aided Reasoning: An Approach. Advances in Formal Methods, vol. 3. Kluwer, Dordrecht (2000)

[KZ88] Kapur, D., Zhang, H.: RRL: A rewrite rule laboratory. In: Lusk, E., Overbeek, R. (eds.) CADE 1988. LNCS, vol. 310, pp. 768–769. Springer, Heidelberg (1988)

[Meh03] Mehlhorn, K.: The reliable algorithmic software challenge RASC. In: Jansen, K., Margraf, M., Mastrolli, M., Rolim, J.D.P. (eds.) WEA 2003. LNCS, vol. 2647, p. 222. Springer, Heidelberg (2003)

[Mil90] Miller, D.: An extension to ML to handle bound variables in data structures: Preliminary report. In: Informal Proceedings of the Logical Frameworks BRA Workshop, Nice, France, Available as UPenn CIS technical report MS-CIS-90-59 (June 1990)

[Muñ99] Muñoz, C.: PBS: Support for the B-method in PVS. Technical Report SRI-CSL-99-1, Computer Science Laboratory, SRI International, Menlo Park, CA (February 1999)

[ORSvH95] Owre, S., Rushby, J., Shankar, N., von Henke, F.: Formal verification for fault-tolerant architectures: Prolegomena to the design of PVS. IEEE Transactions on Software Engineering 21(2), 107–125 (1995), PVS home page, `http://pvs.csl.sri.com`

[Pau84] Paulson, L.C.: Verifying the unification algorithm in LCF. Technical Report 50, University of Cambridge Computer Laboratory (1984)

[Pau94] Paulson, L.C. (ed.): Isabelle: A Generic Theorem Prover. LNCS, vol. 828. Springer, Heidelberg (1994), Isabelle home page, `http://www.cl.cam.ac.uk/research/hvg/Isabelle/`

[PSS98] Pnueli, A., Siegel, M., Singerman, E.: Translation validation. In: Steffen, B. (ed.) TACAS 1998. LNCS, vol. 1384, pp. 151–166. Springer, Heidelberg (1998)

[RELM03] Rosu, G., Eker, S., Lincoln, P., Meseguer, J.: Certifying and synthesizing membership equational proofs. In: Araki, K., Gnesi, S., Mandrioli, D. (eds.) FME 2003. LNCS, vol. 2805, pp. 359–380. Springer, Heidelberg (2003)

[RS01] Rueß, H., Shankar, N.: Deconstructing Shostak. In: 16th Annual IEEE Symposium on Logic in Computer Science, Boston, MA, pp. 19–28. IEEE Computer Society, Los Alamitos (July 2001)

[SC98] Stringer-Calvert, D.W.J.: Mechanical Verification of Compiler Correctness. PhD thesis, University of York, Department of Computer Science, York, England (March 1998), `http://www.csl.sri.com/~dave_sc/papers/thesis.html`

[Sha05] Shankar, N.: Inference systems for logical algorithms. In: Ramanujam, R., Sen, S. (eds.) FSTTCS 2005. LNCS, vol. 3821, pp. 60–78. Springer, Heidelberg (2005)

[Sha08] Shankar, N.: Trust and automation in verification tools. In: Cha, S.(S.), Choi, J.-Y., Kim, M., Lee, I., Viswanathan, M. (eds.) ATVA 2008. LNCS, vol. 5311, pp. 4–17. Springer, Heidelberg (2008)

[Sha09] Shankar, N.: Automated deduction for verification. ACM Comput. Surv. 41(4), 20 (2009)

[Sha10] Shankar, N.: Fixpoint and search in pvs. In: Müller, P. (ed.) LASER 2010. LNCS, vol. 6029, pp. 140–161. Springer, Heidelberg (2010)

[Sho67] Shoenfield, J.R.: Mathematical Logic. Addison-Wesley, Reading (1967)

[Sho84] Shostak, R.E.: Deciding combinations of theories. Journal of the ACM 31(1), 1–12 (1984)

[SORSC99] Shankar, N., Owre, S., Rushby, J.M., Stringer-Calvert, D.W.J.: PVS Prover Guide. Computer Science Laboratory, SRI International, Menlo Park, CA (September 1999)

[SR02] Shankar, N., Rueß, H.: Combining Shostak Theories. In: Tison, S. (ed.) RTA 2002. LNCS, vol. 2378, pp. 1–18. Springer, Heidelberg (2002)

[The09] The Coq Development Team. The Coq proof assistant reference manual version 8.2. Technical report, INRIA (February 2009)

[ZM03] Zhang, L., Malik, S.: Validating SAT solvers using an independent resolution-based checker: Practical implementations and other applications. In: DATE, pp. 10880–10885. IEEE Computer Society, Los Alamitos (2003)

Twenty Years of Rewriting Logic

José Meseguer

Department of Computer Science,
University of Illinois at Urbana-Champaign
meseguer@cs.uiuc.edu

The first three papers on rewriting logic were published in 1990 [4,3,2]; they were then expanded in [5,6]. Since that time, many researchers around the world have made important contributions to its foundations, tools, and applications. Since 1996, the WRLA workshop has met biennially, with the 2010 Paphos meeting being its eighth edition, and many hundreds of papers have been published on the subject (for a bibliography up to 2002 see [1]). This growth makes it desirable to reflect from time to time upon the advances made, survey such advances, and perhaps get some glimpses and make some guesses about future directions. I thank the organizers of WRLA 2010 for giving me the opportunity and the stimulus to do some reflecting, surveying, and guessing about rewriting logic at this point, when twenty years have passed since the first papers were published. It is somewhat like taking a snapshot of a person at age twenty. I have taken some similar, total or partial pictures at earlier ages, as a child [7,9,8], and as a teenager [1] (with Narciso Martí-Oliet) and [10]. It seems appropriate to attempt taking a coming-of-age picture, and to ask some questions about rewriting logic such as the following:

- How well-developed are its mathematical foundations?
- To what extent have its goals as a semantic framework for concurrency, and as a logical framework, been achieved?
- Which languages and tools supporting rewriting logic programming, specication, and verification have been developed?
- In which application areas has it been shown useful?
- How do its future prospects look like?

In this short abstract (as opposed to the talk itself) I cannot give details on my answers to these questions: I hope to provide such details in a projected survey. I can however outline the different headings covered in the talk as a way of addressing the above questions:

1. **Foundations**, including:
   - Generalized Rewrite Theories
   - Coherence and Computability
   - Termination
   - Narrowing and Reachability Analysis
   - Reflection

P.C. Ölveczky (Ed.): WRLA 2010, LNCS 6381, pp. 15–17, 2010.

   - Strategies
   - Temporal Logic Properties
   - Simulation and Abstraction
   - Real-Time Rewrite Theories
   - Probabilistic Rewrite Theories
2. **Rewriting Logic as a Logical and Semantic Framework**, including:
   - Representing Logics
   - Representing Models of Concurrency
   - Representing Modeling Languages
   - Rewriting Logic Semantics of Programming Languages
3. **Rewriting Logic Languages**, including:
   - Cafe-OBJ
   - ELAN
   - Maude
4. **Applications and Tools**, including applications to:
   - Software/Hardware Specification and Verification
   - Network Systems
   - Real-Time and Embedded Systems
   - Probabilistic Systems
   - Security
   - Bioinformatics

   and tools such as rewriting logic interpreters and model checkers, formal tool environments for rewriting logic specifications, and domain-specific formal tools.
5. **Future Directions**, where, besides future developments in all the above-mentioned areas, the following areas seem particularly promising:
   - Deductive and Symbolic Reachability Verification Methods for Rewrite Theories
   - Verification Methods and Tools for Probabilistic Rewrite Theories
   - Parallel and Distributed Computing
   - Cyber-Physical Systems
   - Further Advances in the Rewriting Logic Semantics Project.

Acknowledgments. In a talk of this kind, one feels a great sense of gratitude to the many efforts of gifted researchers who have made important contributions. They are too many to mention here by name; but in the planned survey I hope to do justice to many of their contributions. This work has been supported in part by NSF Grants CNS 07-16638, CNS 08-34709, CNS 08-31064, CNS 09-04749, and CCF 09-05584.

References

1. Martí-Oliet, N., Meseguer, J.: Rewriting logic: roadmap and bibliography. Theoretical Computer Science 285, 121–154 (2002)
2. Meseguer, J.: Conditional rewriting logic: deduction, models and concurrency. In: Okada, M., Kaplan, S. (eds.) CTRS 1990. LNCS, vol. 516, pp. 64–91. Springer, Heidelberg (1991)

3. Meseguer, J.: A logical theory of concurrent objects. In: ECOOP-OOPSLA 1990 Conference on Object-Oriented Programming, Ottawa, Canada, pp. 101–115. ACM, New York (October 1990)
4. Meseguer, J.: Rewriting as a unified model of concurrency. In: Baeten, J.C.M., Klop, J.W. (eds.) CONCUR 1990. LNCS, vol. 458, pp. 384–400. Springer, Heidelberg (1990)
5. Meseguer, J.: Conditional rewriting logic as a unified model of concurrency. Theoretical Computer Science 96(1), 73–155 (1992)
6. Meseguer, J.: A logical theory of concurrent objects and its realization in the Maude language. In: Agha, G., Wegner, P., Yonezawa, A. (eds.) Research Directions in Concurrent Object-Oriented Programming, pp. 314–390. MIT Press, Cambridge (1993)
7. Meseguer, J.: Rewriting logic as a semantic framework for concurrency: a progress report. In: Sassone, V., Montanari, U. (eds.) CONCUR 1996. LNCS, vol. 1119, pp. 331–372. Springer, Heidelberg (1996)
8. Meseguer, J.: Rewriting logic and Maude: a wide-spectrum semantic framework for object-based distributed systems. In: Smith, S., Talcott, C.L. (eds.) Formal Methods for Open Object-Based Distributed Systems, FMOODS 2000, pp. 89–117. Kluwer, Dordrecht (2000)
9. Meseguer, J.: Rewriting logic and Maude: Concepts and applications. In: Bachmair, L. (ed.) RTA 2000. LNCS, vol. 1833, pp. 1–26. Springer, Heidelberg (2000)
10. Meseguer, J.: A rewriting logic sampler. In: Van Hung, D., Wirsing, M. (eds.) ICTAC 2005. LNCS, vol. 3722, pp. 1–28. Springer, Heidelberg (2005)

Proving Termination in the Context-Sensitive Dependency Pair Framework*

Raúl Gutiérrez and Salvador Lucas

ELP Group, DSIC, Universitat Politècnica de València
Camí de Vera s/n, 46022 València, Spain

Abstract. Termination of *context-sensitive rewriting* (CSR) is an interesting problem with several applications in the fields of term rewriting and in the analysis of programming languages like CafeOBJ, Maude, OBJ, etc. The dependency pair approach, one of the most powerful techniques for proving termination of rewriting, has been adapted to be used for proving termination of CSR. The corresponding notion of *context-sensitive dependency pair* (CSDP) is different from the standard one in that *collapsing pairs* (i.e., rules whose right-hand side is a variable) are considered. Although the implementation and practical use of CSDPs lead to a powerful framework for proving termination of CSR, handling collapsing pairs is not easy and often leads to impose heavy requirements over the base orderings which are used to achieve the proofs. A recent proposal removes collapsing pairs by transforming them into sets of new (standard) pairs. In this way, though, the role of collapsing pairs for modeling context-sensitive computations gets lost. This leads to a less intuitive and accurate description of the termination behavior of the system. In this paper, we show how to get the best of the two approaches, thus obtaining a powerful *context-sensitive dependency pair framework* which satisfies all practical and theoretical expectations.

1 Introduction

In Context-Sensitive Rewriting (CSR, [1]), a *replacement map* μ satisfying $\mu(\mathsf{f}) \subseteq \{1, \ldots, \mathsf{ar}(\mathsf{f})\}$ for every function symbol f of arity ar(f) in the signature $\mathcal{F}$ is used to discriminate the argument positions on which the rewriting steps are allowed. In this way, a terminating behavior of (context-sensitive) computations with Term Rewriting Systems (TRSs) can be obtained. CSR has shown useful to model evaluation strategies in programming languages. In particular, it is an essential ingredient to analyze the *termination behavior* of programs in programming languages (like CafeOBJ, Maude, OBJ, etc.) which implement recent presentations of rewriting logic like the *Generalized Rewrite Theories* [2], see [3,4,5].

* Partially supported by the EU (FEDER) and the Spanish MEC/MICINN, under grant TIN 2007-68093-C02-02.

P.C. Ölveczky (Ed.): WRLA 2010, LNCS 6381, pp. 18–34, 2010.

Example 1. Consider the following TRS in [6]:

$$\begin{aligned} \mathsf{gt}(\mathsf{0},y) &\to \mathsf{false} & \mathsf{p}(\mathsf{0}) &\to \mathsf{0}\\ \mathsf{gt}(\mathsf{s}(x),\mathsf{0}) &\to \mathsf{true} & \mathsf{p}(\mathsf{s}(x)) &\to x\\ \mathsf{gt}(\mathsf{s}(x),\mathsf{s}(y)) &\to \mathsf{gt}(x,y) & \mathsf{minus}(x,y) &\to \mathsf{if}(\mathsf{gt}(y,\mathsf{0}),\mathsf{minus}(\mathsf{p}(x),\mathsf{p}(y)),x)\\ \mathsf{if}(\mathsf{true},x,y) &\to x & \mathsf{div}(\mathsf{0},\mathsf{s}(y)) &\to \mathsf{0}\\ \mathsf{if}(\mathsf{false},x,y) &\to y & \mathsf{div}(\mathsf{s}(x),\mathsf{s}(y)) &\to \mathsf{s}(\mathsf{div}(\mathsf{minus}(x,y),\mathsf{s}(y))) \end{aligned}$$

with $\mu(\mathsf{if}) = \{1\}$ and $\mu(\mathsf{f}) = \{1,\ldots,\mathsf{ar}(\mathsf{f})\}$ for all other symbols f. *Note that, if no replacement restriction is considered, then the following sequence is possible and the system would be nonterminating:*

$$\underline{\mathsf{minus}(\mathsf{0},\mathsf{0})} \to^*_{\mathcal{R}} \mathsf{if}(\mathsf{gt}(\mathsf{0},\mathsf{0}),\underline{\mathsf{minus}(\mathsf{0},\mathsf{0})},\mathsf{0}) \to^*_{\mathcal{R}} \ldots,\underline{\mathsf{minus}(\mathsf{0},\mathsf{0})},\ldots \to^*_{\mathcal{R}} \cdots$$

In CSR, *though, this sequence is not possible because reductions on the second argument of the* if*-operator are disallowed due to $\mu(\mathsf{if}) = \{1\}$.*

In [7], Arts and Giesl's dependency pair approach [8], a powerful technique for proving termination of rewriting, was adapted to CSR (see [9] for a more recent presentation). Regarding proofs of termination of rewriting, the dependency pair technique focuses on the following idea: since a TRS $\mathcal{R}$ is terminating if there is no infinite rewrite sequence starting from any term, the rules that are really able to produce such infinite sequences are those rules $\ell \to r$ such that r contains some *defined* symbol[1] g. Intuitively, we can think of these rules as representing possible (direct or indirect) recursive calls. Recursion paths associated to each rule $\ell \to r$ are represented as new rules $u \to v$ (called *dependency pairs*) where $u = \mathsf{f}^\sharp(\ell_1,\ldots,\ell_k)$ if $\ell = \mathsf{f}(\ell_1,\ldots,\ell_k)$ and $v = \mathsf{g}^\sharp(s_1,\ldots,s_m)$ if $s = \mathsf{g}(s_1,\ldots,s_m)$ is a subterm of r and g is a defined symbol. The notation $\mathsf{f}^\sharp$ for a given symbol f means that f is *marked*. In practice, we often capitalize f and use F instead of $\mathsf{f}^\sharp$ in our examples. For this reason, the dependency pair technique starts by considering a new TRS $\mathsf{DP}(\mathcal{R})$ which contains all these dependency pairs for each $\ell \to r \in \mathcal{R}$. The rules in $\mathcal{R}$ and $\mathsf{DP}(\mathcal{R})$ determine the so-called *dependency chains* whose finiteness characterizes termination of $\mathcal{R}$ [8]. Furthermore, the dependency pairs can be presented as a *dependency graph*, where the infinite chains are captured by the *cycles* in the graph.

These intuitions are valid for CSR, but the subterms s of the right-hand sides r of the rules $\ell \to r$ which are considered to build the *context-sensitive dependency pairs* $\ell^\sharp \to s^\sharp$ must be *μ-replacing* terms. In sharp contrast with the dependency pair approach, though, we also need *collapsing dependency pairs* $u \to x$ where u is obtained from the left-hand side ℓ of a rule $\ell \to r$ in the usual way, i.e., $u = \ell^\sharp$ but x is a *migrating variable* which is μ-replacing in r but which only occurs at *non-μ-replacing positions* in ℓ [7,9]. Collapsing pairs are essential in our approach. They express that infinite context-sensitive rewrite sequences can involve not only the kind of recursion which is represented by the *usual* dependency pairs but also a new kind of recursion which is *hidden* inside

[1] A symbol $\mathsf{g} \in \mathcal{F}$ is defined in $\mathcal{R}$ if there is a rule $\ell \to r$ in $\mathcal{R}$ whose left-hand side ℓ is of the form $\mathsf{g}(\ell_1,\ldots,\ell_k)$ for some $k \geq 0$.

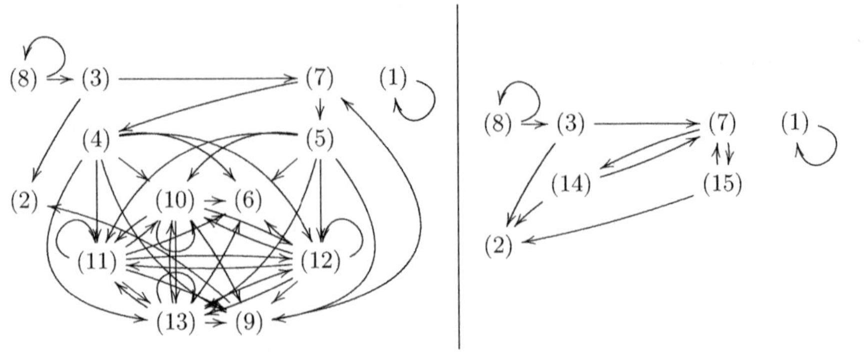

Fig. 1. Dependency graph for Example 1 following [6] (left) and [9] (right)

the non-μ-replacing parts of the terms involved in the infinite sequence until a *migrating* variable within a rule $\ell \to r$ shows them up.

In [6], a transformation that replaces the *collapsing pairs* by a new set of pairs that simulate their behavior was introduced. This new set of pairs is used to simplify the definition of context-sensitive dependency chain; but, on the other hand, we loose the intuition of what collapsing pairs mean in a context-sensitive rewriting chain. And understanding the new dependency graph is harder.

Example 2. (Continuing Example 1) If we follow the transformational definition in [6], we have the following dependency pairs (a new symbol U *is introduced):*

$$
\begin{array}{rll}
\mathsf{GT}(\mathsf{s}(x),\mathsf{s}(y)) \to \mathsf{GT}(x,y) & (1) \\
\mathsf{M}(x,y) \to \mathsf{GT}(y,\mathsf{0}) & (2) \\
\mathsf{D}(\mathsf{s}(x),\mathsf{s}(y)) \to \mathsf{M}(x,y) & (3) \\
\mathsf{IF}(\mathsf{true},x,y) \to \mathsf{U}(x) & (4) \\
\mathsf{IF}(\mathsf{false},x,y) \to \mathsf{U}(y) & (5) \\
\mathsf{U}(\mathsf{p}(x)) \to \mathsf{P}(x) & (6)
\end{array}
\qquad
\begin{array}{rll}
\mathsf{M}(x,y) \to \mathsf{IF}(\mathsf{gt}(y,\mathsf{0}),\mathsf{minus}(\mathsf{p}(x),\mathsf{p}(y)),x) & (7) \\
\mathsf{D}(\mathsf{s}(x),\mathsf{s}(y)) \to \mathsf{D}(\mathsf{minus}(x,y),\mathsf{s}(y)) & (8) \\
\mathsf{U}(\mathsf{minus}(\mathsf{p}(x),\mathsf{p}(y))) \to \mathsf{M}(\mathsf{p}(x),\mathsf{p}(y)) & (9) \\
\mathsf{U}(\mathsf{p}(x)) \to \mathsf{U}(x) & (10) \\
\mathsf{U}(\mathsf{p}(y)) \to \mathsf{U}(y) & (11) \\
\mathsf{U}(\mathsf{minus}(x,y)) \to \mathsf{U}(x) & (12) \\
\mathsf{U}(\mathsf{minus}(x,y)) \to \mathsf{U}(y) & (13)
\end{array}
$$

and the dependency graph has the unreadable aspect shown in Figure 1 (left). In contrast, if we consider the original definition of CSDPs and CSDG in [7,9], our set of dependency pairs is the following:

$$
\begin{array}{rll}
\mathsf{GT}(\mathsf{s}(x),\mathsf{s}(y)) \to \mathsf{GT}(x,y) & (1) \\
\mathsf{M}(x,y) \to \mathsf{GT}(y,\mathsf{0}) & (2) \\
\mathsf{D}(\mathsf{s}(x),\mathsf{s}(y)) \to \mathsf{M}(x,y) & (3)
\end{array}
\qquad
\begin{array}{rll}
\mathsf{M}(x,y) \to \mathsf{IF}(\mathsf{gt}(y,\mathsf{0}),\mathsf{minus}(\mathsf{p}(x),\mathsf{p}(y)),x) & (7) \\
\mathsf{D}(\mathsf{s}(x),\mathsf{s}(y)) \to \mathsf{D}(\mathsf{minus}(x,y),\mathsf{s}(y)) & (8) \\
\mathsf{IF}(\mathsf{true},x,y) \to x & (14) \\
\mathsf{IF}(\mathsf{false},x,y) \to y & (15)
\end{array}
$$

and the dependency graph is much more clear, see Figure 1 (right).

The work in [6] was motivated by the fact that mechanizing proofs of termination of CSR according to the results in [7] can be difficult due to the presence of collapsing dependency pairs. The problem is that [7] imposes hard restrictions on the orderings which are used in proofs of termination of CSR when collapsing dependency pairs are present. In this paper we address this problem in a different way. We keep collapsing CSDPs (and their descriptive power and simplicity) while the practical problems for handling them are overcome.

After some preliminaries in Section 2, in Section 3 we introduce the notion of *hidden term* and *hiding context* and discuss their role in infinite μ-rewrite sequences. In Section 4 we introduce a new notion of CSDP chain which is well-suited for mechanizing proofs of termination of CSR with CSDPs. In Section 5 we introduce our dependency pair framework for proving termination of CSR. Furthermore, we show that with the new definition we can also use all the existing processors from the two previous approaches and we can define new powerful processors. Section 6 shows an specific example of the power of this framework. Section 7 shows our experimental results. Section 8 discusses the differences between our approach and the one in [6]. Section 9 concludes. Proofs can be found in [10].

2 Preliminaries

We assume a basic knowledge about standard definitions and notations for term rewriting as given in, e.g., [11]. Positions $p, q, \ldots$ are represented by chains of positive natural numbers used to address subterms of t. Given positions p, q, we denote its concatenation as $p.q$. If p is a position, and Q is a set of positions, then $p.Q = \{p.q \mid q \in Q\}$. We denote the root or top position by Λ. The set of positions of a term t is $\mathcal{P}os(t)$. Positions of nonvariable symbols $\mathsf{f} \in \mathcal{F}$ in $t \in \mathcal{T}(\mathcal{F},\mathcal{X})$ are denoted as $\mathcal{P}os_{\mathcal{F}}(t)$. The *subterm* at position p of t is denoted as $t|_p$ and $t[s]_p$ is the term t with the subterm at position p replaced by s. We write $t \trianglerighteq s$ if $s = t|_p$ for some $p \in \mathcal{P}os(t)$ and $t \triangleright s$ if $t \trianglerighteq s$ and $t \neq s$. The symbol labeling the root of t is denoted as $\mathsf{root}(t)$. A *substitution* is a mapping $\sigma : \mathcal{X} \to \mathcal{T}(\mathcal{F},\mathcal{X})$ from a set of variables $\mathcal{X}$ into the set $\mathcal{T}(\mathcal{F},\mathcal{X})$ of terms built from the symbols in the *signature* $\mathcal{F}$ and the variables in $\mathcal{X}$. A *context* is a term $C \in \mathcal{T}(\mathcal{F} \cup \{\Box\},\mathcal{X})$ with a 'hole' $\Box$ (a fresh constant symbol). A *rewrite rule* is an ordered pair (ℓ, r), written $\ell \to r$, with $\ell, r \in \mathcal{T}(\mathcal{F},\mathcal{X})$, $\ell \notin \mathcal{X}$ and $\mathcal{V}ar(r) \subseteq \mathcal{V}ar(\ell)$. The left-hand side (*lhs*) of the rule is ℓ and r is the right-hand side (*rhs*). A *TRS* is a pair $\mathcal{R} = (\mathcal{F}, R)$ where $\mathcal{F}$ is a signature and R is a set of rewrite rules over terms in $\mathcal{T}(\mathcal{F},\mathcal{X})$. Given $\mathcal{R} = (\mathcal{F}, R)$, we consider $\mathcal{F}$ as the disjoint union $\mathcal{F} = \mathcal{C} \uplus \mathcal{D}$ of symbols $\mathsf{c} \in \mathcal{C}$, called *constructors* and symbols $\mathsf{f} \in \mathcal{D}$, called *defined symbols*, where $\mathcal{D} = \{\mathsf{root}(\ell) \mid \ell \to r \in R\}$ and $\mathcal{C} = \mathcal{F} \setminus \mathcal{D}$.

In the following, we introduce some notions and notation about CSR [1]. A mapping $\mu : \mathcal{F} \to \wp(\mathbb{N})$ is a *replacement map* if $\forall \mathsf{f} \in \mathcal{F},\ \mu(\mathsf{f}) \subseteq \{1, \ldots, \mathsf{ar}(\mathsf{f})\}$. Let $M_{\mathcal{F}}$ be the set of all replacement maps (or $M_{\mathcal{R}}$ for the replacement maps of a TRS $\mathcal{R} = (\mathcal{F}, R)$). The set of *$\mu$-replacing positions* $\mathcal{P}os^{\mu}(t)$ of $t \in \mathcal{T}(\mathcal{F},\mathcal{X})$ is: $\mathcal{P}os^{\mu}(t) = \{\Lambda\}$, if $t \in \mathcal{X}$ and $\mathcal{P}os^{\mu}(t) = \{\Lambda\} \cup \bigcup_{i \in \mu(root(t))} i.\mathcal{P}os^{\mu}(t|_i)$, if $t \notin \mathcal{X}$. The set of *μ-replacing variables* of t is $\mathcal{V}ar^{\mu}(t) = \{x \in \mathcal{V}ar(t) \mid \exists p \in \mathcal{P}os^{\mu}(t), t|_p = x\}$ and $\mathcal{V}ar^{\not\mu}(t) = \{x \in \mathcal{V}ar(t) \mid \exists p \in \mathcal{P}os(t) \setminus \mathcal{P}os^{\mu}(t), t|_p = x\}$ is the set of *non-μ-replacing variables* of t. Note that $\mathcal{V}ar^{\mu}(t)$ and $\mathcal{V}ar^{\not\mu}(t)$ do not need to be disjoint. The *μ-replacing subterm* relation $\trianglerighteq_{\mu}$ is given by $t \trianglerighteq_{\mu} s$ if there is $p \in \mathcal{P}os^{\mu}(t)$ such that $s = t|_p$. We write $t \triangleright_{\mu} s$ if $t \trianglerighteq_{\mu} s$ and $t \neq s$. We write

$t \rhd_{\not\mu} s$ to denote that s is a *non-μ-replacing strict subterm* of t, i.e., there is a *non-μ-replacing position* $p \in \mathcal{P}os(t) \setminus \mathcal{P}os^{\mu}(t)$ such that $s = t|_p$. In CSR, we (only) contract *μ-replacing redexes*: t μ-rewrites to s, written $t \hookrightarrow_{\mathcal{R},\mu} s$ (or $t \stackrel{p}{\hookrightarrow}_{\mathcal{R},\mu} s$ to make position p explicit), iff there are $\ell \to r \in R$, $p \in \mathcal{P}os^{\mu}(t)$ and a substitution σ such that $t|_p = \sigma(\ell)$ and $s = t[\sigma(r)]_p$; $t \stackrel{>q}{\hookrightarrow}_{\mathcal{R},\mu} s$ means that the μ-rewrite step is applied below position q, i.e., $p > q$. We say that a variable x is *migrating* in $\ell \to r \in R$ if $x \in \mathcal{V}ar^{\mu}(r) \setminus \mathcal{V}ar^{\mu}(\ell)$. A term t is *μ-terminating* if there is no infinite μ-rewrite sequence $t = t_1 \hookrightarrow_{\mathcal{R},\mu} t_2 \hookrightarrow_{\mathcal{R},\mu} \cdots \hookrightarrow_{\mathcal{R},\mu} t_n \hookrightarrow_{\mathcal{R},\mu} \cdots$ starting from t. A TRS $\mathcal{R} = (\mathcal{F}, R)$ is *μ-terminating* if $\hookrightarrow_{\mathcal{R},\mu}$ is terminating. A pair $(\mathcal{R}, \mu)$ where $\mathcal{R}$ is a TRS and $\mu \in M_{\mathcal{R}}$ is often called a *CS-TRS*.

3 Infinite μ-Rewrite Sequences

Let $\mathcal{M}_{\infty,\mu}$ be a set of *minimal non-μ-terminating terms* in the following sense: t belongs to $\mathcal{M}_{\infty,\mu}$ if t is non-μ-terminating and every strict *μ-replacing* subterm s of t (i.e., $t \rhd_{\mu} s$) is μ-terminating [7]. Minimal terms allow us to characterize infinite μ-rewrite sequences [9]. In [9], we show that if we have migrating variables x that "unhide" infinite computations starting from terms u which are introduced by the binding $\sigma(x)$ of the variable, then we can obtain information about the "incoming" term u if this term does not occur in the initial term of the sequence. In order to formalize this, we need a restricted notion of minimality.

Definition 1 (Strongly Minimal Terms [9]). *Let $\mathcal{T}_{\infty,\mu}$ be a set of* strongly minimal non-μ-terminating terms *in the following sense: t belongs to $\mathcal{T}_{\infty,\mu}$ if t is non-μ-terminating and every strict subterm u (i.e., $t \rhd u$) is μ-terminating. It is obvious that* $\mathsf{root}(t) \in \mathcal{D}$ *for all $t \in \mathcal{T}_{\infty,\mu}$.*

Every non-μ-terminating term has a subterm that is strongly minimal. Then, given a non-μ-terminating term t we can always find a subterm $t_0 \in \mathcal{T}_{\infty,\mu}$ of t which starts a *minimal infinite μ-rewrite sequence* of the form $t_0 \stackrel{>\Lambda}{\hookrightarrow}^*_{\mathcal{R},\mu} \sigma_1(\ell_1) \stackrel{\Lambda}{\hookrightarrow}_{\mathcal{R},\mu} \sigma_1(r_1) \trianglerighteq_{\mu} t_1 \stackrel{>\Lambda}{\hookrightarrow}^*_{\mathcal{R},\mu} \sigma_2(\ell_2) \stackrel{\Lambda}{\hookrightarrow}_{\mathcal{R},\mu} \sigma_2(r_2) \trianglerighteq_{\mu} t_2 \stackrel{>\Lambda}{\hookrightarrow}^*_{\mathcal{R},\mu} \cdots$ where $t_i, \sigma_i(\ell_i) \in \mathcal{M}_{\infty,\mu}$ for all $i > 0$ [9]. Theorem 1 below tells us that we have two possibilities:

- The minimal non-μ-terminating terms $t_i \in \mathcal{M}_{\infty,\mu}$ in the sequence are partially introduced by a μ-replacing nonvariable subterm of the right-hand sides r_i of the rules $\ell_i \to r_i$.
- The minimal non-μ-terminating terms $t_i \in \mathcal{M}_{\infty,\mu}$ in the sequence are introduced by instantiated *migrating variables* x_i of (the respective) rules $\ell_i \to r_i$, i.e., $x_i \in \mathcal{V}ar^{\mu}(r_i) \setminus \mathcal{V}ar^{\mu}(\ell_i)$. Then, t_i is partially introduced by terms occurring at non-μ-replacing positions in the right-hand sides of the rules (*hidden terms*) within a given (*hiding*) context.

We use the following functions [7,9]: $\textsc{Ren}^{\mu}(t)$, which *independently* renames all *occurrences* of μ-replacing variables by using new fresh variables which are not in $\mathcal{V}ar(t)$, and $\text{N}\text{ARR}^{\mu}_{\mathcal{R}}(t)$, which indicates whether t is μ-narrowable[2] (w.r.t. the intended TRS $\mathcal{R}$).

A nonvariable term $t \in \mathcal{T}(\mathcal{F},\mathcal{X}) \setminus \mathcal{X}$ is a *hidden term* [6,9] if there is a rule $\ell \to r \in R$ such that t is a non-μ-replacing subterm of r. In the following, $\mathcal{HT}(\mathcal{R},\mu)$ is the set of all hidden terms in $(\mathcal{R},\mu)$ and $\mathcal{NHT}(\mathcal{R},\mu)$ the set of μ-narrowable hidden terms headed by a defined symbol:

$$\mathcal{NHT}(\mathcal{R},\mu) = \{t \in \mathcal{HT}(\mathcal{R},\mu) \mid \mathsf{root}(t) \in \mathcal{D} \text{ and } \text{NARR}^{\mu}_{\mathcal{R}}(\text{REN}^{\mu}(t))\}$$

Definition 2 (Hiding Context). *Let $\mathcal{R}$ be a TRS and $\mu \in M_{\mathcal{R}}$. A function symbol* f *hides position* i *in the rule* $\ell \to r \in \mathcal{R}$ *if* $r \rhd_{\not\mu} \mathsf{f}(r_1,\ldots,r_n)$ *for some terms* $r_1,\ldots,r_n$*, and there is* $i \in \mu(\mathsf{f})$ *such that* r_i *contains a μ-replacing defined symbol (i.e.,* $\mathcal{P}os^{\mu}_{\mathcal{D}}(r_i) \neq \varnothing$*) or a variable* $x \in (\mathcal{V}ar^{\not\mu}(\ell) \cap \mathcal{V}ar^{\not\mu}(r)) \setminus (\mathcal{V}ar^{\mu}(\ell) \cup \mathcal{V}ar^{\mu}(r))$ *which is μ-replacing in* r_i *(i.e.,* $x \in \mathcal{V}ar^{\mu}(r_i)$*). A context* $C[\Box]$ *is* hiding *[6] if* $C[\Box] = \Box$*, or* $C[\Box] = \mathsf{f}(t_1,\ldots,t_{i-1},C'[\Box],t_{i+1},\ldots,t_k)$*, where* f *hides position* i *and* $C'[\Box]$ *is a hiding context.*

Definition 2 is a refinement of [6, Definition 7], where the new condition $x \in (\mathcal{V}ar^{\not\mu}(\ell) \cap \mathcal{V}ar^{\not\mu}(r)) \setminus (\mathcal{V}ar^{\mu}(\ell) \cup \mathcal{V}ar^{\mu}(r))$ is useful to discard contexts that are not valid when minimality is considered.

Example 3. The hidden terms in Example 1 are $\mathsf{minus}(\mathsf{p}(x),\mathsf{p}(y))$, $\mathsf{p}(x)$ and $\mathsf{p}(y)$. Symbol minus hides positions 1 and 2, but p hides no position. Without the new condition in Definition 2, p would hide position 1.

These notions are used and combined to model infinite context-sensitive rewrite sequences starting from strongly minimal non-μ-terminating terms as follows.

Theorem 1 (Minimal Sequence). *Let $\mathcal{R}$ be a TRS and $\mu \in M_{\mathcal{R}}$. For all* $t \in \mathcal{T}_{\infty,\mu}$*, there is an infinite sequence*

$$t = t_0 \overset{>\Lambda}{\hookrightarrow}^*_{\mathcal{R},\mu} \sigma_1(\ell_1) \overset{\Lambda}{\hookrightarrow}_{\mathcal{R},\mu} \sigma_1(r_1) \trianglerighteq_{\mu} t_1 \overset{>\Lambda}{\hookrightarrow}^*_{\mathcal{R},\mu} \sigma_2(\ell_2) \overset{\Lambda}{\hookrightarrow}_{\mathcal{R},\mu} \cdots$$

where, for all $i \geq 1$, $\ell_i \to r_i \in R$ *are rewrite rules,* σ_i *are substitutions, and terms* $t_i \in \mathcal{M}_{\infty,\mu}$ *are minimal non-μ-terminating terms such that either*

1. $t_i = \sigma_i(s_i)$ *for some nonvariable term* s_i *such that* $r_i \trianglerighteq_{\mu} s_i$*, or*
2. $\sigma_i(x_i) = \theta_i(C_i[t'_i])$ *and* $t_i = \theta_i(t'_i)$ *for some variable* $x_i \in \mathcal{V}ar^{\mu}(r_i) \setminus \mathcal{V}ar^{\mu}(\ell_i)$*,* $t'_i \in \mathcal{NHT}(\mathcal{R},\mu)$*, hiding context* $C_i[\Box]$*, and substitution* θ_i*.*

4 Chains of Context-Sensitive Dependency Pairs

In this section, we revise the definition of chain of context-sensitive dependency pairs given in [9]. First, we recall the notion of context-sensitive dependency pair.

[2] A term s *μ-narrows* to the term t if there is a nonvariable position $p \in \mathcal{P}os^{\mu}_{\mathcal{F}}(s)$ and a rule $\ell \to r$ such that $s|_p$ and ℓ *unify* with *mgu* σ, and $t = \sigma(s[r]_p)$.

Definition 3 (Context-Sensitive Dependency Pairs [9]). *Let $\mathcal{R} = (\mathcal{F}, R) = (\mathcal{C} \uplus \mathcal{D}, R)$ be a TRS and $\mu \in M_{\mathcal{F}}$. We define $\mathsf{DP}(\mathcal{R},\mu) = \mathsf{DP}_{\mathcal{F}}(\mathcal{R},\mu) \cup \mathsf{DP}_{\mathcal{X}}(\mathcal{R},\mu)$ to be set of* context-sensitive dependency pairs *(CSDPs) where:*

$$\begin{aligned}\mathsf{DP}_{\mathcal{F}}(\mathcal{R},\mu) &= \{\ell^\sharp \to s^\sharp \mid \ell \to r \in R, r \trianglerighteq_\mu s, \mathsf{root}(s) \in \mathcal{D}, \ell \not\trianglerighteq_\mu s, \textsc{Narr}^{\mu}_{\mathcal{R}}(\textsc{Ren}^{\mu}(s))\}\\ \mathsf{DP}_{\mathcal{X}}(\mathcal{R},\mu) &= \{\ell^\sharp \to x \mid \ell \to r \in R, x \in \mathcal{V}ar^{\mu}(r) \setminus \mathcal{V}ar^{\mu}(\ell)\}\end{aligned}$$

We extend $\mu \in M_{\mathcal{F}}$ into $\mu^\sharp \in M_{\mathcal{F}\cup\mathcal{D}^\sharp}$ by $\mu^\sharp(\mathsf{f}) = \mu(\mathsf{f})$ if $\mathsf{f} \in \mathcal{F}$ and $\mu^\sharp(\mathsf{f}^\sharp) = \mu(\mathsf{f})$ if $\mathsf{f} \in \mathcal{D}$.

Now, we provide a new notion of *chain* of CSDPs. In contrast to [6], we store the information about hidden terms and hiding contexts which is relevant to model infinite minimal μ-rewrite sequences as a new *unhiding TRS* instead of introducing them as new (transformed) pairs.

Definition 4 (Unhiding TRS). *Let $\mathcal{R}$ be a TRS and $\mu \in M_{\mathcal{R}}$. We define $\mathsf{unh}(\mathcal{R},\mu)$ as the TRS consisting of the following rules:*

1. *$\mathsf{f}(x_1,\ldots,x_i,\ldots,x_k) \to x_i$ for all function symbols f of arity k, distinct variables $x_1,\ldots,x_k$, and $1 \leq i \leq k$ such that f* hides position *i in $\ell \to r \in R$, and*
2. *$t \to t^\sharp$ for every $t \in \mathcal{NHT}(\mathcal{R},\mu)$.*

Example 4. The unhiding TRS $\mathsf{unh}(\mathcal{R},\mu)$ for $\mathcal{R}$ and μ in Example 1 is:

$$\begin{aligned}\mathsf{minus}(\mathsf{p}(x),\mathsf{p}(y)) &\to \mathsf{M}(\mathsf{p}(x),\mathsf{p}(y)) \quad (16) \qquad & \mathsf{minus}(x,y) &\to y \quad (18)\\ \mathsf{p}(x) &\to \mathsf{P}(x) \quad (17) & \mathsf{minus}(x,y) &\to x \quad (19)\end{aligned}$$

Definitions 3 and 4 lead to a suitable notion of *chain* which captures minimal infinite μ-rewrite sequences according to the description in Theorem 1. In the following, given a TRS $\mathcal{S}$, we let $\mathcal{S}_{\rhd_\mu}$ be the rules from $\mathcal{S}$ of the form $s \to t \in \mathcal{S}$ and $s \rhd_\mu t$; and $\mathcal{S}_\sharp = \mathcal{S} \setminus \mathcal{S}_{\rhd_\mu}$.

Definition 5 (Chain of Pairs - Minimal Chain). *Let $\mathcal{R}$, $\mathcal{P}$ and $\mathcal{S}$ be TRSs and $\mu \in M_{\mathcal{R}\cup\mathcal{P}\cup\mathcal{S}}$. A $(\mathcal{P},\mathcal{R},\mathcal{S},\mu)$-chain is a finite or infinite sequence of pairs $u_i \to v_i \in \mathcal{P}$, together with a substitution σ satisfying that, for all $i \geq 1$,*

1. *if $v_i \notin \mathcal{V}ar(u_i) \setminus \mathcal{V}ar^{\mu}(u_i)$, then $\sigma(v_i) = t_i \hookrightarrow^*_{\mathcal{R},\mu} \sigma(u_{i+1})$, and*
2. *if $v_i \in \mathcal{V}ar(u_i) \setminus \mathcal{V}ar^{\mu}(u_i)$, then $\sigma(v_i) \overset{\Lambda}{\hookrightarrow}{}^*_{\mathcal{S}_{\rhd_\mu},\mu} \circ \overset{\Lambda}{\hookrightarrow}_{\mathcal{S}_\sharp,\mu} t_i \hookrightarrow^*_{\mathcal{R},\mu} \sigma(u_{i+1})$.*

A $(\mathcal{P},\mathcal{R},\mathcal{S},\mu)$-chain is called minimal *if for all $i \geq 1$, t_i is $(\mathcal{R},\mu)$-terminating.*

Notice that if rules $\mathsf{f}(x_1,\ldots,x_k) \to x_i$ for all $\mathsf{f} \in \mathcal{D}$ and $i \in \mu(\mathsf{f})$ (where $x_1,\ldots,x_k$ are variables) are used in Item 1 of Definition 4, then Definition 5 yields the notion of chain in [9]; and if, additionally, rules $\mathsf{f}(x_1,\ldots,x_k) \to \mathsf{f}^\sharp(x_1,\ldots,x_k)$ for all $\mathsf{f} \in \mathcal{D}$ are used in Item 2 of Definition 4, then we have the original notion of chain in [7]. Thus, the new definition covers all previous ones.

Theorem 2 (Soundness and Completeness of CSDPs). *Let $\mathcal{R}$ be a TRS and $\mu \in M_{\mathcal{R}}$. A CS-TRS $(\mathcal{R},\mu)$ is terminating if and only if there is no infinite $(\mathsf{DP}(\mathcal{R},\mu),\mathcal{R},\mathsf{unh}(\mathcal{R},\mu),\mu^\sharp)$-chain.*

5 Context-Sensitive Dependency Pair Framework

In the DP framework [12], proofs of termination are handled as *termination problems* involving two TRSs $\mathcal{P}$ and $\mathcal{R}$ instead of just the 'target' TRS $\mathcal{R}$. In our setting we start with the following definition (see also [6,9]).

Definition 6 (CS Problem and CS Processor). *A* CS problem τ *is a tuple* $\tau = (\mathcal{P}, \mathcal{R}, \mathcal{S}, \mu)$*, where* $\mathcal{R}$*,* $\mathcal{P}$ *and* $\mathcal{S}$ *are TRSs, and* $\mu \in M_{\mathcal{R}\cup\mathcal{P}\cup\mathcal{S}}$*. The* CS *problem* $(\mathcal{P}, \mathcal{R}, \mathcal{S}, \mu)$ *is* finite *if there is no infinite* $(\mathcal{P}, \mathcal{R}, \mathcal{S}, \mu)$*-chain. The* CS *problem* $(\mathcal{P}, \mathcal{R}, \mathcal{S}, \mu)$ *is* infinite *if* $\mathcal{R}$ *is non-*μ*-terminating or there is an infinite minimal* $(\mathcal{P}, \mathcal{R}, \mathcal{S}, \mu)$*-chain.*

A CS *processor* Proc *is a mapping from CS problems into sets of CS problems. Alternatively, it can also return "*no*". A CS processor* Proc *is* sound *if for all CS problems* τ*,* τ *is finite whenever* $\mathsf{Proc}(\tau) \neq \mathsf{no}$ *and* $\forall \tau' \in \mathsf{Proc}(\tau)$*,* τ' *is finite. A CS processor* Proc *is* complete *if for all CS problems* τ*,* τ *is infinite whenever* $\mathsf{Proc}(\tau) = \mathsf{no}$ *or* $\exists \tau' \in \mathsf{Proc}(\tau)$ *such that* τ' *is infinite.*

In order to prove the μ-termination of a TRS $\mathcal{R}$, we adapt the result from [12] to CSR.

Theorem 3 (CSDP Framework). *Let* $\mathcal{R}$ *be a TRS and* $\mu \in M_{\mathcal{R}}$*. We construct a tree whose nodes are labeled with CS problems or "*yes*" or "*no*", and whose root is labeled with* $(\mathsf{DP}(\mathcal{R}, \mu), \mathcal{R}, \mathsf{unh}(\mathcal{R}, \mu), \mu^\sharp)$*. For every inner node labeled with* τ*, there is a sound processor* Proc *satisfying one of the following conditions:*

1. $\mathsf{Proc}(\tau) = \mathsf{no}$ *and the node has just one child, labeled with "*no*".*
2. $\mathsf{Proc}(\tau) = \varnothing$ *and the node has just one child, labeled with "*yes*".*
3. $\mathsf{Proc}(\tau) \neq \mathsf{no}$*,* $\mathsf{Proc}(\tau) \neq \varnothing$*, and the children of the node are labeled with the CS problems in* $\mathsf{Proc}(\tau)$*.*

If all leaves of the tree are labeled with "yes*", then* $\mathcal{R}$ *is* μ*-terminating. Otherwise, if there is a leaf labeled with "*no*" and if all processors used on the path from the root to this leaf are complete, then* $\mathcal{R}$ *is non-*μ*-terminating.*

In the following subsections we describe a number of sound and complete CS processors.

5.1 Collapsing Pair Processors

The following processor integrates the transformation of [6] into our framework. The pairs in a CS-TRS $(\mathcal{P}, \mu)$, where $\mathcal{P} = (\mathcal{G}, P)$, are partitioned as follows: $P_{\mathcal{X}} = \{u \to v \in P \mid v \in \mathcal{V}ar(u) \setminus \mathcal{V}ar^{\mu}(u)\}$ and $P_{\mathcal{G}} = P \setminus P_{\mathcal{X}}$.

Theorem 4 (Collapsing Pair Transformation). *Let* $\tau = (\mathcal{P}, \mathcal{R}, \mathcal{S}, \mu)$ *be a CS problem where* $\mathcal{P} = (\mathcal{G}, P)$ *and* P_{U} *be given by the following rules:*

- $u \to \mathsf{U}(x)$ *for every* $u \to x \in \mathcal{P}_{\mathcal{X}}$*,*
- $\mathsf{U}(s) \to \mathsf{U}(t)$ *for every* $s \to t \in \mathcal{S}_{\rhd_\mu}$*, and*
- $\mathsf{U}(s) \to t$ *for every* $s \to t \in \mathcal{S}_{\sharp}$*.*

Here, U *is a new fresh symbol. Let* $\mathcal{P}' = (\mathcal{G} \cup \{\mathsf{U}\}, P')$ *where* $P' = (P \setminus P_{\mathcal{X}}) \cup P_{\mathsf{U}}$*, and* μ' *extends* μ *by* $\mu'(\mathsf{U}) = \varnothing$*. The processor* Proc_{eColl} *given by* $\mathsf{Proc}_{eColl}(\tau) = \{(\mathcal{P}', \mathcal{R}, \varnothing, \mu')\}$ *is sound and complete.*

Now, we can apply all CS processors from [6] and [9] which did not consider any $\mathcal{S}$ component in CS problems.

In our framework, we can also apply specific processors for collapsing pairs that are very useful, but these only are used if we have collapsing pairs in the chains (as in [9]). For instance, we can use the processor in Theorem 5 below, which is often applied in proofs of termination of CSR with MU-TERM [13,14]. The subTRS of $\mathcal{P}_{\mathcal{X}}$ containing the rules whose migrating variables occur on non-μ-replacing immediate subterms in the left-hand side is $\mathcal{P}^1_{\mathcal{X}} = \{\mathsf{f}(u_1, \ldots, u_k) \to x \in \mathcal{P}_{\mathcal{X}} \mid \exists i, 1 \leq i \leq k, i \notin \mu(\mathsf{f}), x \in \mathcal{V}ar(u_i)\}$.

Theorem 5 (Basic CS Processor for Collapsing Pairs). *Let* $\tau = (\mathcal{P}, \mathcal{R}, \mathcal{S}, \mu)$ *be a CS problem where* $\mathcal{R} = (\mathcal{C} \uplus \mathcal{D}, R)$ *and* $\mathcal{S} = (\mathcal{H}, S)$*. Assume that (1) all the rules in* $\mathcal{S}_\sharp$ *are noncollapsing, i.e., for all* $s \to t \in \mathcal{S}_\sharp$*,* $t \notin \mathcal{X}$ *(2)* $\{\mathsf{root}(t) \mid s \to t \in \mathcal{S}_\sharp\} \cap \mathcal{D} = \varnothing$ *and (3) for all* $s \to t \in \mathcal{S}_\sharp$*, we have that* $s = \mathsf{f}(s_1, \ldots, s_k)$ *and* $t = \mathsf{g}(s_1, \ldots, s_k)$ *for some* $k \in \mathbb{N}$*, funtion symbols* $\mathsf{f}, \mathsf{g} \in \mathcal{H}$*, and terms* $s_1, \ldots, s_k$*. Then, the processors* Proc_{Coll1} *given by*

$$\mathsf{Proc}_{Coll1}(\tau) = \begin{cases} \varnothing & \text{if } \mathcal{P} = \mathcal{P}^1_{\mathcal{X}} \text{ and} \\ \{(\mathcal{P}, \mathcal{R}, \mathcal{S}, \mu)\} & \text{otherwise} \end{cases}$$

is sound and complete.

Example 5. (Continuing Example 1) Consider the CS problem $\tau = (\mathcal{P}_4, \mathcal{R}, \mathcal{S}_3, \mu^\sharp)$ *where* $\mathcal{P}_4 = \{(14), (15)\}$ *and* $\mathcal{S}_3 = \{(16), (18), (19)\}$*. We can apply* $\mathsf{Proc}_{Coll1}(\tau)$ *to conclude that the CS problem* τ *is finite.*

5.2 Context-Sensitive Dependency Graph

In the DP-approach [8,12], a *dependency graph* is associated to the TRS $\mathcal{R}$. The nodes of the graph are the dependency pairs in $\mathsf{DP}(\mathcal{R})$ and there is an arc from a dependency pair $u \to v$ to a dependency pair $u' \to v'$ if there are substitutions θ and θ' such that $\theta(v) \to^*_{\mathcal{R}} \theta'(u')$. In our setting, we have the following.

Definition 7 (Context-Sensitive Graph of Pairs). *Let* $\mathcal{R}, \mathcal{P}$ *and* $\mathcal{S}$ *be TRSs and* $\mu \in M_{\mathcal{R} \cup \mathcal{P} \cup \mathcal{S}}$*. The* context-sensitive (CS) graph $\mathsf{G}(\mathcal{P}, \mathcal{R}, \mathcal{S}, \mu)$ *has* $\mathcal{P}$ *as the set of nodes. Given* $u \to v, u' \to v' \in \mathcal{P}$*, there is an arc from* $u \to v$ *to* $u' \to v'$ *if* $u \to v, u' \to v'$ *is a minimal* $(\mathcal{P}, \mathcal{R}, \mathcal{S}, \mu)$*-chain for some substitution* σ*.*

In termination proofs, we are concerned with the so-called *strongly connected components* (SCCs) of the dependency graph, rather than with the cycles themselves (which are exponentially many) [15]. The following result formalizes the use of SCCs for dealing with CS problems.

Theorem 6 (SCC Processor). *Let* $\tau = (\mathcal{P}, \mathcal{R}, \mathcal{S}, \mu)$ *be a CS problem. Then, the processor* Proc_{SCC} *given by*

$$\mathsf{Proc}_{SCC}(\mathcal{P}, \mathcal{R}, \mathcal{S}, \mu) = \{(\mathcal{Q}, \mathcal{R}, \mathcal{S}_{\mathcal{Q}}, \mu) \mid \mathcal{Q} \text{ are the pairs of an SCC in } \mathsf{G}(\mathcal{P}, \mathcal{R}, \mathcal{S}, \mu)\}$$

(where $\mathcal{S}_{\mathcal{Q}}$ are the rules from $\mathcal{S}$ involving a possible $(\mathcal{Q}, \mathcal{R}, \mathcal{S}, \mu)$-chain) is sound and complete.

The CS graph is not computable. Thus, we have to use an over-approximation of it. In the following definition, we use the function $\text{TCAP}^{\mu}_{\mathcal{R}}(t)$, which renames all subterms headed by a 'defined' symbol in $\mathcal{R}$ by new fresh variables if it can be rewritten:

Definition 8 ($\text{TCAP}^{\mu}_{\mathcal{R}}$ [9]). *Given a TRS $\mathcal{R}$ and a replacement map μ, we let $\text{TCAP}^{\mu}_{\mathcal{R}}$ be as follows:*

$$\text{TCAP}^{\mu}_{\mathcal{R}}(x) = y \quad \text{if } x \text{ is a variable, and}$$

$$\text{TCAP}^{\mu}_{\mathcal{R}}(\mathsf{f}(t_1, \ldots, t_k)) = \begin{cases} \mathsf{f}([t_1]^{\mathsf{f}}_1, \ldots, [t_k]^{\mathsf{f}}_k) & \text{if } \mathsf{f}([t_1]^{\mathsf{f}}_1, \ldots, [t_k]^{\mathsf{f}}_k) \text{ does not unify} \\ & \text{with } \ell \text{ for any } \ell \to r \text{ in } \mathcal{R} \\ y & \text{otherwise} \end{cases}$$

where y is a new fresh variable, $[s]^{\mathsf{f}}_i = \text{TCAP}^{\mu}_{\mathcal{R}}(s)$ if $i \in \mu(\mathsf{f})$ and $[s]^{\mathsf{f}}_i = s$ if $i \notin \mu(\mathsf{f})$. We assume that ℓ shares no variable with $\mathsf{f}([t_1]^{\mathsf{f}}_1, \ldots, [t_k]^{\mathsf{f}}_k)$ when the unification is attempted.

Definition 9 (Estimated CS Graph of Pairs). *Let $\tau = (\mathcal{P}, \mathcal{R}, \mathcal{S}, \mu)$ be a CS problem. The* estimated CS graph *associated to $\mathcal{R}$, $\mathcal{P}$ and $\mathcal{S}$ (denoted $\mathsf{EG}(\mathcal{P}, \mathcal{R}, \mathcal{S}, \mu)$) has $\mathcal{P}$ as the set of nodes and arcs which connect them as follows:*

1. *there is an arc from $u \to v \in \mathcal{P}_{\mathcal{G}}$ to $u' \to v' \in \mathcal{P}$ if $\text{TCAP}^{\mu}_{\mathcal{R}}(v)$ and u' unify, and*
2. *there is an arc from $u \to v \in \mathcal{P}_{\mathcal{X}}$ to $u' \to v' \in \mathcal{P}$ if there is $s \to t \in \mathcal{S}_{\sharp}$ such that $\text{TCAP}^{\mu}_{\mathcal{R}}(t)$ and u' unify.*

We have the following.

Theorem 7 (Approximation of the CS Graph). *Let $\mathcal{R}$, $\mathcal{P}$ and $\mathcal{S}$ be TRSs and $\mu \in M_{\mathcal{R} \cup \mathcal{P} \cup \mathcal{S}}$. The* estimated *CS graph $\mathsf{EG}(\mathcal{P}, \mathcal{R}, \mathcal{S}, \mu)$ contains the CS graph $\mathsf{G}(\mathcal{P}, \mathcal{R}, \mathcal{S}, \mu)$.*

We also provide a computable definition of the SCC processor in Theorem 8.

Theorem 8 (SCC Processor using $\text{TCAP}^{\mu}_{\mathcal{R}}$). *Let $\tau = (\mathcal{P}, \mathcal{R}, \mathcal{S}, \mu)$ be a CS problem. The CS processor Proc_{SCC} given by*

$$\mathsf{Proc}_{SCC}(\tau) = \{(\mathcal{Q}, \mathcal{R}, \mathcal{S}_{\mathcal{Q}}, \mu) \mid \mathcal{Q} \text{ contains the pairs of an SCC in } \mathsf{EG}(\mathcal{P}, \mathcal{R}, \mathcal{S}, \mu)\}$$

where

- $\mathcal{S}_{\mathcal{Q}} = \varnothing$ *if* $\mathcal{Q}_{\mathcal{X}} = \varnothing$.
- $\mathcal{S}_{\mathcal{Q}} = \mathcal{S}_{\rhd_\mu} \cup \{s \to t \mid s \to t \in \mathcal{S}_{\sharp}, \text{TCAP}^{\mu}_{\mathcal{R}}(t)$ *and u' unify for some* $u' \to v' \in \mathcal{Q}\}$ *if* $\mathcal{Q}_{\mathcal{X}} \neq \varnothing$.

is sound and complete.

Example 6. In Figure 1 (right) we show $\mathsf{EG}(\mathsf{DP}(\mathcal{R}, \mu), \mathcal{R}, \mathsf{unh}(\mathcal{R}, \mu), \mu^{\sharp})$ for $\mathcal{R}$ in Example 1. The graph has three SCCs $\mathcal{P}_1 = \{(1)\}$, $\mathcal{P}_2 = \{(8)\}$, and $\mathcal{P}_3 = \{(7), (14), (15)\}$. If we apply the CS processor Proc_{SCC} to the initial CS problem $(\mathsf{DP}(\mathcal{R}, \mu), \mathcal{R}, \mathsf{unh}(\mathcal{R}, \mu), \mu^{\sharp})$ for $(\mathcal{R}, \mu)$ in Example 1, then we obtain the problems: $(\mathcal{P}_1, \mathcal{R}, \varnothing, \mu^{\sharp})$, $(\mathcal{P}_2, \mathcal{R}, \varnothing, \mu^{\sharp})$, $(\mathcal{P}_3, \mathcal{R}, \mathcal{S}_3, \mu^{\sharp})$, where $\mathcal{S}_3 = \{(16), (18), (19)\}$.

5.3 Reduction Triple Processor

A *μ-reduction pair* $(\gtrsim, \sqsupset)$ consists of a stable and μ-monotonic[3] quasi-ordering $\gtrsim$, and a well-founded stable relation $\sqsupset$ on terms in $\mathcal{T}(\mathcal{F}, \mathcal{X})$ which are compatible, i.e., $\gtrsim \circ \sqsupset \subseteq \sqsupset$ or $\sqsupset \circ \gtrsim \subseteq \sqsupset$ [7].

In [7,9], when a collapsing pair $u \to x$ occurs in a chain, we have to *look inside* the instantiated right-hand side $\sigma(x)$ for a μ-replacing subterm that, after marking it, does μ-rewrite to the (instantiated) left-hand side of another pair. For this reason, the quasi-orderings $\gtrsim$ of reduction pairs $(\gtrsim, \sqsupset)$ which are used in [7,9] are required to have the μ-subterm property, i.e. $\trianglerighteq_\mu \subseteq \gtrsim$. This is equivalent to impose $\mathsf{f}(x_1, \ldots, x_k) \gtrsim x_i$ for all projection rules $\mathsf{f}(x_1, \ldots, x_k) \to x_i$ with $\mathsf{f} \in \mathcal{F}$ and $i \in \mu(\mathsf{f})$. This is similar for markings: in [7] we have to ensure that $\mathsf{f}(x_1, \ldots, x_k) \gtrsim \mathsf{f}^\sharp(x_1, \ldots, x_k)$ for all defined symbols f in the signature. In [9], thanks to the notion of hidden term, we relaxed the last condition: we require $t \gtrsim t^\sharp$ for all (narrowable) *hidden terms* t. In [6], thanks to the notion of hiding context, we only require that $\gtrsim$ is compatible with the projections $\mathsf{f}(x_1, \ldots, x_k) \to x_i$ for those symbols f and positions i such that f *hides position* i. However, this information is implicitly encoded as (new) pairs $\mathsf{U}(\mathsf{f}(x_1, \ldots, x_k)) \to \mathsf{U}(x_i)$ in the set $\mathcal{P}$. The strict component $\sqsupset$ of the reduction pair $(\gtrsim, \sqsupset)$ is used with these new pairs now.

In this paper, since the rules in $\mathcal{S}$ are not considered as ordinary pairs (in the sense of [6,9]) we can relax the conditions imposed to the orderings dealing with these rules. Furthermore, since rules in $\mathcal{S}$ are applied only once to the root of the terms, we only have to impose stability to the relation which is compatible with these rules (no transitivity, reflexivity, well-foundedness or μ-monotonicity is required).

Therefore, we can use *μ-reduction triples* $(\gtrsim, \sqsupset, \succeq)$ now, where $(\gtrsim, \sqsupset)$ is a μ-reduction pair and $\succeq$ is a stable relation on terms which is compatible with $\gtrsim$ or $\sqsupset$, i.e., $\succeq \circ \gtrsim \subseteq \gtrsim$ or $\sqsupset \circ \succeq \subseteq \sqsupset$.

Theorem 9 (μ-Reduction Triple Processor). *Let $\tau = (\mathcal{P}, \mathcal{R}, \mathcal{S}, \mu)$ be a CS problem. Let $(\gtrsim, \sqsupset, \succeq)$ be a μ-reduction triple such that*

1. *$\mathcal{P} \subseteq \gtrsim \cup \sqsupset$, $\mathcal{R} \subseteq \gtrsim$, and*
2. *whenever $\mathcal{P}_\mathcal{X} \neq \varnothing$ we have that $\mathcal{S} \subseteq \gtrsim \cup \sqsupset \cup \succeq$.*

Let $\mathcal{P}_\sqsupset = \{u \to v \in \mathcal{P} \mid u \sqsupset v\}$ and $\mathcal{S}_\sqsupset = \{s \to t \in \mathcal{S} \mid s \sqsupset t\}$. Then, the processor Proc_{RT} given by

$$\mathsf{Proc}_{RT}(\tau) = \begin{cases} \{(\mathcal{P} \setminus \mathcal{P}_\sqsupset, \mathcal{R}, \mathcal{S} \setminus \mathcal{S}_\sqsupset, \mu)\} & \textit{if (1) and (2) hold} \\ \{(\mathcal{P}, \mathcal{R}, \mathcal{S}, \mu)\} & \textit{otherwise} \end{cases}$$

is sound and complete.

---

[3] A binary relation R on terms is μ-monotonic if for all terms $s, t, t_1, \ldots, t_k$, and k-ary symbols f, whenever $s \mathrel{\mathsf{R}} t$ and $i \in \mu(\mathsf{f})$ we have $\mathsf{f}(t_1, \ldots, t_{i-1}, s, \ldots, t_k) \mathrel{\mathsf{R}} \mathsf{f}(t_1, \ldots, t_{i-1}, t, \ldots, t_k)$.

Since rules from $\mathcal{S}$ are only applied after using a collapsing pair, we only need to make them compatible with some component of the triple if $\mathcal{P}$ contains collapsing pairs, i.e., if $\mathcal{P}_{\mathcal{X}} \neq \varnothing$. Another advantage is that we can now *remove rules from* $\mathcal{S}$. Furthermore, we can increase the power of this definition by considering the *usable rules* corresponding to $\mathcal{P}$, instead of $\mathcal{R}$ as a whole (see [6,16]), and also by using argument filterings [9].

Example 7. (Continuing Example 6) Consider the CS problem $\tau = (\mathcal{P}_3, \mathcal{R}, \mathcal{S}_3, \mu^\sharp)$ *where* $\mathcal{P}_3 = \{(7), (14), (15)\}$, $\mathcal{S}_3 = \{(16), (18), (19)\}$ *and* $\mathcal{R}$ *is the TRS in Example 1. If we apply* Proc_{RT} *to the CS problem* τ *by using the* μ*-reduction triple* $(\geq, >, \geq)$ *where* $\geq$ *and* $>$ *are the orderings induced by the following polynomial interpretation (see [17] for missing notation and definitions):*

$$
\begin{array}{llll}
[\mathsf{if}](x,y,z) = (1/2 \times x) + y + z & & [\mathsf{minus}](x,y) = (2 \times x) + (2 \times y) + 1/2 \\
[\mathsf{p}](x) = (1/2 \times x) & & [\mathsf{0}] = 0 \\
[\mathsf{false}] = 0 & & [\mathsf{s}](x) = (2 \times x) + 2 \\
[\mathsf{true}] = 2 & & [\mathsf{gt}](x,y) = (2 \times x) + (1/2 \times y) \\
[\mathsf{M}](x,y) = (2 \times x) + (2 \times y) + 1/2 & & [\mathsf{IF}](x,y,z) = (1/2 \times x) + y + z
\end{array}
$$

then, we have $[\ell] \geq [r]$ *for all (usable) rules in* $\mathcal{R}$ *and, for the rules in* $\mathcal{P}_3$ *and* $\mathcal{S}_3$*, we have*

$$
\begin{array}{rclrcl}
[\mathsf{M}(x,y)] & \geq & [\mathsf{IF}(\mathsf{gt}(y,\mathsf{0}), \mathsf{minus}(\mathsf{p}(x),\mathsf{p}(y)), x)] & [\mathsf{minus}(\mathsf{p}(x),\mathsf{p}(y))] & \geq & [\mathsf{M}(\mathsf{p}(x),\mathsf{p}(y))] \\
[\mathsf{IF}(\mathsf{true},x,y)] & > & [x] & [\mathsf{minus}(x,y)] & > & [y] \\
[\mathsf{IF}(\mathsf{false},x,y)] & \geq & [y] & [\mathsf{minus}(x,y)] & > & [x]
\end{array}
$$

Then, we get $\mathsf{Proc}_{RT}(\tau) = \{(\{(7), (15)\}, \mathcal{R}, \{(16)\}, \mu^\sharp)\}$.

5.4 Subterm Processor

The subterm criterion was adapted to CSR in [7], but its use was restricted to noncollapsing pairs [7, Theorem 5]. In [9], a new version for collapsing pairs was defined, but in this version you can only remove all collapsing pairs and the projection π is restricted to μ-replacing positions. Our new version is fully general and able to remove collapsing and noncollapsing pairs at the same time. Furthermore, we are also able to remove rules in $\mathcal{S}$. Before introducing it, we need the following definition.

Definition 10 (Root Symbols of a TRS [9]). *Let* $\mathcal{R} = (\mathcal{F}, R)$ *be a TRS. The set of* root symbols *associated to* $\mathcal{R}$ *is:*

$$\mathsf{Root}(\mathcal{R}) = \{\mathsf{root}(\ell) \mid \ell \to r \in R\} \cup \{\mathsf{root}(r) \mid \ell \to r \in R, r \notin \mathcal{X}\}$$

Definition 11 (Simple Projection). *Let* $\mathcal{R}$ *be a TRS. A* simple projection *for* $\mathcal{R}$ *is a mapping* π *that assigns to every* k*-ary symbol* $\mathsf{f} \in \mathsf{Root}(\mathcal{R})$ *an argument position* $i \in \{1, \ldots, k\}$. *This mapping is extended to terms by*

$$\pi(t) = \begin{cases} t|_{\pi(\mathsf{f})} & \text{if } t = \mathsf{f}(t_1, \ldots, t_k) \text{ and } \mathsf{f} \in \mathsf{Root}(\mathcal{R}) \\ t & \text{otherwise} \end{cases}$$

Theorem 10 (Subterm Processor). *Let* $\tau = (\mathcal{P}, \mathcal{R}, \mathcal{S}, \mu)$ *be a CS problem where* $\mathcal{R} = (\mathcal{F}, R) = (\mathcal{C} \uplus \mathcal{D}, R)$, $\mathcal{P} = (\mathcal{G}, P)$ *and* $\mathcal{S} = (\mathcal{H}, S)$*. Assume that (1)* $\mathsf{Root}(\mathcal{P}) \cap \mathcal{D} = \varnothing$*, (2) the rules in* $\mathcal{P}_\mathcal{G} \cup \mathcal{S}_\sharp$ *are noncollapsing, (3) for all* $s_i \to t_i \in \mathcal{S}_{\rhd_\mu}$, $\mathsf{root}(s_i), \mathsf{root}(t_i) \notin \mathsf{Root}(\mathcal{P})$ *and (4) for all* $s_i \to t_i \in \mathcal{S}_\sharp$, $\mathsf{root}(s_i) \notin \mathsf{Root}(\mathcal{P})$ *and* $\mathsf{root}(t_i) \in \mathsf{Root}(\mathcal{P})$*. Let* π *be a simple projection for* $\mathcal{P}$*. Let* $\mathcal{P}_{\pi,\rhd_\mu} = \{u \to v \in \mathcal{P} \mid \pi(u) \rhd_\mu \pi(v)\}$ *and* $\mathcal{S}_{\pi,\rhd_\mu} = \{s \to t \in \mathcal{S} \mid \pi(s) \rhd_\mu \pi(t)\}$*. Then,* $\mathsf{Proc}_{subterm}$ *given by*

$$\mathsf{Proc}_{subterm}(\tau) = \begin{cases} \{(\mathcal{P} \setminus \mathcal{P}_{\pi,\rhd_\mu}, \mathcal{R}, \mathcal{S} \setminus \mathcal{S}_{\pi,\rhd_\mu}, \mu)\} & \textit{if } \pi(\mathcal{P}) \subseteq \unrhd_\mu \\ & \textit{and whenever } \mathcal{P}_\mathcal{X} \neq \varnothing, \\ & \textit{then } \pi(\mathcal{S}) \subseteq \unrhd_\mu \\ \{(\mathcal{P}, \mathcal{R}, \mathcal{S}, \mu)\} & \textit{otherwise} \end{cases}$$

is sound and complete.

Notice that the conditions in Theorem 10 are not harmful in practice because the CS problems which are obtained from CS-TRSs normally satisfy those conditions.

Example 8. (Continuing Example 7) We have the CS problem $(\mathcal{P}_5, \mathcal{R}, \mathcal{S}_5, \mu^\sharp)$ *where* $\mathcal{P}_5 = \{(7), (15)\}$ *and* $\mathcal{S}_5 = \{(16)\}$*. We can apply the subterm processor* $\mathsf{Proc}_{subterm}$ *by using the projection* $\pi(\mathsf{IF}) = 3$ *and* $\pi(\mathsf{M}) = 1$*:*

$$\begin{aligned} \pi(\mathsf{M}(x,y)) = x &\unrhd_\mu x = \pi(\mathsf{IF}(\mathsf{gt}(y,0), \mathsf{minus}(\mathsf{p}(x),\mathsf{p}(y)), x)) \\ \pi(\mathsf{IF}(\mathsf{false},x,y)) = y &\unrhd_\mu y = \pi(y) \\ \pi(\mathsf{minus}(\mathsf{p}(x),\mathsf{p}(y))) = \mathsf{minus}(\mathsf{p}(x),\mathsf{p}(y)) &\rhd_\mu \mathsf{p}(x) = \pi(\mathsf{M}(\mathsf{p}(x),\mathsf{p}(y))) \end{aligned}$$

We obtain the CS problem $\tau' = (\{(7), (15)\}, \mathcal{R}, \varnothing, \mu)$ *for which we can use* Proc_{SCC} *to conclude that there is no cycle, i.e.,* $\mathsf{Proc}_{SCC}(\tau') = \varnothing$*.*

6 Using the CSDP Framework in Maude

Proving termination of programs in sophisticated equational languages like OBJ, CafeOBJ or Maude is difficult because these programs combine different features that are not supported by state-of-the-art termination tools. For instance, the following Maude program combines the use of an evaluation strategy and types given as sorts in the specification [3].

```
fmod LengthOfFiniteLists is
   sorts Nat NatList NatIList .
   subsort NatList < NatIList .
   op 0 : -> Nat .
   op s : Nat -> Nat .
   op zeros : -> NatIList .
   op nil : -> NatList .
   op cons : Nat NatIList -> NatIList [strat (1 0)] .
   op cons : Nat NatList -> NatList [strat (1 0)] .
   op length : NatList -> Nat .
   vars M N : Nat .
```

```
  var IL : NatIList .
  var L : NatList .
  eq zeros = cons(0,zeros) .
  eq length(nil) = 0 .
  eq length(cons(N, L)) = s(length(L)) .
endfm
```

Nowadays, MU-TERM [14,13] can separately prove termination of order-sorted rewriting [18] and CSR, but it is not able to handle programs which combine both of them. Then, we use the transformation developed in [3] to transform this system into a CS-TRS (without sorts). Such a CS-TRS can be found in the Termination Problems Data Base[4] (TPDB): `TRS/CSR_Maude/LengthOfFiniteLists_complete.trs`. As far as we know, MU-TERM is the only tool that can prove termination of this system thanks to the CSDP framework presented in this paper[5].

7 Experimental Evaluation

From Friday to Saturday, December 18-19, 2009, the 2009 International Termination Competition took place and a CSR termination category was included. In the termination competition, the benchmarks are executed in a completely automatic way with a timeout of 60 seconds over a subset of 37 systems[6] of the complete collection of the 109 CS-TRSs of the TPDB 7.0.

The results in this paper have been implemented as part of the termination tool MU-TERM. Our tool MU-TERM participated in the aforementioned CSR category of the 2009 Termination Competition. The results of the competition are summarized in Table 1. Tools AProVE [19] and VMTL [20] implement the context-sensitive dependency pairs using the transformational approach in [6]. The techniques implemented by Jambox [21] to prove termination of CSR are not documented yet, to our knowledge. As showed in Table 1, we are able to prove the same number of systems than AProVE, but MU-TERM is almost two and a half times faster. Furthermore, we prove termination of 95 of the 109 examples. To our knowledge, there is no tool that can prove more than those 95 examples from this collection of problems. And, as remarked in Section 6, there are interesting examples which can be handled by MU-TERM only.

We have also executed the complete collection of systems of the CSR category[7], where we compare the 2009 and 2007 competition versions of MU-TERM. In the

[4] `http://www.lri.fr/~marche/tpdb/`

[5] On May 12, 2010, we introduced this system in the online version of AProVE `http://aprove.informatik.rwth-aachen.de/`, and a timeout occurred after 120 seconds (maximum timeout). MU-TERM proof can be found in `http://zenon.dsic.upv.es/muterm/benchmarks/benchmarks-csr/benchmarks.html`

[6] See `http://termcomp.uibk.ac.at/termcomp/competition/competitionResults.seam?category=10230&competitionId=101722&actionMethod=competition%2FcategoryList.xhtml%3AcompetitionCategories.forward&conversationPropagation=begin`

[7] A complete report of our experiments can be found in `http://zenon.dsic.upv.es/muterm/benchmarks/`

Table 1. 2009 Termination Competition Results (Context-Sensitive Rewriting)

Tool Version	Proved	Average time
AProVE	34/37	3.084s
Jambox	28/37	2.292s
MU-TERM	34/37	1.277s
VMTL	29/37	6.708s

2007 version, the CSDP framework was not available. Now, we can prove 15 more examples and, when comparing the execution times which they took over the 80 examples where both tools succeeded (84, 48 seconds vs. 15, 073 seconds), we are more than 5, 5 times faster now.

8 Related Work

In [6], a transformation of collapsing pairs into 'ordinary' (i.e., noncollapsing) pairs is introduced by using the new notion of *hiding context* [6, Definition 7]. We easily and naturally included such a transformation as a new processor Proc_{eColl} in our framework (see Theorem 4). The claimed advantage of [6] is that the notion of chain is simplified to Item 1 in Definition 5. But, although the definition of chain in [6] is apparently closer to the standard one [12, Definition 3], this does *not* mean that we can use or easily 'translate' existing DP-processors (see [12]) to be used with CSR. Besides the narrowing processor in [9, Theorem 16], the reduction pair processor with usable rules in [6, Theorem 21] is a clear example, because the avoidance of collapsing pairs does not improve the previous results about usable rules for CSR investigated in [16].

As we have seen in this paper, collapsing pairs are an essential part of the theoretical description of termination of CSR. Actually, the transformational approach in [6] *explicitly* uses them for introducing the new unhiding pairs in [6, Definition 9]. This shows that the most basic notion when *modeling* the termination behavior of CSR is that of collapsing pair and that unhiding pairs should be better considered as an ingredient for handling collapsing pairs in proofs of termination (as implemented by processor Proc_{eColl} above). Furthermore, the application of such a transformation in the very beginning of the termination analysis of CS-TRSs (as done in [6]) typically leads to obtain a more complex dependency graph (see in Figure 1 (left)) which, as witnessed by our experimental analysis in Section 7, can be more difficult to analyze when proving termination in practice.

Our approach clarifies the role of collapsing pairs to model the termination behavior of CSR. Furthermore, the new notions introduced in this paper lead to a more 'robust' framework. For instance, in order to integrate in [6] the new improvement in the notion of hiding context (see Definition 2), one has to *redefine* the notion of context-sensitive dependency pair in [6]. In our approach, the context-sensitive dependency pairs are always the same.

9 Conclusions

When proofs of termination of CSR are mechanized following the context-sensitive dependency pair approach [7], handling collapsing pairs is difficult. In [6] this problem is solved by a transformation which disregards collapsing pairs (so we loose their descriptive power), adds a new fresh symbol U which has nothing to do with the original CS-TRS, and makes the dependency graph harder to understand.

We have shown a different way to mechanize the context-sensitive dependency pair approach. The idea is adding a new TRS, the *unhiding TRS*, which avoids the extra requirements in [7]. Thanks to the flexibility of our framework, we can use all existing processors in the literature, improve the existing ones by taking advantage of having collapsing pairs, and define new processors. Furthermore, we have improved the notion of *hide* given in [6]. Our experimental evaluation shows that our techniques lead to an implementation which offers the best performance in terms of solved problems and efficiency.

References

1. Lucas, S.: Context-Sensitive Computations in Functional and Functional Logic Programs. Journal of Functional and Logic Programming (1), 1–61 (1998)
2. Bruni, R., Meseguer, J.: Semantic Foundations for Generalized Rewrite Theories. Theoretical Computer Science 360(1), 386–414 (2006)
3. Durán, F., Lucas, S., Marché, C., Meseguer, J., Urbain, X.: Proving Operational Termination of Membership Equational Programs. Higher-Order and Symbolic Computation 21(1-2), 59–88 (2008)
4. Endrullis, J., Hendriks, D.: From Outermost to Context-Sensitive Rewriting. In: Treinen, R. (ed.) RTA 2009. LNCS, vol. 5595, pp. 305–319. Springer, Heidelberg (2009)
5. Fernández, M.L.: Relaxing Monotonicity for Innermostt Termination. Information Processing Letters 93(3), 117–123 (2005)
6. Alarcón, B., Emmes, F., Fuhs, C., Giesl, J., Gutiérrez, R., Lucas, S., Schneider-Kamp, P., Thiemann, R.: Improving Context-Sensitive Dependency Pairs. In: Cervesato, I., Veith, H., Voronkov, A. (eds.) LPAR 2008. LNCS (LNAI), vol. 5330, pp. 636–651. Springer, Heidelberg (2008)
7. Alarcón, B., Gutiérrez, R., Lucas, S.: Context-Sensitive Dependency Pairs. In: Arun-Kumar, S., Garg, N. (eds.) FSTTCS 2006. LNCS, vol. 4337, pp. 297–308. Springer, Heidelberg (2006)
8. Arts, T., Giesl, J.: Termination of Term Rewriting Using Dependency Pairs. Theoretical Computer Science 236(1-2), 133–178 (2000)
9. Alarcón, B., Gutiérrez, R., Lucas, S.: Context-Sensitive Dependency Pairs. Information and Computation (2010) (to appear)
10. Gutiérrez, R., Lucas, S.: Proving Termination in the Context-Sensitive Dependency Pairs Framework. Technical report, Universidad Politécnica de Valencia (February 2010) Available as Technical Report DSIC-II/02/10
11. Ohlebusch, E.: Advanced Topics in Term Rewriting. Springer, Heidelberg (2002)
12. Giesl, J., Thiemann, R., Schneider-Kamp, P., Falke, S.: Mechanizing and Improving Dependency Pairs. Journal of Automatic Reasoning 37(3), 155–203 (2006)

13. Lucas, S.: MU-TERM: A Tool for Proving Termination of Context-Sensitive Rewriting. In: van Oostrom, V. (ed.) RTA 2004. LNCS, vol. 3091, pp. 200–209. Springer, Heidelberg (2004), http://zenon.dsic.upv.es/muterm/
14. Alarcón, B., Gutiérrez, R., Iborra, J., Lucas, S.: Proving Termination of Context-Sensitive Rewriting with MU-TERM. Electronic Notes in Theoretical Computer Science 188, 105–115 (2007)
15. Hirokawa, N., Middeldorp, A.: Automating the Dependency Pair Method. Information and Computation 199, 172–199 (2005)
16. Gutiérrez, R., Lucas, S., Urbain, X.: Usable Rules for Context-Sensitive Rewrite Systems. In: Voronkov, A. (ed.) RTA 2008. LNCS, vol. 5117, pp. 126–141. Springer, Heidelberg (2008)
17. Lucas, S.: Polynomials over the Reals in Proofs of Termination: from Theory to Practice. RAIRO Theoretical Informatics and Applications 39(3), 547–586 (2005)
18. Lucas, S., Meseguer, J.: Order-Sorted Dependency Pairs. In: Antoy, S., Albert, E. (eds.) Proc. of 10th International ACM SIGPLAN Sympsium on Principles and Practice of Declarative Programming PPDP 2008, pp. 108–119. ACM Press, New York (2008)
19. Giesl, J., Schneider-Kamp, P., Thiemann, R.: AProVE 1.2: Automatic Termination Proofs in the Dependency Pair Framework. In: Furbach, U., Shankar, N. (eds.) IJCAR 2006. LNCS (LNAI), vol. 4130, pp. 281–286. Springer, Heidelberg (2006), http://www-i2.informatik.rwth-aachen.de/AProVE
20. Schernhammer, F., Gramlich, B.: VMTL - A Modular Termination Laboratory. In: Treinen, R. (ed.) RTA 2009. LNCS, vol. 5595, pp. 285–294. Springer, Heidelberg (2009)
21. Endrullis, J.: Jambox, Automated Termination Proofs For String and Term Rewriting (2009), http://joerg.endrullis.de/jambox.html

A Dependency Pair Framework for $A \lor C$-Termination*

Beatriz Alarcón[1], Salvador Lucas[1], and José Meseguer[2]

[1] ELP group, DSIC, Universidad Politécnica de Valencia,
Camino de Vera s/n, 46022 Valencia, Spain
{balarcon,slucas}@dsic.upv.es

[2] CS Dept. University of Illinois at Urbana-Champaign, IL, USA

Abstract. The development of powerful techniques for proving termination of rewriting modulo a set of equations is essential when dealing with rewriting logic-based programming languages like CafeOBJ, Maude, OBJ, etc. One of the most important techniques for proving termination over a wide range of variants of rewriting (strategies) is the *dependency pair approach*. Several works have tried to adapt it to rewriting modulo *associative and commutative* (AC) equational theories, and even to more general theories. However, as we discuss in this paper, no appropriate notion of minimality (and minimal chain of dependency pairs) which is well-suited to develop a *dependency pair framework* has been proposed to date. In this paper we carefully analyze the structure of infinite rewrite sequences for rewrite theories whose equational part is a (free) *combination* of associative and commutative axioms which we call $A \lor C$*-rewrite theories*. Our analysis leads to a more accurate and optimized notion of dependency pairs through the new notion of *stably minimal term*. Then, we have developed a suitable dependency pair framework for proving termination of $A \lor C$-rewrite theories.

Keywords: equational rewriting, termination, dependency pairs.

1 Introduction

Rewriting with rules R modulo axioms E is a widely used technique in both rule-based programming languages and in automated deduction. Consequently, termination of rewriting modulo specific axioms E (e.g., associativity-commutativity, AC) has been studied. Methods for proving termination of rewriting systems modulo AC-axioms are known and even implemented. Several works have tried to adapt the *dependency pair approach* (DP-approach [1]) to rewriting modulo *associative and commutative* (AC) theories [13,9,10,11,14]. The corresponding proof methods, though, cannot be applied to commonly occurring combinations

* Partially supported by EU (FEDER) and MICINN grant TIN 2007-68093-C02-02. José Meseguer has been partially supported by NSF Grants CCF-0905584, CNS-07-16038, and CNS-08-34709. Beatriz Alarcón was partially supported by the Spanish MEC/MICINN under FPU grant AP2005-3399.

P.C. Ölveczky (Ed.): WRLA 2010, LNCS 6381, pp. 35–51, 2010.

```
fmod LIST&SET is
  sorts Bool Nat List Set .
  subsorts Nat < List Set .
  ops true false : -> Bool .
  ops _and_ _or_ : Bool Bool -> Bool [assoc comm] .
  op 0 : -> Nat .
  op s_ : Nat -> Nat .
  op _;_ : List List -> List [assoc] .
  op null : -> Set .
  op __ : Set Set -> Set [assoc comm] .
  op _in_ : Nat Set -> Bool .
  op _==_ : List List -> Bool [comm] .
  op list2set : List -> Set .
  var  B : Bool .                    vars N M : Nat .
  vars L L' : List .                 var  S : Set .
  eq N N = N .
  eq true and B = B .                eq false and B = false .
  eq true or B = true .              eq false or B = B .
  eq 0 == s N = false .              eq s N == s M = N == M .
  eq N ; L == M = false .            eq N ; L == M ; L' = (N == M) and L == L' .
  eq L == L = true .
  eq list2set(N) = N .               eq list2set(N ; L) = N list2set(L) .
  eq N in null = false .             eq N in M S = (N == M) or N in S .
endfm
```

Fig. 1. Example in Maude syntax [3]

of axioms that fall outside their scope. For instance, they could not be applied to prove termination of the TRS in Figure 1, (specified in Maude with self-explanatory syntax; we would not care about sort information here) where we have a (free) *combination* of associative and commutative axioms which we call an *A∨C-rewrite theory* in this paper. Furthermore, the *Dependency Pair Framework* (DP-framework [6]), which is the basis of state-of-the-art tools for proving termination of (different variants of) term rewriting has not yet been adapted to the AC case.

In this paper, we address these two problems. Giesl and Kapur generalized the previous works on AC-termination with dependency pairs to deal with more general kinds of equational theories E satisfying some restrictions [5]. In principle, the *A∨C*-theories that we are going to investigate here fit Giesl and Kapur's approach. However, as we discuss below, they did not provide any definition of *minimal chain* needed for further developments in the DP-framework. In the DP-framework, the central notion regarding termination proofs is that of *DP problem*: the goal is checking the absence (or presence) of the so-called *infinite minimal chains*, where the notion of minimal chain can be thought as an abstraction of the infinite rewrite sequences starting from *minimal non-terminating terms*. The most important notion regarding mechanization of the proofs is that of *processor.* A (correct) processor basically transforms DP problems into (hopefully) *simpler* ones, in such a way that the existence of an infinite chain in the original DP problem implies the existence of an infinite chain in the transformed one. Here 'simpler' usually means that fewer pairs are involved. Processors are used in a pipe (more precissely, a *tree*) to incrementally simplify the original DP problem as much as possible, possibly decomposing it into smaller pieces which are then independently treated in the very same way. This is the crucial new feature of the DP-framework w.r.t. the DP-approach of [1]. This makes it so powerful as a basis for implementing termination provers.

Before being able to adapt the DP-framework to deal with $A \vee C$-theories, we start by giving a more refined notion of minimality. In fact, the notion of minimality which is used in [5] is the straightforward extension of the one which is used to prove termination of standard rewriting but without dealing with *equivalence preservation* which, as we show below, is essential to provide an appropriate notion of minimal non-E-terminating term for $A \vee C$-theories E which can be used to define a suitable $A \vee C$-DP-framework. We carefully analyze the structure of infinite rewrite sequences for $A \vee C$-rewrite theories. This leads to appropriate definitions of $A \vee C$-dependency pair and minimal chain.

After some technical preliminaries, in Section 3 we investigate the drawbacks of previous notions of minimal term when modeling infinite $A \vee C$-rewrite sequences. Then, we introduce the notion of *stably minimal* non-E-terminating term which is the basis of our development. Section 4 investigates the structure of infinite sequences starting from such stably minimal terms. Section 5 uses these results to formalize our notion of $A \vee C$-dependency pairs and minimal chains. Section 6 introduces an $A \vee C$-DP-framework for proving $A \vee C$-termination using $A \vee C$-DPs. in particular, we introduce the notion of $A \vee C$-dependency graph and a first processor for proving termination in the $A \vee C$-DP-framework. We also show how to use orderings for defining a second processor. Section 7 compares our approach with the related work and concludes.

2 Rewriting Modulo Equational Theories

Given a rewrite theory $\mathcal{R} = (\Sigma, E, R)$, we write $s \to_{R/E} t$ if there exist u, v such that $s \sim_E u$, $u \to_R v$, and $v \sim_E t$. We say that a rewrite theory $\mathcal{R} = (\Sigma, E, R)$ is *E-terminating*, iff $\to_{R/E}$ is terminating. In general, given terms s and t, the problem of whether $s \to_{R/E} t$ holds is undecidable: in order to check whether $s \to_{R/E} t$ we have to search through the possibly infinite equivalence classes $[s]_E$ and $[t]_E$ to see whether a matching is found for a subterm of some $u \in [s]_E$ and the result of rewriting u belongs to the equivalence class $[t]_E$. For this reason, a much simpler relation $\to_{R,E}$ is defined, which becomes decidable if an E-matching algorithm exists. For any terms s, t, $s \to_{R,E} t$ holds iff there is a position p in s, a rule $l \to r$ in R, and a substitution σ such that $s|_p \sim_E \sigma(l)$ and $t = s[\sigma(r)]_p$ (see [15]). We say that a rewrite theory $\mathcal{R} = (\Sigma, E, R)$ is *(R, E)-terminating*, if $\to_{R,E}$ is terminating.

Regarding E-termination analysis using *dependency pairs* (DPs), Kusakari and Toyama observed that there is no simple extension of DPs to directly deal with $\rightarrow_{R/E}$-computations [11,9]. In contrast, several approaches have been developed for $\rightarrow_{R,E}$-computations [5,11,13]. Since $\rightarrow_{R,E}\subseteq\rightarrow_{R/E}$ (but the opposite inclusion does not hold, in general), E-termination cannot be concluded from (R,E)-termination. Actually, Marché and Urbain showed that there are (R,E)-terminating rewrite theories $\mathcal{R}$ which are *not* E-terminating.

Example 1. Consider the following rewrite theory $\mathcal{R} = (\Sigma, E, R)$, where '+' is an AC symbol [13]: $a + b \rightarrow a + (b + c)$. Note that $t = a + (b + c)$ is an $\rightarrow_{R,E}$-normal form (hence (R,E)-terminating). However, $t \sim_{AC} (a + b) + c$ which is non-E-terminating.

Giesl and Kapur [5] proved the equivalence of both notions of termination with respect to a notion of *extension completion* $\mathcal{E}xt_E(R)$ of a rewrite theory $\mathcal{R} = (\Sigma, E, R)$ which, for E being a set containing associative or commutative axioms, goes back to Peterson and Stickel [15].

Theorem 1. [5, Theorem 11] *Let $\mathcal{R} = (\Sigma, E, R)$ be a rewrite theory with E a regular and linear equational theory and $t \in \mathcal{T}(\Sigma, \mathcal{X})$. Then, t starts an infinite $\rightarrow_{R/E}$-reduction if and only if t starts an infinite $\rightarrow_{\mathcal{E}xt_E(R),E}$-reduction. Therefore, $\mathcal{R}$ is E-terminating if and only if $\rightarrow_{\mathcal{E}xt_E(R),E}$ is terminating.*

2.1 Combination of Associative and Commutative Theories

Let E be a set of equations that has the modular decomposition $E = \bigcup_{f\in\Sigma} E_f$, where if $k = ar(f) \neq 2$, then $E_f = \varnothing$, and if $k = 2$, then $E_f \subseteq \{A_f, C_f\}$, where:

- A_f is the associativity axiom $f(f(x,y),z) = f(x, f(y,z))$,
- C_f is the commutativity axiom $f(x,y) = f(y,x)$.

We also define $\Sigma = \Sigma_A \uplus \Sigma_C \uplus \Sigma_{AC} \uplus \Sigma_\varnothing$ where $f \in \Sigma_A \Leftrightarrow E_f = \{A_f\}$, $f \in \Sigma_C \Leftrightarrow E_f = \{C_f\}$, $f \in \Sigma_{AC} \Leftrightarrow E_f = \{A_f, C_f\}$, $f \in \Sigma_\varnothing \Leftrightarrow E_f = \varnothing$. In the following, we often say that a symbol $f \in \Sigma$ *is associative* if $f \in \Sigma_A \cup \Sigma_{AC}$.

Definition 1 (**$A{\vee}C$-rewrite theory**). *An equational theory $E = \bigcup_{f\in\Sigma} E_f$, where if $k = ar(f) \neq 2$, then $E_f = \varnothing$, and if $k = 2$, then $E_f \subseteq \{A_f, C_f\}$ is called an $A{\vee}C$-*theory. *A rewrite theory $\mathcal{R} = (\Sigma, E, R)$ such that E is an $A{\vee}C$-theory, is called an $A{\vee}C$-*rewrite theory.

To deal with rewriting modulo $A{\vee}C$-theories by using (R,E)-rewriting we have to extend R by following [15, Definition 10.4]:

$$\begin{aligned}
\mathcal{E}xt_{AC}(R) &= R \cup \{f(l,w) \rightarrow f(r,w) \mid l \rightarrow r \in R, f = root(l) \in \Sigma_{AC}\}\\
\mathcal{E}xt_{A}(R) &= R \cup \{f(l,w) \rightarrow f(r,w), f(w,l) \rightarrow f(w,r), f(z, f(l,w)) \rightarrow f(z, f(r,w))\\
&\qquad \mid l \rightarrow r \in R, f = root(l) \in \Sigma_A\}\\
\mathcal{E}xt_{C}(R) &= R
\end{aligned}$$

where w and z are fresh variables which do not occur in the original rule of R. Therefore, given an $A{\vee}C$ theory E, we let: $\mathcal{E}xt_E(R) = \mathcal{E}xt_{AC}(R) \cup \mathcal{E}xt_A(R) \cup \mathcal{E}xt_C(R)$. Note that $R \subseteq \mathcal{E}xt_E(R)$.

2.2 Minimal Terms and Infinite Rewrite Sequences

Given a TRS $\mathcal{R} = (\mathcal{C} \uplus \mathcal{D}, R)$, with $\mathcal{C}$ a subsignature of constructors and $\mathcal{D}$ a subsignature of defined symbols, the *minimal* nonterminating terms associated to $\mathcal{R}$ are nonterminating terms t whose proper subterms u (i.e., $t \rhd u$) are terminating; $\mathcal{T}_\infty$ is the set of minimal nonterminating terms associated to $\mathcal{R}$ [7]. Minimal nonterminating terms have two important properties:

1. Every nonterminating term s contains a minimal nonterminating term $t \in \mathcal{T}_\infty$ (i.e., $s \unrhd t$), and
2. minimal nonterminating terms t are always rooted by a *defined* symbol $f \in \mathcal{D}$: $\forall t \in \mathcal{T}_\infty, root(t) \in \mathcal{D}$.

Considering the structure of the infinite rewrite sequences starting from a minimal nonterminating term $t = f(t_1, \ldots, t_k) \in \mathcal{T}_\infty$ is helpful to arrive at the notion of dependency pair. Such sequences proceed as follows (see, e.g., [7]):

1. a finite number of reductions can be performed *below* the root of t, thus rewriting t into t'; then
2. a rule $f(l_1, \ldots, l_k) \to r$ applies *at the root* of t' (i.e., $t' = \sigma(f(l_1, \ldots, l_k))$ for some substitution σ); and
3. there is a minimal nonterminating term $u \in \mathcal{T}_\infty$ (hence $root(u) \in \mathcal{D}$) at some position p of $\sigma(r)$ which is a *nonvariable position of* r which 'continues' the infinite sequence initiated by t in a similar way.

This means that considering the occurrences of defined symbols in the right-hand sides of the rewrite rules suffices to 'catch' *every possible infinite rewrite sequence starting from* $\sigma(r)$. In particular, no infinite sequence can be issued from t' *below* the variables of r (more precisely: all bindings $\sigma(x)$ are *terminating* terms). The standard definition of dependency pair [1] and (minimal) chain of dependency pairs [6] relies on (1)–(3) above [7]. These facts are formalized as follows:

Proposition 1. [7, Lemma 1] *Let $\mathcal{R} = (\mathcal{C} \uplus \mathcal{D}, R)$ be a TRS. For all $t \in \mathcal{T}_\infty$, there exist $l \to r \in R$, a substitution σ and a term $u \in \mathcal{T}_\infty$ such that $root(u) \in \mathcal{D}$, $t \xrightarrow{>\Lambda}{}^{*} \sigma(l) \xrightarrow{\Lambda} \sigma(r) \unrhd u$ and there is a nonvariable subterm v of r, $r \unrhd v$, such that $u = \sigma(v)$.*

In the following section we begin the analysis of infinite E-rewrite sequences according to this schema. We aim at providing an appropriate notion of minimal non-E-terminating term (for $A \vee C$-theories E) which allows us to reach a result similar to Proposition 1.

3 Stably Minimal Non-E-Terminating Terms

In the dependency pair approach [1,7,6], the analysis of infinite rewrite sequences is restricted to those starting from *minimal nonterminating terms* $t \in \mathcal{T}_\infty$. The following notion of minimal non-E-terminating term is implicit in [5, proof of Theorem 16]. Similar definitions can be found in [10,11,9,14].

Definition 2 (Minimal non-E-terminating term [5]). *Let* $\mathcal{R} = (\Sigma, E, R)$ *be a rewrite theory. A non-E-terminating term* $t \in \mathcal{T}(\Sigma, \mathcal{X})$ *is said to be* minimal *(written* $t \in \mathcal{T}_{\infty,R,E}$*) if every strict subterm* s *of* t *(i.e.,* $t \rhd s$*) is* $(\mathcal{E}xt_E(R), E)$*-terminating.*

Remark 1. In Definition 2, if we assume that E is linear and regular (like $A{\vee}C$-theories), then, by Theorem 1, we could equivalently start by saying that t is non-$(\mathcal{E}xt_E(R), E)$-terminating. This leads to a more symmetric definition which we often use in the following without further comment.

Every non-E-terminating term s contains a minimal non-E-terminating term $t \in \mathcal{T}_{\infty,R,E}$ (this is stated without proof in [5, proof of Theorem 16]).

Remark 2 (Root symbols of minimal terms). Note that, if E is an $A{\vee}C$-equational theory, then $root(t) \in \mathcal{D}$ whenever $t \in \mathcal{T}_{\infty,R,E}$. As remarked by Giesl and Kapur (see also Example 5 below) this is not true for arbitrary equational theories.

The problem with Giesl and Kapur's Definition 2 is that minimality is *not* preserved under E-equivalence.

Example 2. Consider the following TRS $\mathcal{R}$:

$$f(x, x) \to f(0, f(1, 2)) \tag{1}$$

where $f \in \Sigma_{AC}$. Hence, $\mathcal{E}xt_{AC}(R)$ only adds the following rule to $\mathcal{R}$:

$$f(f(x, x), y) \to f(f(0, f(1, 2)), y) \tag{2}$$

Note that $t = f(f(0, 1), f(0, f(1, 2)))$ is non-$(\mathcal{E}xt_{AC}(R), AC)$-terminating:

$$\underline{f(f(0,1), f(0, f(1,2)))} \sim_A f(0, \underline{f(1, f(0, f(1,2)))}) \sim_A f(0, f(\underline{f(1,0)}, f(1,2))) \sim_C f(0, \underline{f(f(0,1), f(1,2))}) \sim_A \underline{f(0, f(0, f(1, f(1,2))))} \sim_A \underline{f(f(0,0), f(1, f(1,2)))} \xrightarrow{\Lambda}_{\mathcal{E}xt_{AC}(R)} f(f(0, f(1,2)), f(1, f(1,2))) \to_{\mathcal{E}xt_{AC}(R),AC} \cdots$$

Since $f(0, 1)$ and $f(0, f(1, 2))$ are in $(\mathcal{E}xt_{AC}(R), AC)$-normal form, we have that $t \in \mathcal{T}_{\infty,R,AC}$. However, $t' = f(f(0, 0), f(1, f(1, 2)))$, which is AC-equivalent to t (i.e., $t \sim_{AC} t'$), is non-AC-terminating but it is *not* minimal because its strict subterm $f(1, f(1, 2)))$ is non-$(\mathcal{E}xt_{AC}(R), AC)$-terminating:

$$\underline{f(1, f(1,2)))} \sim_A \underline{f(f(1,1), 2)} \xrightarrow{\Lambda}_{\mathcal{E}xt_{AC}(R)} \underline{f(f(0, f(1,2)), 2)} \sim_A f(0, \underline{f(f(1,2), 2)}) \sim_A f(0, f(1, f(2,2))) \sim_A \underline{f(f(0,1), f(2,2))} \sim_C \underline{f(f(2,2), f(0,1))} \xrightarrow{\Lambda}_{\mathcal{E}xt_{AC}(R)} f(f(0, f(1,2)), f(0,1)) \to_{\mathcal{E}xt_{AC}(R),AC} \cdots$$

Example 2 shows that an essential property of minimal terms when considered as part of infinite $(\mathcal{E}xt_E(R), E)$-rewriting sequences for $A{\vee}C$-theories E gets lost: the application of $(\mathcal{E}xt_E(R), E)$-rewrite steps *at the root* of a minimal term s by means of a rule $l \to r$ (i.e., $s \sim_{AC} \sigma(l) \xrightarrow{\Lambda}_{\mathcal{E}xt_E(R)} \sigma(r)$) does *not* guarantee that there is a *nonvariable subterm* v of the right-hand side r which is a prefix of the 'next' minimal term in the infinite sequence.

Example 3. Term t in Example 2 can be rewritten at the root *only* by rule (2) of $\mathcal{E}xt_{AC}(R)$. We can apply this rule to t' in Example 2 (for instance) to obtain $s' = \sigma(r) = f(f(0,f(1,2)),f(1,f(1,2)))$ (where $r = f(f(0,f(1,2)),y))$, which is non-$(\mathcal{E}xt_{AC}(R),AC)$-terminating. Note that s' contains a minimal term $u \in \mathcal{T}_{\infty,R,E}$. Since $s'|_2 = f(1,f(1,2))$ is non-$(\mathcal{E}xt_{AC}(R),AC)$-terminating, it follows that s' is *not* minimal. Since $s'|_1 = f(0,f(1,2))$ is $(\mathcal{E}xt_{AC}(R),AC)$-terminating, the only possibility is that u occurs in $s'|_2$. Actually, $s'|_2$ is minimal already; hence, $u = s'|_2$. But note the absence of any nonvariable position $p \in \mathcal{P}os(r)$ in the right-hand side of the considered rule such that $\sigma(r|_p) = u = f(1,f(1,2))$.

This is in sharp contrast with the situation of the DP-approach for ordinary rewriting. Furthermore, it is not difficult to see that for all $t'' \sim_{AC} t$ such $t'' = \sigma'(l)$ for some substitution σ', we have a similar situation. Thus, the problem illustrated here cannot be solved by using a different $\sim_{AC}$ sequence before performing the $\mathcal{E}xt_{AC}(R)$-root-step.

In the following we introduce a new notion of minimality which solves these problems.

3.1 A New Notion of Minimal Non-E-Terminating Terms

The following definition solves the problems discussed above by explicitly requiring that the condition defining minimality is preserved under E-equivalence.

Definition 3 (Stably minimal non-E-terminating term). *Let $\mathcal{R} = (\Sigma, E, R)$ be a rewrite theory. Let $\mathcal{M}_{\infty,R,E}$ be a set of stably minimal non-E-terminating terms in the following sense: $t \in \mathcal{T}(\Sigma,\mathcal{X})$ belongs to $\mathcal{M}_{\infty,R,E}$ if t is non-E-terminating, and for all $t' \sim_E t$ and every proper subterm s' of t' (i.e., $t' \rhd s'$), s' is $(\mathcal{E}xt_E(R),E)$-terminating.*

We have the following useful characterization of minimality.

Proposition 2 (Characterization of stably minimal terms). *Let $\mathcal{R} = (\Sigma, R, E)$ be a rewrite theory and $t \in \mathcal{T}(\Sigma,\mathcal{X})$. Then, $t \in \mathcal{M}_{\infty,R,E}$ if and only if $[t]_E \subseteq \mathcal{T}_{\infty,R,E}$. Therefore, $\mathcal{M}_{\infty,R,E} = \{t \in \mathcal{T}(\Sigma,\mathcal{X}) \mid [t]_E \subseteq \mathcal{T}_{\infty,R,E}\}$.*

The problem in Example 2 disappears now: t is *not* minimal according to Definition 3. The following result shows how to *find* stably minimal non-E-terminating terms associated to a given non-E-terminating term. This is essential in our development.

Proposition 3. *Let $\mathcal{R} = (\Sigma, E, R)$ be a rewrite theory such that $[t]_E = \{t\}$ for all constant and variable terms t. Let $s \in \mathcal{T}(\Sigma,\mathcal{X})$. If s is non-E-terminating, then there is a subterm t of some $s' \sim_E s$ ($s' \unrhd t$) such that $t \in \mathcal{M}_{\infty,R,E}$.*

Clearly, Proposition 3 holds whenever $\mathcal{R}$ is an $A\vee C$-rewrite theory.

Example 4. Consider the term t in Example 2. Although $t \in \mathcal{T}_{\infty,R,E}$, $t \notin \mathcal{M}_{\infty,R,E}$: the term $t' = f(f(0,0),f(1,f(1,2)))$, which is AC-equivalent to t, contains a subterm $u = f(1,f(1,2))$ which is non-E-terminating. It is not difficult to see that actually $u \in \mathcal{M}_{\infty,R,E}$.

In general, Proposition 3 does *not* hold for arbitrary sets of equations E.

Example 5. Consider the following example [5, Example 13]:

$$R : f(x) \to x \qquad E : f(a) = a$$

Note that $a \in \mathcal{T}_{\infty,R,E}$. However, a is *not* stably minimal because $a \sim_E f(a)$ but $f(a) \notin \mathcal{T}_{\infty,R,E}$. Thus, Proposition 3 does not hold.

Now we provide a more precise result about where we can find stably minimal subterms within a non-E-terminating term for $A{\vee}C$-rewrite theories $\mathcal{R} = (\Sigma, E, R)$. In the following theorem, given a term s and a symbol f, by an f-subterm t of s (written $s \trianglerighteq_f t$) we mean a subterm t of s such that $t = s|_p$ and for all $q < p$, $root(s|_q) = f$. We also write $s \triangleright_f t$ if $s \trianglerighteq_f t$ and $s \neq t$.

Theorem 2. *Let $\mathcal{R} = (\Sigma, E, R)$ be an $A{\vee}C$-rewrite theory. If s is non-E-terminating, then there is a subterm $t \in \mathcal{T}_{\infty,R,E}$ of s ($s \trianglerighteq t$) and*

1. *If (1) $A_{root(t)} \notin E_{root(t)}$ or (2) $t = f(t_1, t_2)$, $A_f \in E_f$, $root(t_1) \neq f$, and $root(t_2) \neq f$, then $t \in \mathcal{M}_{\infty,R,E}$.*
2. *If $t = f(t_1, t_2)$, $A_f \in E_f$, and $root(t_1) = f$ or $root(t_2) = f$, and $t \notin \mathcal{M}_{\infty,R,E}$, then there is $s' \sim_E t$ and a strict f-subterm u of s' ($s' \triangleright_f u$) such that $root(u) = f$ and $u \in \mathcal{M}_{\infty,R,E}$.*

The following result is just a convenient reformulation of the previous one.

Corollary 1. *Let $\mathcal{R} = (\Sigma, E, R)$ be an $A{\vee}C$-rewrite theory. If s is non-E-terminating, then either there is a subterm $t \in \mathcal{M}_{\infty,R,E}$ of s ($s \trianglerighteq t$), or there is a subterm $t \in \mathcal{T}_{\infty,R,E}$ of s satisfying that $t = f(t_1, t_2)$, $A_f \in E_f$, and $root(t_1) = f$ or $root(t_2) = f$, and such that there is $s' \sim_E t$ and a strict f-subterm u of s' ($s' \triangleright_f u$) such that $root(u) = f$ and $u \in \mathcal{M}_{\infty,R,E}$.*

4 Structure of (Stably) Minimal Infinite $A{\vee}C$-Rewrite Sequences

Now we analyze $A{\vee}C$-rewrite sequences starting from stably minimal non-$A{\vee}C$-terminating terms. First we consider a restricted case.

Proposition 4. *Let $\mathcal{R} = (\Sigma, E, R) = (\mathcal{C} \uplus \mathcal{D}, E, R)$ be an $A{\vee}C$-rewrite theory. Let $s \in \mathcal{M}_{\infty,R,E}$ be such that $f = root(s)$ and either (1) $A_f \notin E_f$, or (2) $s = f(s_1, s_2)$, $A_f \in E_f$, and $root(s_1), root(s_2) \in \mathcal{C}$. Assume that for all $l \to r \in R$ such that $root(l) = f$ and all subterms v of r ($r \trianglerighteq v$) such that $v = g(v_1, v_2)$ for some associative symbol g, we have that $root(v_1), root(v_2) \notin \mathcal{X} \cup \{g\}$. Then, there exist $l \to r \in R$, a substitution σ and terms $t \in \mathcal{T}(\Sigma, \mathcal{X})$ and $u \in \mathcal{M}_{\infty,R,E}$ such that*

$$s \xrightarrow{>\Lambda}{}^{*}_{\mathcal{E}xt_E(R),E} t \sim_E \sigma(l) \xrightarrow{\Lambda}_R \sigma(r) \trianglerighteq u$$

and there is a nonvariable subterm v of r, $r \trianglerighteq v$, such that $u = \sigma(v)$.

Unfortunately, stable minimality of (arbitrary) non-E-terminating terms s for $A\vee C$-theories E is not preserved under inner $(\mathcal{E}xt_E(R), E)$-rewritings.

Example 6. Term $u = f(f(1,1),2)$ in Example 2 is stably minimal: $u \in \mathcal{M}_{\infty,R,E}$. We have that $f(\underline{f(1,1)},2) \xrightarrow{>\Lambda}_R f(f(0,f(1,2)),2)$. Note that $f(f(0,f(1,2)),2) \notin \mathcal{M}_{\infty,R,E}$: we have $\underline{f(f(0,f(1,2)),2)} \sim_A f(0,\underline{f(f(1,2),2)}) \sim_A f(0,f(1,f(2,2)))$ where $f(0,f(1,\overline{f(2,2)}))$ contains a subterm $\overline{f(1,f(2,2))}$ which is non-$(\mathcal{E}xt_E(R),E)$-terminating.

In the following, we show how to avoid this problem. We define *deep reduction* as a restriction $\xrightarrow{>1,2}_{\mathcal{E}xt_E(R),E}$ of inner $(\mathcal{E}xt_E(R), E)$-rewriting which restricts reductions on terms like u above. We will show that deep reduction preserves stable minimality of non-E-terminating terms for $A\vee C$-rewrite theories $\mathcal{R} = (\Sigma, E, R)$.

Definition 4 (Deep reduction). *Let $\mathcal{R} = (\Sigma, E, R)$ be an $A\vee C$-rewrite theory. Given $t \in \mathcal{T}(\Sigma,\mathcal{X})$, $t \xrightarrow{>1,2}_{\mathcal{E}xt_E(R),E} s$ if $t \xrightarrow{q}_{\mathcal{E}xt_E(R),E} s$ for some position $q \in \mathcal{P}os(t)$ such that $q > p$ for $p \in \{1,2\}$ if $t = \sigma(u)$ for some $u = v \in E$ or $v = u \in E$ and $u|_p \notin \mathcal{X}$; otherwise, $q > \Lambda$.*

Obviously, $\xrightarrow{>1,2}_{\mathcal{E}xt_E(R),E} \subseteq \xrightarrow{>\Lambda}_{\mathcal{E}xt_E(R),E}$. The following proposition shows that *deep reduction* preserves stable minimality.

Proposition 5. *Let $\mathcal{R} = (\Sigma, E, R)$ be an $A\vee C$-rewrite theory and $t \in \mathcal{M}_{\infty,R,E}$. If $t \xrightarrow{>1,2}{}^{*}_{\mathcal{E}xt(\mathcal{R}),E} s$ and s is non-E-terminating, then $s \in \mathcal{M}_{\infty,R,E}$.*

As a consequence, the following theorem establishes the desired property for stable minimal non-$A\vee C$-terminating terms.

Theorem 3. *Let $\mathcal{R} = (\Sigma, E, R)$ be an $A\vee C$-rewrite theory. For all $s \in \mathcal{M}_{\infty,R,E}$, there exist $l \to r \in \mathcal{E}xt_E(R)$ and a substitution σ such that*

$$s\ (\sim_E \circ \xrightarrow{>1,2}_{\mathcal{E}xt_E(R),E})^* t \sim_E \sigma(l) \xrightarrow{\Lambda}_{\mathcal{E}xt_E(R)} \sigma(r)$$

and there is a nonvariable subterm v of r $(r \trianglerighteq v)$, such that either

1. *$v = f(v_1, v_2)$ for some associative symbol f, $root(v_1) \in \mathcal{X} \cup \{f\}$ or $root(v_2) \in \mathcal{X} \cup \{f\}$, $root(\sigma(v_1)) = f$ or $root(\sigma(v_2)) = f$, $\sigma(v) \in \mathcal{T}_{\infty,R,E}$ and there is a term $t' \sim_E \sigma(v)$ containing a strict f-subterm $u = f(u_1,u_2)$ $(t' \rhd_f u)$ such that $u \in \mathcal{M}_{\infty,R,E}$, or*
2. *$\sigma(v) \in \mathcal{M}_{\infty,R,E}$ otherwise.*

Example 2 shows that Theorem 3 does not hold for Giesl and Kapur's minimal terms $s \in \mathcal{T}_{\infty,R,E}$.

5 $A\vee C$-Dependency Pairs and Chains

Propositions 3 and 4 together with Theorem 3 are the basis for our definition of *$A\vee C$-Dependency Pairs* and the corresponding *chains*. Together, they show that given an $A\vee C$-rewrite theory $\mathcal{R} = (\Sigma, E, R)$, every non-$E$-terminating term s has an associated infinite $(\mathcal{E}xt_E(R), E)$-rewrite sequence starting from a stably minimal subterm $t \in \mathcal{M}_{\infty,R,E}$. Such a sequence proceeds as described in Proposition 4 and Theorem 3, depending on the shape of t.

This process is abstracted in the following definition of $A\vee C$-dependency pairs (Definition 5) and in the definition of chain below (Definition 6).

Given a signature Σ and $f \in \Sigma$, we let $f^\sharp$ denote a fresh new symbol (often called *tuple* symbol or DP-symbol) associated to a symbol f [1]. Let $\Sigma^\sharp$ be the set of tuple symbols associated to symbols in Σ. As usual, for $t = f(t_1, \ldots, t_k) \in \mathcal{T}(\Sigma, \mathcal{X})$, we write $t^\sharp$ to denote the *marked* term $f^\sharp(t_1, \ldots, t_k)$ (written sometimes $F(t_1, \ldots, t_k)$). Given a set of rules R and a symbol $f \in \Sigma$, we let $R_f = \{l \to r \in R \mid root(l) = f\}$. Given a set of rules R, the set $\mathsf{DP}(R)$ of dependency pairs associated to R is [1]: $\mathsf{DP}(R) = \{l^\sharp \to s^\sharp \mid l \to r \in R, r \trianglerighteq s, root(s) \in \mathcal{D}\}$.

Definition 5 ($A\vee C$-Dependency Pairs). *Let $\mathcal{R} = (\Sigma, E, R) = (\mathcal{C} \uplus \mathcal{D}, E, R)$ be an $A\vee C$-rewrite theory. Then, $\mathsf{DP}_E(R) = \mathsf{DP}(\mathcal{E}xt_E(R))$ is the set of $A\vee C$-dependency pairs ($A\vee C$-DPs) of $\mathcal{R}$.*

In general, the set of $A\vee C$-DPs which is obtained from Definition 5 is a subset of those which are obtained by particularizing Giesl and Kapur's definitions to the $A\vee C$ case [5].

Example 7. Consider the AC-rewrite theory $\mathcal{R} = (\Sigma, E, R)$ in Example 2. The set $DP_E(R)$ consists of the following pairs:

$$F(x,x) \to F(0, f(1,2)) \tag{3}$$
$$F(x,x) \to F(1,2) \tag{4}$$
$$F(f(x,x),y) \to F(f(0,f(1,2)),y) \tag{5}$$
$$F(f(x,x),y) \to F(0,f(1,2)) \tag{6}$$
$$F(f(x,x),y) \to F(1,2) \tag{7}$$

5.1 Chains of $A\vee C$-DPs

An essential property of the dependency pair method is that it provides a *characterization* of termination of TRSs $\mathcal{R}$ as the absence of infinite (minimal) *chains of dependency pairs* [1,6]. If we want to prove the same for $A\vee C$-rewrite theories, we have to introduce a suitable notion of chain which can be used with $A\vee C$-DPs. As in the DP-framework, where the origin of *pairs* does not matter, we should rather think of another rewrite theory $\mathcal{P} = (\Gamma, F, P)$ which is used together with $\mathcal{R}$ to build the chains. According to the usual terminology [6], we often call *pairs* to the rules $u \to v \in P$.

Definition 6 (Chain of pairs - Minimal chain). *Let $\mathcal{P} = (\Gamma, F, P)$ and $\mathcal{R} = (\Sigma, E, R)$ be rewrite theories, and $\mathcal{S} = (\mathcal{F}, S)$ be a TRS. An (F, P, E, R, S)-chain is a finite or infinite sequence of pairs $u_i \to v_i \in P$, together with substitutions σ and θ_i satisfying that, for all $i \geq 1$:*

1. *If* $\sigma(v_i) = f_i(v_{i1}, v_{i2})$ *satisfies* $\sigma(v_i) = \theta_i(u'_i)$ *for some* $u'_i = v'_i \in F$ *or* $v'_i = u'_i \in F$ *such that* $u'_i = f_i(u'_{i1}, u'_{i2})$ *satisfies* $u'_{i1} \notin \mathcal{X}$ *or* $u'_{i2} \notin \mathcal{X}$, *then*

$$\sigma(v_i) \sim_F \circ \xrightarrow{\Lambda}{}^{+}_{\mathcal{S}_{f_i}} t_i \,(\sim_F \circ \xrightarrow{>1,2}_{\mathcal{E}xt_E(R),E})^* \circ \sim_F \sigma(u_{i+1})$$

2. *and* $\sigma(v_i) = t_i \to^*_{\mathcal{E}xt_E(R),E} \circ \sim_F \sigma(u_{i+1})$, *otherwise.*

An (F, P, E, R, S)*-chain is called* minimal *if for all* $i \geq 1$, *and* $t'_i \in [t_i]_F$, t'_i *is* $(\mathcal{E}xt_E(R), E)$*-terminating.*

As usual, in Definition 6 we assume that different occurrences of dependency pairs do not share any variable (renaming substitutions are used if necessary).

This more abstract notion of chain can be particularized to be used with $A\vee C$-DPs, by just taking

1. $P = \mathsf{DP}_E(R)$,
2. $F = E \cup E^\sharp$, where $E^\sharp = \{s^\sharp = t^\sharp \mid s = t \in E\}$, and
3. $\mathcal{S} = \{f^\sharp(f(x,y),z) \to f^\sharp(x,y), f^\sharp(x, f(y,z)) \to f^\sharp(y,z) \mid f \in \Sigma_A \cup \Sigma_{AC}\}$.

Theorem 4 (Characterization of $A\vee C$-termination). *Let* $\mathcal{R} = (\Sigma, E, R)$ *be an* $A\vee C$*-rewrite theory. Let* $\mathcal{S} = (\Sigma \cup \mathcal{D}^\sharp, S)$ *be a TRS such that* $S = \{f^\sharp(f(x,y),z) \to f^\sharp(x,y), f^\sharp(x, f(y,z)) \to f^\sharp(y,z) \mid f \in \Sigma_A \cup \Sigma_{AC}\}$. *Then,* $\mathcal{R}$ *is* $(\mathcal{E}xt_E(R), E)$*-terminating if and only if there is no infinite minimal* $(E^\sharp \cup E, \mathsf{DP}_E(R), E, R, S)$*-chain.*

6 An $A\vee C$-Dependency Pair Framework

In the following, we adapt Giesl et al. DP-framework to provide a suitable framework for mechanizing proofs of $A\vee C$-termination using $A\vee C$-DPs.

Definition 7 ($A\vee C$ problem). *An $A\vee C$ problem* τ *is a tuple* $\tau = (F, P, E, R, S)$, *where* $\mathcal{R} = (\Sigma, E, R)$ *is an $A\vee C$-rewrite theory,* $\mathcal{P} = (\Gamma, F, P)$ *is a rewrite theory, and* $\mathcal{S} = (\mathcal{F}, S)$ *is a TRS. An $A\vee C$ problem is finite if there is no infinite minimal* (F, P, E, R, S)*-chain. An $A\vee C$ problem* τ *is infinite if* $\mathcal{R}$ *is non-$A\vee C$-terminating or there is an infinite minimal* (F, P, E, R, S)*-chain.*

The following definition adapts the notion of *processor* [6] to prove termination of $A\vee C$-rewrite theories.

Definition 8 ($A\vee C$ processor). *An $A\vee C$ processor* Proc *is a mapping from $A\vee C$ problems into sets of $A\vee C$ problems. Alternatively, it can also return* "no". *An $A\vee C$ processor* Proc *is*

- sound *if for all $A\vee C$ problems* τ, τ *is finite whenever* $\mathsf{Proc}(\tau) \neq$ no *and* $\forall \tau' \in \mathsf{Proc}(\tau)$, τ' *is finite.*
- complete *if for all $A\vee C$ problems* τ, τ *is infinite whenever* $\mathsf{Proc}(\tau) =$ no *or* $\exists \tau' \in \mathsf{Proc}(\tau)$ *such that* τ' *is infinite.*

Similar to [6] for the DP-framework, we construct a tree whose nodes are labeled with $A\vee C$ problems or "yes" or "no", and whose root is labeled with $(E^\sharp \cup E, \mathsf{DP}_E(R), E, R, S)$. Now we have the following result which adapts [6, Corollary 5] to $A\vee C$-rewrite theories.

Theorem 5 ($A\vee C$-DP framework). *Let $\mathcal{R} = (\Sigma, E, R)$ be an $A\vee C$-theory. We construct a tree whose nodes are labeled with $A\vee C$ problems or "yes" or "no", and whose root is labeled with $(E^\sharp \cup E, \mathsf{DP}_E(R), E, R, S)$, where $S = \{f^\sharp(f(x,y),z) \to f^\sharp(x,y), f^\sharp(x, f(y,z)) \to f^\sharp(y,z) \mid f \in \Sigma_A \cup \Sigma_{AC}\}$. For every inner node labeled with τ, there is a sound processor* Proc *satisfying one of the following conditions:*

1. $\mathsf{Proc}(\tau) = \mathsf{no}$ *and the node has just one child, labeled with* "no".
2. $\mathsf{Proc}(\tau) = \varnothing$ *and the node has just one child, labeled with* "yes".
3. $\mathsf{Proc}(\tau) \neq \mathsf{no}$, $\mathsf{Proc}(\tau) \neq \varnothing$, *and the children of the node are labeled with the $A\vee C$ problems in* $\mathsf{Proc}(\tau)$.

If all leaves of the tree are labeled with "yes", then $\mathcal{R}$ is E-terminating. Otherwise, if there is a leaf labeled with "no" and if all processors used on the path from the root to this leaf are complete, then $\mathcal{R}$ is not E-terminating.

6.1 $A\vee C$-Dependency Graph

$A\vee C$ problems focus our attention on the analysis of *infinite minimal chains*. Our aim here is obtaining a notion of graph which is able to represent all infinite *minimal* chains of pairs as given in Definition 6.

Definition 9 ($A\vee C$-Graph of Pairs). *Let $\mathcal{R} = (\Sigma, E, R)$ and $\mathcal{P} = (\Gamma, F, P)$ be rewrite theories and $\mathcal{S} = (\mathcal{F}, S)$ be a TRS. The $A\vee C$-graph associated to them (denoted $\mathsf{G}(F, P, E, R, S)$) has P as the set of nodes. There is an arc from $u \to v \in P$ to $u' \to v' \in P$ if $u \to v, u' \to v'$ is an (F, P, E, R, S)-chain.*

In termination proofs, we are concerned with the so-called *strongly connected components* (SCCs) of the dependency graph, rather than with the cycles themselves (which are exponentially many) [8]. A strongly connected component in a graph is a *maximal cycle*, i.e., a cycle which is not contained in any other cycle. In the following result, given two sets of rules S and Q, we let S_Q be the least subset of S satisfying that whenever there is a rule $u \to v \in Q$, such that v unifies with s for some $s = t \in F$ or $t = s \in F$ such that $s = f(s_1, s_2)$ and $s_1 \notin \mathcal{X}$ or $s_2 \notin \mathcal{X}$, then $S_f \subseteq S_Q$.

Theorem 6 (SCC processor). *Let $\mathcal{R} = (\Sigma, E, R)$ and $\mathcal{P} = (\Gamma, F, P)$ be rewrite theories and $\mathcal{S} = (\mathcal{F}, S)$ be a TRS. Then, the processor* Proc_{SCC} *given by*

$$\mathsf{Proc}_{SCC}(F, P, E, R, S) = \{(F, Q, E, R, S_Q) \mid Q \text{ are the pairs of an SCC in } \mathsf{G}(F, P, E, R, S)\}$$

is sound and complete.

As a consequence, we can *separately* work with the strongly connected components of $\mathsf{G}(F, P, E, R, S)$, disregarding other parts of the graph. Now we can use these notions to introduce the $A\vee C$-dependency graph, i.e., the $A\vee C$-graph whose nodes are the $A\vee C$-DPs instead of an arbitrary set of pairs.

Definition 10 ($A\vee C$-Dependency Graph). *Let $\mathcal{R} = (\Sigma, E, R)$ be an $A\vee C$-rewrite theory with $\Sigma = \mathcal{C} \uplus \mathcal{D}$. Let $\mathcal{S} = (\Sigma \cup \mathcal{D}^\sharp, S)$ be a TRS such that $S = \{f^\sharp(f(x, y), z) \to f^\sharp(x, y), f^\sharp(x, f(y, z)) \to f^\sharp(y, z) \mid f \in \Sigma_A \cup \Sigma_{AC}\}$. The* $A\vee C$-Dependency Graph *associated to $\mathcal{R}$ is: $\mathsf{DG}(\mathcal{R}) = \mathsf{G}(E^\sharp \cup E, \mathsf{DP}_E(R), E, R, S)$.*

6.2 Estimating the $A\vee C$-Dependency Graph

As in standard rewriting, the $A\vee C$-dependency graph of an $A\vee C$-rewrite theory is in general *not* computable. So, we need to use some approximation of it. For any term $t \in \mathcal{T}(\Sigma, \mathcal{X})$ let $\textsc{Cap}(t)$ result from replacing all proper subterms rooted by a defined symbol by fresh variables and let $\textsc{Ren}(t)$ which *independently* renames all *occurrences* of variables in t by using new fresh variables [1].

As usual, we do not have to talk about *mgu* when dealing with rewriting modulo equations. Instead, it is used the notion of complete set of E-unifiers. However, although in theory, all these unifiers have to be considered, for our results of reachability it is enough to check the existence of one.

Proposition 6. *Let $\mathcal{R} = (\Sigma, E, R)$ be an $A\vee C$-rewrite theory with $\Sigma = \mathcal{C} \uplus \mathcal{D}$. Let $u, t \in \mathcal{T}(\Sigma, \mathcal{X})$ be such that $\mathcal{V}ar(u) \cap \mathcal{V}ar(t) = \varnothing$ and θ, θ' be substitutions. If $\theta(t) \to^*_{\mathcal{E}xt_E(R),E} \circ \sim_E \theta'(u)$, then $\textsc{Ren}(\textsc{Cap}(t))$ and u E-unify.*

Now, we are ready to provide a correct estimation of our graph of pairs. Correctness of our definition relies on Proposition 6.

Definition 11 (Estimated $A\vee C$-Graph of Pairs). *Let $\mathcal{R} = (\Sigma, E, R)$ and $\mathcal{P} = (\Gamma, F, P)$ be rewrite theories and $\mathcal{S} = (\mathcal{F}, S)$ be a TRS. The* estimated $A\vee C$-graph *associated to them (denoted $\mathsf{EG}(F, P, E, R, S)$) has P as the set of nodes and arcs which connect them as follows:*

1. *If v unifies with s for some $s = t \in F$ or $t = s \in F$ such that $s = f(s_1, s_2)$ and $s_1 \notin \mathcal{X}$ or $s_2 \notin \mathcal{X}$, then, there is an arc from $u \to v \in P$ to $u' \to v' \in P$ if $root(u') = f$.*
2. *Otherwise, there is an arc from $u \to v \in P$ to $u' \to v' \in P$ if $\textsc{Ren}(\textsc{Cap}(v))$ and u' E-unify.*

According to Definition 9, we would have the corresponding one for the *estimated* $A\vee C$-DG: $\mathsf{EDG}(\mathcal{R}) = \mathsf{EG}(E^\sharp \cup E, \mathsf{DP}_E(R), E, R, S)$, where

$$S = \{f^\sharp(f(x, y), z) \to f^\sharp(x, y), f^\sharp(x, f(y, z)) \to f^\sharp(y, z) \mid f \in \Sigma_A \cup \Sigma_{AC}\}.$$

Example 8. For the $A\vee C$-rewrite theory in Figure 1, the set $\mathsf{DP}_E(R)$ is[1]:

$$LIST2SET(cons(N,L)) \to UNION(N, list2set(L)) \tag{8}$$
$$LIST2SET(cons(N,L)) \to LIST2SET(L) \tag{9}$$
$$IN(N, union(M,S)) \to EQ(N,M) \tag{10}$$
$$IN(N, union(M,S)) \to OR(eq(N,M), in(N,S)) \tag{11}$$
$$IN(N, union(M,S)) \to IN(N,S) \tag{12}$$
$$UNION(union(N,N),Z) \to UNION(N,Z) \tag{13}$$
$$AND(and(true,B),Z) \to AND(B,Z) \tag{14}$$
$$AND(and(false,B),Z) \to AND(false,Z) \tag{15}$$
$$OR(or(true,B),Z) \to OR(true,Z) \tag{16}$$
$$OR(or(false,B),Z) \to OR(B,Z) \tag{17}$$
$$EQ(s(N),s(M)) \to EQ(N,M) \tag{18}$$
$$EQ(cons(N,L), cons(M,L')) \to EQ(N,M) \tag{19}$$
$$EQ(cons(N,L), cons(M,L')) \to EQ(L,L') \tag{20}$$
$$EQ(cons(N,L), cons(M,L')) \to AND(eq(N,M), eq(L,L')) \tag{21}$$

The (estimated) $A\vee C$-DG is:

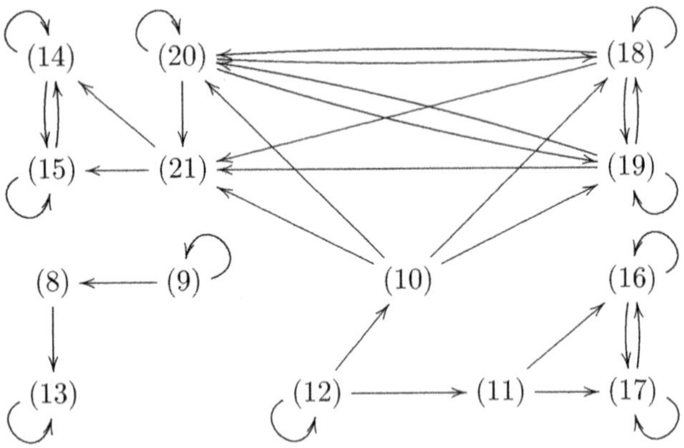

By Theorem 6 we transform the $A\vee C$ problem $(E\cup E^\sharp, \mathsf{DP}(R), E, R, S)$ into a set $\mathsf{Proc}_{SCC}(E\cup E^\sharp, \mathsf{DP}(R), E, R, S)$ given by

$$\{(E\cup E^\sharp, \{(9)\}, E, R, \varnothing), (E\cup E^\sharp, \{(12)\}, E, R, \varnothing), (E\cup E^\sharp, \{(13)\}, E, R, S_{union}),$$

$$(E\cup E^\sharp, \{(14),(15)\}, E, R, S_{and}), (E\cup E^\sharp, \{(16),(17)\}, E, R, S_{or}), (E\cup E^\sharp, \{(18),(19),(20)\}, E, R, \varnothing)\}$$

which contains six new (but simpler) $A\vee C$ problems.

6.3 Use of Reduction Pairs

A reduction pair $(\gtrsim, \sqsupset)$ consists of a stable and monotonic quasi-ordering $\gtrsim$, and a stable and well-founded ordering $\sqsupset$ satisfying either $\gtrsim \circ \sqsupset \subseteq \sqsupset$ or $\sqsupset \circ \gtrsim \subseteq \sqsupset$. In the dependency pair framework reduction pairs are used to obtain *smaller*

[1] We have introduced new 'prefix' symbols *eq*, *cons* and *union* instead of the original 'infix' ones `_==_`, `_;_`, `__`.

sets of pairs $\mathcal{P}' \subseteq \mathcal{P}$ by removing the *strict* pairs, i.e., those pairs $u \to v \in \mathcal{P}$ such that $u \sqsupset v$. Stability is required both for $\gtrsim$ and $\sqsupset$ because, although we only check the left- and right-hand sides of the rewrite rules $l \to r$ (with $\gtrsim$) and pairs $u \to v$ (with $\gtrsim$ or $\sqsupset$), the chains of pairs involve *instances* $\sigma(l)$, $\sigma(r)$, $\sigma(u)$, and $\sigma(v)$ of rules and pairs and we aim at concluding $\sigma(l) \gtrsim \sigma(r)$, and $\sigma(u) \gtrsim \sigma(v)$ or $\sigma(u) \sqsupset \sigma(v)$, respectively. Monotonicity is required for $\gtrsim$ to deal with the application of rules $l \to r$ to an arbitrary depth in terms. Since the pairs are 'applied' only at the root level, no monotonicity is required for $\sqsupset$ (but, for this reason, we cannot compare the rules in $\mathcal{R}$ using $\sqsupset$). Dealing with associative-commutative axioms, we will compare them with the equivalence relation defined by the stable, reflexive, transitive, and symmetric equivalence $\sim$ induced by $\gtrsim$, i.e., $\sim = \gtrsim \cap \lesssim$, since we need to impose compatibility with the equational theories E and F. The following theorem formalizes a generic processor to remove pairs from $\mathcal{P}$ by using reduction pairs.

Theorem 7 (Reduction pair processor). *Let $\mathcal{P} = (\Gamma, F, P)$ be a rewrite theory, $\mathcal{R} = (\Sigma, E, R)$ be an $A \vee C$-rewrite theory, and $\mathcal{S} = (\mathcal{F}, S)$ be a TRS. Let $(\gtrsim, \sqsupset)$ be a reduction pair such that*

1. $R \subseteq \gtrsim$,
2. $P \cup S \subseteq \gtrsim \cup \sqsupset$, *and*
3. $E \cup F \subseteq \sim$.

Let $P_{\sqsupset} = \{u \to v \in P \mid u \sqsupset v\}$. Then, the processor Proc_{RP} *given by*

$$\mathsf{Proc}_{RP}(F, P, E, R, S) = \begin{cases} \{(F, P - P_{\sqsupset}, E, R, S)\} & \text{if (1), (2), and (3) hold} \\ \{(F, P, E, R, S)\} & \text{otherwise} \end{cases}$$

is sound and complete.

7 Related Work and Conclusions

As remarked in the introduction, this is not the first work which tries to use dependency pairs for proving termination of rewriting modulo an equational theory, see [5,10,11,9,13,14]. Our work, however, is, as far as the authors know, the first one which provides a correct notion of minimal non-terminating term for an $A \vee C$-rewrite theory $\mathcal{R} = (\Sigma, E, R)$ which can be used to provide a suitable definition of minimal chain of dependency pairs which can be used to characterize $A \vee C$-termination (Theorem 4). In order to substantiate this claim, consider the AC-rewrite theory $\mathcal{R} = (\Sigma, E, R)$ in Example 2 again. The $A \vee C$-DPs for $\mathcal{R}$ are enumerated in Example 7. Such dependency pairs coincide with the ones which would be computed by, e.g., [5,10,11]. Remember that t in Example 2 is *minimal* in Giesl and Kapur's sense (Definition 2). We should, then, be able to find an infinite *minimal* chain of DPs starting from $t^\sharp$. According to [5,10,11], 'minimal' means that $\sigma(v_i)$ is $(\mathcal{E}xt_E(R), E)$-terminating for all pairs $u_i \to v_i \in \mathsf{DP}_E(R)$ in the chain of dependency pairs induced by the substitution σ. However, this is *not* possible: the marked version $t^\sharp$ of t is $F(f(0,1), f(0, f(1,2)))$, which is an

$(\mathcal{E}xt_E(R), E)$-terminating term. After some $E^\sharp \cup E$-equivalence steps we would be able to apply one of the rules in $\mathsf{DP}_E(R)$. Note, however, that *no rule* $u \to v \in \mathsf{DP}_E(R)$ except (5) has a right-hand side v which can be rewritten (after instantiation into $\sigma(v)$) into an instance $\sigma(u')$ of the left-hand side u' of any other pair in $\mathsf{DP}_E(R)$ by means of $(\mathcal{E}xt_E(R), E^\sharp \cup E)$-rewriting steps. This means that only the dependency pair (5) could be used in any infinite minimal chain of dependency pairs starting from $t^\sharp$. But such a chain would start as follows:

$$F(f(0,1), f(0, f(1,2))) \sim_{E^\sharp \cup E} F(f(0,0), f(1, f(1,2))) \to_{(5)} F(f(0, f(1,2)), f(1, f(1,2)))$$

where $F(f(0, f(1,2)), f(1, f(1,2)))$ contains a subterm $f(1, f(1,2))$ which, as showed in Example 2, is non-$(\mathcal{E}xt_E(R), E)$-terminating. Therefore, this chain of dependency pairs is *not* minimal. We conclude that, according to the notion of minimal chain in the aforementioned papers, *there is no minimal chain of pairs starting from* $t^\sharp$. This means that no *sound* approach to proving AC-termination on the basis of such notion of minimal chain is possible. In this paper we have introduced the notion of *stably minimal term* (Definition 3) which overcomes these problems (Proposition 4 and Theorem 3) and leads to an appropriate characterization of $A\vee C$-termination as the absence of infinite minimal chains of $A\vee C$-DPs (Definitions 5 and 6, and Theorem 4).

Furthermore, we note that [10,11] deal with *AC-rewrite theories* only, and that [5], which considers more general rewrite theories E including $A\vee C$-theories do not cover our work in a second respect: when purely associative theories are considered (i.e., rewrite theories $\mathcal{R} = (\Sigma, E, R)$ such that $E_f \subseteq \{A_f\}$ for all $f \in \Sigma$), then Giesl and Kapur's technique requires the computation of *instances* of the rules in $\mathcal{E}xt_E(R)$ for which the computation of *all* the E-unifiers $uni_E(v,l)$ of v and l for the rules $l \to r$ in $\mathcal{E}xt_E(R)$ and equations $u = v \in E$ or $v = u \in E$ is required. It is well-known, however, that the E-unification problem for associative theories E is *infinitary*, which means that $uni_E(v,l)$ is not guaranteed to be finite, in general. In sharp contrast, we do not have to do that for dealing with purely associative rewrite theories $\mathcal{R}$.

Our second main (and novel) contribution is the formalization of an $A\vee C$-dependency pair framework (Definitions 7 and 8) which, on the basis of the previously developed theory, can be used to develop automatic tools for proving termination of $A\vee C$-rewrite theories (Theorem 5). Two important processors have been adapted as well: the SCC processor (Theorem 6) and the reduction pair processor (Theorem 7).

Much work remains ahead both in terms of further developing the new $A\vee C$-dependency pair framework and in tool support. Appropriate reduction orderings which are well-suited for being used in the reduction pair processor should be investigated. It would also be very useful to explore how the requirements on E can be relaxed to handle even more general sets of axioms. Regarding tool support for the method we have presented, we plan to integrate it within the tool MU-TERM [2]. In this way, our termination technique modulo combinations of associative and commutative axioms will become applicable to an even wider range of rewrite theories, that can be transformed into $A\vee C$-theories by non-termination-preserving transformations [3,4,12].

References

1. Arts, T., Giesl, J.: Termination of Term Rewriting Using Dependency Pairs. Theoretical Computer Science 236(1-2), 133–178 (2000)
2. Alarcón, B., Gutiérrez, R., Iborra, J., Lucas, S.: Proving Termination of Context-Sensitive Rewriting with MU-TERM. Electronic Notes in Theoretical Computer Science 188, 105–115 (2007)
3. Durán, F., Lucas, S., Meseguer, J.: Termination Modulo Combinations of Equational Theories. In: Ghilardi, S., Sebastiani, R. (eds.) FroCoS 2009. LNCS (LNAI), vol. 5749, pp. 246–262. Springer, Heidelberg (2009)
4. Durán, F., Lucas, S., Marché, C., Meseguer, J., Urbain, X.: Proving Operational Termination of Membership Equational Programs. Higher-Order and Symbolic Computation 21(1-2), 59–88 (2008)
5. Giesl, J., Kapur, D.: Dependency Pairs for Equational Rewriting. In: Middeldorp, A. (ed.) RTA 2001. LNCS, vol. 2051, pp. 93–108. Springer, Heidelberg (2001)
6. Giesl, J., Thiemann, R., Schneider-Kamp, P., Falke, S.: Mechanizing and Improving Dependency Pairs. Journal of Automatic Reasoning 37(3), 155–203 (2006)
7. Hirokawa, N., Middeldorp, A.: Dependency Pairs Revisited. In: van Oostrom, V. (ed.) RTA 2004. LNCS, vol. 3091, pp. 249–268. Springer, Heidelberg (2004)
8. Hirokawa, N., Middeldorp, A.: Automating the Dependency Pair Method. Information and Computation 199, 172–199 (2005)
9. Kusakari, K.: Termination, AC-Termination and Dependency Pairs of Term Rewriting Systems. PhD. Thesis, School of Information Science, JAIST (2000)
10. Kusakari, K., Nakamura, M., Toyama, Y.: Elimination Transformations for Associative-Commutative Rewriting Systems. Journal of Automated Reasoning 37, 205–229 (2006)
11. Kusakari, K., Toyama, Y.: On Proving AC-Termination by AC-Dependency Pairs. IEICE Transactions on Information and Systems, E84-D, 604–612 (2001)
12. Lucas, S., Meseguer, J.: Operational Termination of Membership Equational Programs: the Order-Sorted Way. Electronic Notes in Theoretical Computer Science 238(3), 207–225 (2009)
13. Marché, C., Urbain, X.: Termination of associative-commutative rewriting by dependency pairs. In: Nipkow, T. (ed.) RTA 1998. LNCS, vol. 1379, pp. 241–255. Springer, Heidelberg (1998)
14. Marché, C., Urbain, X.: Modular and incremental proofs of AC-termination. Journal of Symbolic Computation 38, 873–897 (2004)
15. Peterson, G.E., Stickel, M.E.: Complete Sets of Reductions for Some Equational Theories. Journal of the ACM 28(2), 233–264 (1981)

Folding Variant Narrowing and Optimal Variant Termination*

Santiago Escobar[1], Ralf Sasse[2], and José Meseguer[2]

[1] DSIC-ELP, Universidad Politécnica de Valencia, Spain
sescobar@dsic.upv.es
[2] University of Illinois at Urbana-Champaign, USA
{rsasse,meseguer}@illinois.edu

Abstract. If a set of equations $E \cup Ax$ is such that E is confluent, terminating, and coherent modulo Ax, narrowing with E modulo Ax provides a complete $E \cup Ax$-unification algorithm. However, except for the hopelessly inefficient case of full narrowing, nothing seems to be known about effective narrowing strategies in the general modulo case beyond the quite depressing observation that basic narrowing is *incomplete* modulo AC. In this work we propose an effective strategy based on the idea of the $E \cup Ax$-variants of a term that we call *folding variant narrowing*. This strategy is *complete*, both for computing $E \cup Ax$-unifiers and for computing a minimal complete set of variants for any input term. And it is *optimally variant terminating* in the sense of terminating for an input term t iff t has a finite, complete set of variants. The applications of folding variant narrowing go beyond providing a complete $E \cup Ax$-unification algorithm: computing the $E \cup Ax$-variants of a term may be just as important as computing $E \cup Ax$-unifiers in recent applications of folding variant narrowing such as termination methods modulo axioms, and checking confluence and coherence of rules modulo axioms.

1 Introduction

Narrowing is a fundamental rewriting technique useful for many purposes, including equational unification and equational theorem proving [15], combinations of functional and logic programming [12,13], partial evaluation [2], symbolic reachability analysis of rewrite theories understood as transition systems [19], and symbolic model checking [7].

Narrowing with confluent and terminating equations E enjoys key completeness results, including the generation of a complete set of E-unifiers and the covering of all rewrite sequences starting at an instance of term t by a normalized substitution, see [15]. However, full narrowing (i.e., narrowing at all non-variable term positions) can be quite inefficient both in space and time. Therefore, much

* S. Escobar has been partially supported by the EU (FEDER) and the Spanish MEC/MICINN under grant TIN 2007-68093-C02-02. J. Meseguer and R. Sasse have been partially supported by NSF Grants CNS 07-16638, CNS 08-31064, and CNS 09-04749.

P.C. Ölveczky (Ed.): WRLA 2010, LNCS 6381, pp. 52–68, 2010.

work has been devoted to *narrowing strategies* that, while remaining complete, can have a much smaller search space. For instance, the *basic narrowing* strategy [15] was shown to be complete w.r.t. a complete set of E-unifiers for confluent and terminating equations E.

Termination aspects are another important potential benefit of narrowing strategies, since they can sometimes *terminate*, generating a finite search tree when narrowing an input term t, while full narrowing may generate an infinite search tree on the same input term. For example, works such as [15,1] investigate conditions under which basic narrowing, one of the most fully studied strategies for termination purposes, terminates. Similarly, so-called lazy narrowing strategies also seek to both reduce the search space and to increase the chances of termination [10], but we are not aware of lazy narrowing strategies for the modulo case.

By decomposing an equational theory $\mathcal{E}$ into a set of rules E and a set of equational axioms Ax for which a finite and complete Ax-unification algorithm exists, and imposing natural requirements such as confluence, termination and coherence of the rules E *modulo* Ax, narrowing can be generalized to narrowing *modulo* axioms Ax. As known since the original study [16], the good completeness properties of standard narrowing extend naturally to similar completeness properties for narrowing modulo Ax. This generalization of narrowing to the modulo case has many applications. It is, to begin with, a key component of theorem proving systems that often reason modulo axioms such as associativity–commutativity, and greatly improves the efficiency of general paramodulation. It is, furthermore, very important for adding functional-logical features to algebraic functional languages supporting rewriting modulo combinations of equational axioms. Yet another recent area with many applications is cryptographic protocol analysis, where there is strong interest in analyzing protocol security *modulo* the algebraic theory $\mathcal{E}$ of a protocol's cryptographic functions, since protocols deemed to be secure under the standard Dolev-Yao model, which treats the underlying cryptography as a black box, can sometimes be broken by clever use of algebraic properties, e.g., [22].

However, very little is known at present about effective narrowing strategies in the modulo case, and some of the known anomalies ring a cautionary note, to the effect that the naive extensions of standard narrowing strategies can fail rather badly in the modulo case. Indeed, except for [16,24], we are not aware of any studies about narrowing strategies in the modulo case. Furthermore, as work in [4,24] shows, narrowing modulo axioms such as associativity-commutativity (AC) can very easily lead to non-terminating behavior and, what is worse, as shown in the Example 1 below, due to Comon-Lundh and Delaune, basic narrowing modulo AC is *not* complete.

Example 1. [4] Consider the equational theory $(\Sigma, E \uplus Ax)$ where E contains the following equations and Ax contains associativity and commutativity for $+$:

$$a + a = 0 \quad (1) \qquad a + a + X = X \ (3) \qquad 0 + X = X \quad (5)$$
$$b + b = 0 \quad (2) \qquad b + b + X = X \ (4)$$

The set E is terminating, AC-convergent, and AC-coherent. Consider now the unification problem $X_1 + X_2 \stackrel{?}{=} 0$ and one of the possible solutions $\sigma = \{X_1 \mapsto a{+}b; X_2 \mapsto a{+}b\}$, which is a normalized solution. It is well-known that in the free case (when $Ax = \emptyset$) basic narrowing is complete for unification in the sense of lifting all innermost rewriting sequences (see [20]). That is, given a term t and a substitution σ, every innermost rewriting sequence starting from $t\sigma$ can be *lifted* to a basic narrowing sequence from t computing a substitution more general than σ. This completeness property fails for basic narrowing modulo AC as shown by the above example when we consider the term $t = X_1 + X_2$ instantiated with σ and the following *innermost rewriting sequence modulo AC* from $t\sigma$ (we underline the redex at each step): $\underline{(a+b)+(a+b)} \rightarrow_{E,AC} \underline{b+b} \rightarrow_{E,AC} 0$. As further explained in Example 3 below, basic narrowing modulo AC, i.e., the extension of basic narrowing to AC where we just replace syntactic unification by AC-unification, cannot lift the above innermost sequence for $t\sigma$, because it is necessary to narrow inside the term generated by instantiation. Therefore, basic narrowing modulo AC is incomplete in the sense of *not* providing a complete $E{\cup}AC$-unification algorithm, even though E may be confluent, terminating, and coherent modulo AC.

It seems clear that full narrowing, although complete, is hopelessly inefficient in the free case, and even more so modulo a set Ax of axioms. The above example shows that known efficient strategies like basic narrowing can totally fail to enjoy the desired completeness properties modulo axioms. What can be done? For equational theories of the form $E{\cup}Ax$, where E is confluent, terminating, and coherent modulo Ax, and such that $E{\cup}Ax$ has the *finite variant property* (FVP) in the sense of [4], we proposed in [9] a narrowing strategy that is complete in the sense of generating a complete set of most general $E \cup Ax$-unifiers, and *terminates* for any input term computing its complete set of variants. And in [8] we gave a method that can be used to check if $E \cup Ax$ is FVP. However, FVP is a quite strong restriction. To the best of our knowledge, except for the hopelessly inefficient case of full narrowing, nothing is known at present about a *general* narrowing strategy that is effective and complete in an adequate sense, including being complete for computing $E \cup Ax$-unifiers, for *any* theory $E \cup Ax$ under the minimum requirements that E is confluent, terminating, and coherent modulo Ax. It turns out that the notion of *variant*, which makes sense for any such theory $E{\cup}Ax$ and does not depend on FVP, provides the key to obtaining a strategy meeting these requirements, and sheds considerable light on the very process of computing $E{\cup}Ax$-unifiers by narrowing.

Our contributions. In this paper, for any theory $E \cup Ax$ with E confluent, terminating, and coherent modulo Ax, we propose *folding variant narrowing* as such a general and effective strategy satisfying the following properties:

1. It is *complete*, both in the sense of computing a complete set of $E \cup Ax$-unifiers, and of computing a minimal and complete set of variants for any input term t.

2. It is *optimal variant terminating*, in the sense that it will terminate for an input term t if and only if t has a finite, complete set of variants (in particular, it will terminate for *any* term t iff $E \cup Ax$ is FVP).

Furthermore, we show that basic narrowing, *both* in the free case ($Ax = \emptyset$) and in the AC case, fails to satisfy properties (1) and/or (2).

The rest of the paper is organized as follows. After some preliminaries in Section 2, we present in Section 3 the notion of variant of a term w.r.t. an order-sorted equational theory and its application to equational unification. Then, we study in Section 4 how to effectively compute the set of variants of a term and provide the folding variant narrowing strategy. In Section 5 we describe future work and conclude the paper.

2 Preliminaries

We follow the classical notation and terminology from [23] for term rewriting and from [18] for rewriting logic and order-sorted notions. We assume an order-sorted signature $\mathbf{\Sigma} = (S, \leq, \Sigma)$ with poset of sorts $(S, \leq)$ and for each sort $s \in S$ where the connected component of s in $(S, \leq)$ has a top sort, denoted $[s]$, and all $f : s_1 \cdots s_n \to s$ with $n \geq 1$ have a top sort overloading $f : [s_1] \cdots [s_n] \to [s]$. We also assume an S-sorted family $\mathcal{X} = \{\mathcal{X}_\mathsf{s}\}_{\mathsf{s} \in \mathsf{S}}$ of disjoint variable sets with each $\mathcal{X}_\mathsf{s}$ countably infinite. $\mathcal{T}_\Sigma(\mathcal{X})_\mathsf{s}$ is the set of terms of sort s, and $\mathcal{T}_{\Sigma,\mathsf{s}}$ is the set of ground terms of sort s. We write $\mathcal{T}_\Sigma(\mathcal{X})$ and $\mathcal{T}_\Sigma$ for the corresponding order-sorted term algebras.

For a term t we write $Var(t)$ for the set of all variables in t. The set of positions of a term t is written $Pos(t)$, and the set of non-variable positions $Pos_\Sigma(t)$. The root position of a term is Λ. The subterm of t at position p is $t|_p$ and $t[u]_p$ is the term t where $t|_p$ is replaced by u.

A *substitution* $\sigma \in Subst(\Sigma, \mathcal{X})$ is a sorted mapping from a finite subset of $\mathcal{X}$, written $Dom(\sigma)$, to $\mathcal{T}_\Sigma(\mathcal{X})$. The set of variables introduced by σ is $Ran(\sigma)$. The identity substitution is *id*. Substitutions are homomorphically extended to $\mathcal{T}_\Sigma(\mathcal{X})$. The application of a substitution σ to a term t is denoted by $t\sigma$. For simplicity, we assume that every substitution is idempotent, i.e., for σ, $Dom(\sigma) \cap Ran(\sigma) = \emptyset$. Substitution idempotency ensures $t\sigma = (t\sigma)\sigma$. The restriction of σ to a set of variables V is $\sigma|_V$; sometimes we write $\sigma|_{t_1,\ldots,t_n}$ to denote $\sigma|_V$ where $V = Var(t_1) \cup \cdots \cup Var(t_n)$. Composition of two substitutions is denoted by $\sigma\sigma'$. We call an idempotent substitution σ a variable *renaming* if there is another substitution σ^{-1} such that $(\sigma\sigma^{-1})|_{Dom(\sigma)} = id$.

A Σ-*equation* is an unoriented pair $t = t'$, where $t, t' \in \mathcal{T}_\Sigma(\mathcal{X})_{[\mathsf{s}]}$ for some sort $\mathsf{s} \in \mathsf{S}$. Given Σ and a set $\mathcal{E}$ of Σ-equations such that $\mathcal{T}_{\Sigma,\mathsf{s}} \neq \emptyset$ for every sort s, order-sorted equational logic induces a congruence relation $=_\mathcal{E}$ on terms $t, t' \in \mathcal{T}_\Sigma(\mathcal{X})$. Throughout this paper we assume that $\mathcal{T}_{\Sigma,\mathsf{s}} \neq \emptyset$ for every sort s. An *equational theory* $(\Sigma, \mathcal{E})$ is a pair with Σ an order-sorted signature and $\mathcal{E}$ a set of Σ-equations.

The $\mathcal{E}$-*subsumption* preorder $\sqsubseteq_\mathcal{E}$ (or $\sqsubseteq$ if $\mathcal{E}$ is understood) holds between $t, t' \in \mathcal{T}_\Sigma(\mathcal{X})$, denoted $t \sqsubseteq_\mathcal{E} t'$ (meaning that t' is *more general* than t modulo $\mathcal{E}$), if there is a substitution σ such that $t =_{\mathcal{E}} t'\sigma$; such a substitution σ is said to be an *$\mathcal{E}$-match* from t to t'. The $\mathcal{E}$-renaming equivalence $t \approx_{\mathcal{E}} t'$, holds if there is a variable renaming θ such that $t\theta =_{\mathcal{E}} t'\theta$. For substitutions σ, ρ and a set of variables V we define $\sigma|_V =_{\mathcal{E}} \rho|_V$ if $x\sigma =_{\mathcal{E}} x\rho$ for all $x \in V$; $\sigma|_V \sqsubseteq_{\mathcal{E}} \rho|_V$ if there is a substitution η such that $\sigma|_V =_{\mathcal{E}} (\rho\eta)|_V$; and $\sigma|_V \approx_{\mathcal{E}} \rho|_V$ if there is a renaming η such that $(\sigma\eta)|_V =_{\mathcal{E}} \rho|_V$.

An *$\mathcal{E}$-unifier* for a Σ-equation $t = t'$ is a substitution σ such that $t\sigma =_{\mathcal{E}} t'\sigma$. For $Var(t) \cup Var(t') \subseteq W$, a set of substitutions $CSU^W_{\mathcal{E}}(t = t')$ is said to be a *complete* set of unifiers of the equation $t =_{\mathcal{E}} t'$ away from W if: (i) each $\sigma \in CSU^W_{\mathcal{E}}(t = t')$ is an $\mathcal{E}$-unifier of $t =_{\mathcal{E}} t'$; (ii) for any $\mathcal{E}$-unifier ρ of $t =_{\mathcal{E}} t'$ there is a $\sigma \in CSU^W_{\mathcal{E}}(t = t')$ such that $\rho|_W \sqsubseteq_{\mathcal{E}} \sigma|_W$; (iii) for all $\sigma \in CSU^W_{\mathcal{E}}(t = t')$, $Dom(\sigma) \subseteq (Var(t) \cup Var(t'))$ and $Ran(\sigma) \cap W = \emptyset$. If the set of variables W is irrelevant or understood from the context, we write $CSU_{\mathcal{E}}(t = t')$ instead of $CSU^W_{\mathcal{E}}(t = t')$. An $\mathcal{E}$-unification algorithm is *complete* if for any equation $t = t'$ it generates a complete set of $\mathcal{E}$-unifiers. Note that this set needs not be finite. A unification algorithm is said to be *finitary* and complete if it always terminates after generating a finite and complete set of solutions. A unification algorithm is said to be *minimal* if it always provides a maximal (w.r.t. $\sqsubseteq_{\mathcal{E}}$) set of unifiers.

A *rewrite rule* is an oriented pair $l \to r$, where $l \notin \mathcal{X}$ and $l, r \in \mathcal{T}_\Sigma(\mathcal{X})_{[\mathsf{s}]}$ for some sort $\mathsf{s} \in \mathsf{S}$. An *(unconditional) order-sorted rewrite theory* is a triple (Σ, Ax, R) with Σ an order-sorted signature, Ax a set of Σ-equations, and R a set of rewrite rules. The rewriting relation on $\mathcal{T}_\Sigma(\mathcal{X})$, written $t \to_R t'$ or $t \to_{p,R} t'$ holds between t and t' iff there exist $p \in Pos_\Sigma(t)$, $l \to r \in R$ and a substitution σ, such that $t|_p = l\sigma$, and $t' = t[r\sigma]_p$. The subterm $t|_p$ is called a *redex*. The relation $\to_{R/Ax}$ on $\mathcal{T}_\Sigma(\mathcal{X})$ is $=_{Ax};\to_R;=_{Ax}$. Note that $\to_{R/Ax}$ on $\mathcal{T}_\Sigma(\mathcal{X})$ induces a relation $\to_{R/Ax}$ on the free (Σ, Ax)-algebra $\mathcal{T}_{\Sigma/Ax}(\mathcal{X})$ by $[t]_{Ax} \to_{R/Ax} [t']_{Ax}$ iff $t \to_{R/Ax} t'$. The transitive closure of $\to_{R/Ax}$ is denoted by $\to^+_{R/Ax}$ and the transitive and reflexive closure of $\to_{R/Ax}$ is denoted by $\to^*_{R/Ax}$. We say that a term t is $\to_{R/Ax}$-irreducible (or just R/Ax-irreducible) if there is no term t' such that $t \to_{R/Ax} t'$.

For substitutions σ, ρ and a set of variables V we define $\sigma|_V \to_{R/Ax} \rho|_V$ if there is $x \in V$ such that $x\sigma \to_{R/Ax} x\rho$ and for all other $y \in V$ we have $y\sigma =_{Ax} y\rho$. A substitution σ is called *R/Ax-normalized* (or normalized) if $x\sigma$ is R/Ax-irreducible for all $x \in V$.

We say that the relation $\to_{R/Ax}$ is *terminating* if there is no infinite sequence $t_1 \to_{R/Ax} t_2 \to_{R/Ax} \cdots t_n \to_{R/Ax} t_{n+1} \cdots$. We say that the relation $\to_{R/Ax}$ is *confluent* if whenever $t \to^*_{R/Ax} t'$ and $t \to^*_{R/Ax} t''$, there exists a term t''' such that $t' \to^*_{R/Ax} t'''$ and $t'' \to^*_{R/Ax} t'''$. An order-sorted rewrite theory (Σ, Ax, R) is confluent (resp. terminating) if the relation $\to_{R/Ax}$ is confluent (resp. terminating). In a confluent, terminating, order-sorted rewrite theory, for each term $t \in \mathcal{T}_\Sigma(\mathcal{X})$, there is a unique (up to Ax-equivalence) R/Ax-irreducible term t' obtained from t by rewriting to canonical form, which is denoted by $t \to^!_{R/Ax} t'$ or $t{\downarrow}_{R/Ax}$ (when t' is not relevant).

2.1 R, Ax-Rewriting

Since Ax-congruence classes can be infinite, $\rightarrow_{R/Ax}$-reducibility is undecidable in general. Therefore, R/Ax-rewriting is usually implemented [16] by R, Ax-rewriting. We assume the following properties on R and Ax:

1. Ax is *regular*, i.e., for each $t = t'$ in Ax, we have $Var(t) = Var(t')$, and *sort-preserving*, i.e., for each substitution σ, we have $t\sigma \in \mathcal{T}_{\Sigma}(\mathcal{X})_{\mathsf{s}}$ iff $t'\sigma \in \mathcal{T}_{\Sigma}(\mathcal{X})_{\mathsf{s}}$; furthermore all variables in $Var(t)$ have a top sort.
2. Ax has a finitary and complete unification algorithm.
3. For each $t \rightarrow t'$ in R we have $Var(t') \subseteq Var(t)$.
4. R is *sort-decreasing*, i.e., for each $t \rightarrow t'$ in R, each $\mathsf{s} \in \mathsf{S}$, and each substitution σ, $t'\sigma \in \mathcal{T}_{\Sigma}(\mathcal{X})_{\mathsf{s}}$ implies $t\sigma \in \mathcal{T}_{\Sigma}(\mathcal{X})_{\mathsf{s}}$.
5. The rewrite rules R are *confluent and terminating modulo* Ax, i.e., the relation $\rightarrow_{R/Ax}$ is confluent and terminating.

Definition 1 (Rewriting modulo [25]). *Let (Σ, Ax, R) be an order-sorted rewrite theory satisfying properties (1)–(5). We define the relation $\rightarrow_{R,Ax}$ on $\mathcal{T}_{\Sigma}(\mathcal{X})$ by $t \rightarrow_{p,R,Ax} t'$ (or just $t \rightarrow_{R,Ax} t'$) iff there is a $p \in Pos_{\Sigma}(t)$, $l \rightarrow r$ in R and substitution σ such that $t|_p =_{Ax} l\sigma$ and $t' = t[r\sigma]_p$.*

Note that, since Ax-matching is decidable, $\rightarrow_{R,Ax}$ is decidable. Notions such as confluence, termination, irreducible terms, and normalized substitution, are defined in a straightforward manner for $\rightarrow_{R,Ax}$. Note that since R is confluent and terminating modulo Ax, the relation $\rightarrow^!_{R,Ax}$ is decidable, i.e., it terminates and produces a unique term (up to Ax-equivalence) for each initial term t, denoted by $t{\downarrow}_{R,Ax}$. Of course $t \rightarrow_{R,Ax} t'$ implies $t \rightarrow_{R/Ax} t'$, but the converse does not need to hold. To prove completeness of $\rightarrow_{R,Ax}$ w.r.t. $\rightarrow_{R/Ax}$ we need the following additional *coherence* assumption; we refer the reader to [11,25,17] for coherence completion algorithms.

6. $\rightarrow_{R,Ax}$ is *Ax-coherent* [16], i.e., $\forall t_1, t_2, t_3$ we have $t_1 \rightarrow_{R,Ax} t_2$ and $t_1 =_{Ax} t_3$ implies $\exists t_4, t_5$ such that $t_2 \rightarrow^*_{R,Ax} t_4$, $t_3 \rightarrow^+_{R,Ax} t_5$, and $t_4 =_{Ax} t_5$.

The following theorem in [16, Proposition 1] that generalizes ideas in [21] and has an easy extension to order-sorted theories, links $\rightarrow_{R/Ax}$ with $\rightarrow_{R,Ax}$.

Theorem 1 (Correspondence [21,16]). *Let (Σ, Ax, R) be an order-sorted rewrite theory satisfying properties (1)–(6). Then $t_1 \rightarrow^!_{R/Ax} t_2$ iff $t_1 \rightarrow^!_{R,Ax} t_3$, where $t_2 =_{Ax} t_3$.*

Finally, we provide the notion of decomposition of an equational theory into rules and axioms.

Definition 2 (Decomposition [9]). *Let $(\Sigma, \mathcal{E})$ be an order-sorted equational theory. We call (Σ, Ax, E) a* decomposition *of $(\Sigma, \mathcal{E})$ if $\mathcal{E} = E \uplus Ax$ and (Σ, Ax, E) is an order-sorted rewrite theory satisfying properties (1)–(6).*

3 Variants and Equational Unification

Suppose that an equational theory $\mathcal{E}$ is decomposed into a set of rules E and a set of equational axioms Ax such that a finite and complete Ax-unification algorithm exists, and the rules E are confluent, terminating, sort-decreasing, and coherent *modulo* Ax. Given a term t, an *E,Ax-variant* of t is a pair (t',θ) with t' an E,Ax-canonical form of the term $t\theta$. That is, the variants of a term intuitively give us all the irreducible *patterns* that instances of t can reduce to. Of course, some variants are *more general* than others, that is, there is a natural preorder $(t',\theta') \sqsubseteq_{E,Ax} (t'',\theta'')$ defining when variant (t'',θ'') is *more general* than variant (t',θ'). This is important, because even though the set of E,Ax-variants of a term t may be infinite, the set of *most general variants* (that is maximal elements in the generalization preorder up to Ax-equivalence and variable renaming) may be finite.

The intimate connection of variants with $\mathcal{E}$-unification is then as follows. Suppose that we add to our theory decomposition $E \uplus Ax$ a binary equality predicate eq, a new constant tt[1] and for each top sort $[s]$ and x of sort $[s]$ an extra rule $eq(x,x) \to tt$. Then, given any two terms t, t', if θ is a $\mathcal{E}$-unifier of t and t', then the E,Ax canonical forms of $t\theta$ and $t'\theta$ must be Ax-equal and therefore the pair (tt,θ) must be a variant of the term $eq(t,t')$. Furthermore, if the term $eq(t,t')$ has a finite set of most general variants, then we are *guaranteed* that the set of most general $\mathcal{E}$-unifiers of t and t' is *finite*.

We characterize a notion of variant semantics for equational theories.

Definition 3 (Variant Semantics). *Let (Σ, Ax, E) be a decomposition of an equational theory and t be a term. We define the set of variants of t as $[\![t]\!]^{\star}_{E,Ax} = \{(t',\theta) \mid \theta \in \mathcal{S}ubst(\Sigma,\mathcal{X}), t\theta \to^{!}_{E,Ax} t'',$ and $t'' =_{Ax} t'\}$.*

Let us make explicit the relation between variants and $\mathcal{E}$-unification.

Proposition 1 (Variant-based Unification). *Let (Σ, Ax, E) be a decomposition of an equational theory $(\Sigma,\mathcal{E})$. Let t_1, t_2 be two terms. Then, ρ is a $\mathcal{E}$-unifier of t_1 and t_2 iff $\exists (t',\rho) \in [\![t_1]\!]^{\star}_{E,Ax} \cap [\![t_2]\!]^{\star}_{E,Ax}$.*

Some variants are more general than others. We write $(t_1,\theta_1) \sqsubseteq_{E,Ax} (t_2,\theta_2)$ to denote that variant (t_2,θ_2) is *more general* than variant (t_1,θ_1). Our notion of being more general takes into account not only the instantiation relation between the two substitutions θ_1 and θ_2 and the two normal forms t_1 and t_2 of a term t, but also whether θ_2 is already an E,Ax-normalized substitution, since, for a substitution, the less E,Ax rewrite steps, the better.

Definition 4 (Variant Preordering). *Let (Σ, Ax, E) be a decomposition of an equational theory and t be a term. Given two variants $(t_1,\theta_1), (t_2,\theta_2) \in [\![t]\!]^{\star}_{E,Ax}$, we write $(t_1,\theta_1) \sqsubseteq_{E,Ax} (t_2,\theta_2)$, meaning (t_2,θ_2) is* more general *than*

[1] We extend Σ to $\widehat{\Sigma}$ by adding a new sort Truth, not related to any sort in Σ, with constant `tt`, and for each top sort of a connected component [s], an operator eq : [s] × [s] → Truth.

(t_1, θ_1), *iff there is a substitution* ρ *such that* $t_1 =_{Ax} t_2\rho$ *and* $\theta_1{\downarrow}_{E,Ax} =_{Ax} \theta_2\rho$. *We write* $(t_1, \theta_1) \sqsubset_{E,Ax} (t_2, \theta_2)$ *if for every substitution* ρ *such that* $t_1 =_{Ax} t_2\rho$ *and* $\theta_1{\downarrow}_{E,Ax} =_{Ax} \theta_2\rho$, *then* ρ *is not a renaming.*

We are, indeed, interested in equivalence classes for variant semantics and provide a notion of semantic equality, written $\simeq_{E,Ax}$, based on $\sqsubseteq_{E,Ax}$.

Definition 5 (Variant Equality). *Let* (Σ, Ax, E) *be a decomposition of an equational theory and* t *be a term. For* $S_1, S_2 \subseteq [\![t]\!]^\star_{E,Ax}$, *we write* $S_1 \sqsubseteq_{E,Ax} S_2$ *iff for each* $(t_1, \theta_1) \in S_1$, *there exists* $(t_2, \theta_2) \in S_2$ *s.t.* $(t_1, \theta_1) \sqsubseteq_{E,Ax} (t_2, \theta_2)$. *We write* $S_1 \simeq_{E,Ax} S_2$ *iff* $S_1 \sqsubseteq_{E,Ax} S_2$ *and* $S_2 \sqsubseteq_{E,Ax} S_1$.

Despite the previous semantic notion of equivalence, the following, more syntactic notion of equality of variants up to renaming is useful.

Definition 6 (Ax-Equality). *Let* (Σ, Ax, E) *be a decomposition of an equational theory and* t *be a term. For* $(t_1, \theta_1), (t_2, \theta_2) \in [\![t]\!]^\star_{E,Ax}$, *we write* $(t_1, \theta_1) \approx_{Ax} (t_2, \theta_2)$ *if there is a renaming* ρ *such that* $t_1\rho =_{Ax} t_2\rho$ *and* $\theta_1\rho =_{Ax} \theta_2\rho$. *For* $S_1, S_2 \subseteq [\![t]\!]^\star_{E,Ax}$, *we write* $S_1 \approx_{Ax} S_2$ *if for each* $(t_1, \theta_1) \in S_1$, *there exists* $(t_2, \theta_2) \in S_2$ *s.t.* $(t_1, \theta_1) \approx_{Ax} (t_2, \theta_2)$, *and for each* $(t_2, \theta_2) \in S_2$, *there exists* $(t_1, \theta_1) \in S_1$ *s.t.* $(t_2, \theta_2) \approx_{Ax} (t_1, \theta_1)$.

The preorder of Definition 4 allows us to provide a most general and complete set of variants that encompasses all the variants for a term t.

Definition 7 (Most General and Complete Variant Semantics). *Let* (Σ, Ax, E) *be a decomposition of an equational theory and* t *be a term. A most general and complete variant semantics of* t, *denoted* $[\![t]\!]_{E,Ax}$, *is a subset* $[\![t]\!]_{E,Ax} \subseteq [\![t]\!]^\star_{E,Ax}$ *such that: (i)* $[\![t]\!]^\star_{E,Ax} \sqsubseteq_{E,Ax} [\![t]\!]_{E,Ax}$, *and (ii) for each* $(t_1, \theta_1) \in [\![t]\!]_{E,Ax}$, *there is no* $(t_2, \theta_2) \in [\![t]\!]_{E,Ax}$ *s.t.* $(t_1, \theta_1) \not\approx_{Ax} (t_2, \theta_2)$ *and* $(t_1, \theta_1) \sqsubseteq_{E,Ax} (t_2, \theta_2)$.

Note that, for any term t, $[\![t]\!]^\star_{E,Ax} \simeq_{E,Ax} [\![t]\!]_{E,Ax}$ but, in general, $[\![t]\!]^\star_{E,Ax} \not\approx_{Ax} [\![t]\!]_{E,Ax}$. Also, by definition, all the substitutions in $[\![t]\!]_{E,Ax}$ are E,Ax-normalized. Moreover, $[\![t]\!]_{E,Ax}$ is unique up to $\approx_{Ax}$ and provides a very succinct description of $[\![t]\!]^\star_{E,Ax}$. Indeed, up to Ax-equality, $[\![t]\!]_{E,Ax}$ characterizes the set of *maximal elements* (therefore, most general variants) of the preorder $([\![t]\!]_{E,Ax}, \sqsubseteq_{E,Ax})$.

Again, let us make explicit the relation between variants and $\mathcal{E}$-unification.

Proposition 2 (Minimal and Complete $\mathcal{E}$-unification). *Let* (Σ, Ax, E) *be a decomposition of an equational theory* $(\Sigma, \mathcal{E})$. *Let* t, t' *be two terms. Then,* $S = \{\theta \mid (\mathtt{tt}, \theta) \in [\![\mathsf{eq}(t, t')]\!]_{\widehat{E},Ax}\}$ *is a minimal and complete set of* $\mathcal{E}$*-unifiers for* $t = t'$, *where* eq *and* $\mathtt{tt}$ *are new symbols defined in Footnote 1 and* $\widehat{E} = E \cup \{\mathsf{eq}(X, X) \to \mathtt{tt}\}$.

Example 2. Let us consider the following equational theory for the exclusive or operator and the cancellation equations for public encryption/decryption, which is actually useful for protocol verification (see [19]). This equational theory is relevant because there are no unification procedures directly applicable to it, e.g.

unification algorithms for exclusive-or such as [3] do not directly apply if extra equations are added. The exclusive or symbol $\oplus$ has associative and commutative (AC) properties with 0 as its unit. The symbol pk is used for public key encryption and the symbol sk for private key encryption. The equational theory $(\Sigma, \mathcal{E})$ has a decomposition into E containing the following oriented equations and Ax containing associativity and commutativity for $\oplus$:

$$X \oplus 0 = X \quad (6) \qquad X \oplus X = 0 \quad (7) \qquad pk(K, sk(K, M)) = M \quad (9)$$

$$X \oplus X \oplus Y = Y \quad (8) \qquad sk(K, pk(K, M)) = M \quad (10)$$

Note that equations (6)–(7) are not AC-coherent, but adding equation (8) is sufficient to recover that property. For $t = M \oplus sk(K, pk(K, M))$ and $s = X \oplus sk(K, pk(K, Y))$, we have that $[\![t]\!]_{E,Ax} = \{(0, id)\}$ and $[\![s]\!]_{E,Ax} = \{(X \oplus Y, id), (Z, \{X \mapsto 0, Y \mapsto Z\}), (Z, \{X \mapsto Z, Y \mapsto 0\}), (Z, \{X \mapsto Z \oplus U, Y \mapsto U\}), (Z, \{X \mapsto U, Y \mapsto Z \oplus U\}), (0, \{X \mapsto U, Y \mapsto U\}), (Z_1 \oplus Z_2, \{X \mapsto U \oplus Z_1, Y \mapsto U \oplus Z_2\})\}$. This set is the most general one w.r.t. $\sqsubseteq_{E,Ax}$.

The *finite variant property* defined by Comon-Lundh and Delaune [4], provides a useful sufficient condition for finitary $\mathcal{E}$-unification. Essentially, it determines whether every term has a finite number of most general variants.

Definition 8 (Finite variant property [4]). *Let (Σ, Ax, E) be a decomposition of an equational theory $(\Sigma, \mathcal{E})$. Then $(\Sigma, \mathcal{E})$, and thus (Σ, Ax, E), has the* finite variant property *iff for each term t, the set $[\![t]\!]_{E,Ax}$ is finite. We will call (Σ, Ax, E) a* finite variant decomposition *of $(\Sigma, \mathcal{E})$ iff (Σ, Ax, E) has the finite variant property.*

In [8] we developed a technique to check whether an equational theory has the finite variant property. Using our technique it is easy to check that Example 2 has the finite variant property, as every right–hand side is a constant symbol or a variable.

Finally, it is clear that when we consider a finite variant decomposition, we have a decidable unification algorithm.

Corollary 1 (Finitary $\mathcal{E}$-unification). *Let (Σ, Ax, E) be a finite variant decomposition of an equational theory $(\Sigma, \mathcal{E})$. Then, for any two given terms t, t', $S = \{\theta \mid (\mathtt{tt}, \theta) \in [\![\mathtt{eq}(t, t')]\!]_{\widehat{E},Ax}\}$ is a* finite, minimal, and complete *set of $\mathcal{E}$-unifiers for $t = t'$, where $\widehat{E}$, $\mathtt{eq}$, and $\mathtt{tt}$ are defined as in Proposition 2.*

Note that the opposite does not hold: given two terms t, t' that have a finite, minimal, and complete set of $\mathcal{E}$-unifiers, the equational theory $(\Sigma, \mathcal{E})$ may not have a finite variant decomposition (Σ, Ax, E). An example is the unification under homomorphism (or one-side distributivity), where there is a finite number of unifiers of two terms but the theory does not satisfy the finite variant property (see [4,8]); the key idea is that the term $\mathtt{eq}(t, t')$ may have an infinite number of variants even though there is only a finite set of most general variants of the form $(\mathtt{tt}, \theta)$.

Once we have clarified the intimate relation between variants and equational unification, we consider in the next section how to compute a complete set of variants of a term.

4 Variants and Narrowing-Based Equational Unification

Narrowing generalizes rewriting by performing unification at non-variable positions instead of the usual matching. The essential idea behind narrowing is to *symbolically* represent the rewriting relation between terms as a narrowing relation between more general terms with variables.

Definition 9 (Narrowing modulo [16,19]). *Let (Σ, Ax, R) be an order-sorted rewrite theory. Let $CSU_{Ax}(u = u')$ provide a finitary and complete set of Ax-unifiers for any pair of terms u, u' with the same top sort. Let t be a term and W be a set of variables such that $Var(t) \subseteq W$. The R, Ax-narrowing relation on $\mathcal{T}_\Sigma(\mathcal{X})$ is defined as $t \leadsto_{p,\sigma,R,Ax} t'$ ($\leadsto_{\sigma,R,Ax}$ if p is understood, and $\leadsto$ if σ, R, Ax are understood) if there is $p \in Pos_\Sigma(t)$, a rule $l \to r \in R$ properly renamed s.t. $Var(l) \cap W = \emptyset$, and $\sigma \in CSU^{W'}_{Ax}(t|_p = l)$ for $W' = W \cup Var(l)$ such that $t' = (t[r]_p)\sigma$.*

For convenience, in each narrowing step $t \leadsto_\sigma t'$ we only provide the part of σ that binds variables of t. The transitive closure of $\leadsto$ is denoted by $\leadsto^+$ and the transitive and reflexive closure by $\leadsto^*$. We may write $t \leadsto^*_\sigma t'$ instead of $t \leadsto^* t'$ if there are $s_1, \ldots, s_{k-1}$ and substitutions $\rho_1, \ldots, \rho_k$ such that $t \leadsto_{\rho_1} s_1 \cdots s_{k-1} \leadsto_{\rho_k} t'$, $k \geq 0$, and $\sigma = \rho_1 \cdots \rho_k$. Several notions of completeness of narrowing w.r.t. rewriting have been given in the literature (e.g. [15,16,19]).

Theorem 2 (Completeness of Full Narrowing Modulo [16]). *Let (Σ, Ax, E) be a decomposition of an equational theory. Let t_1 be a term and θ be an E,Ax-normalized substitution. If $t_1\theta \to^!_{E,Ax} t_2$, then there exists a term t'_2 and two E,Ax-normalized substitutions θ' and ρ s.t. $t_1 \leadsto^*_{\theta',E,Ax} t'_2$, $\theta|_{Var(t_1)} =_{Ax} (\theta'\rho)|_{Var(t_1)}$, and $t_2 =_{Ax} t'_2\rho$. Furthermore, the rewriting sequence and the narrowing sequence have the same number of steps, with the same rules and at the same positions.*

Narrowing completeness ensures complete generation of all the variants of a term and, thus, an $\mathcal{E}$-unification algorithm: if the term $eq(t, t')$ has a finite set of most general variants, then we are *guaranteed* that the set of most general substitutions *computed* by E,Ax-narrowing is *finite* and provides the set of most general $\mathcal{E}$-unifiers of t and t'. However, can we compute the set of most general $\mathcal{E}$-unifiers of t and t' *effectively*? This is not entirely obvious. Full E,Ax-narrowing may never terminate, since it will compute a complete set of variants of the form (tt, θ) for the term $eq(t, t')$, but that set may easily be infinite, even though a finite set of most general elements for it exists. The solution, of course, is that we should look for adequate narrowing *strategies* that have better properties than full E,Ax-narrowing so that, in the end, we can obtain a *terminating* narrowing-based $\mathcal{E}$-unification *algorithm* to unify t and t' whenever any term $eq(t, t')$ has a finite set of most general variants.

4.1 Narrowing Strategies and Their Properties

In order to provide an appropriate narrowing strategy that enjoys better properties than full E,Ax-narrowing, we need to characterize what a narrowing strategy is and which properties it must satisfy. E.g., the notion of variant–completeness rather than the standard full narrowing completeness becomes essential.

First, we define the notion of a narrowing strategy and several useful properties. Given a narrowing sequence $\alpha : (t_0 \leadsto_{\sigma_0,p_0,R,Ax} t_1 \cdots \leadsto_{\sigma_{n-1},p_{n-1},R,Ax} t_n)$, we denote by α_i the narrowing sequence $\alpha_i : (t_0 \leadsto_{\sigma_0,p_0,R,Ax} t_1 \cdots \leadsto_{\sigma_{i-1},p_{i-1},R,Ax} t_i)$ which is a prefix of α. We denote by $\mathit{Full}_{\mathcal{R}}(t)$ the set of all narrowing sequences starting at term t.

Definition 10 (Narrowing Strategy). *A narrowing strategy $\mathcal{S}$ is a function of two arguments, namely, a rewrite theory $\mathcal{R} = (\Sigma, Ax, R)$ and a term $t \in \mathcal{T}_{\Sigma}(\mathcal{X})$, which we denote by $\mathcal{S}_{\mathcal{R}}(t)$, such that $\mathcal{S}_{\mathcal{R}}(t) \subseteq \mathit{Full}_{\mathcal{R}}(t)$. We require $\mathcal{S}_{\mathcal{R}}(t)$ to be prefix closed, i.e., for each narrowing sequence $\alpha \in \mathcal{S}_{\mathcal{R}}(t)$, and each $i \in \{1, \ldots, n\}$, we also have $\alpha_i \in \mathcal{S}_{\mathcal{R}}(t)$.*

We say a narrowing strategy $\mathcal{S}$ is *complete* if it satisfies Theorem 2. In this paper we are interested in a notion of completeness of a narrowing strategy slightly different than previous notions, which we call *variant-completeness.* First, we extend the variant semantics to narrowing and consider only narrowing sequences to normalized terms.

Definition 11 (Narrowing Semantics). *Let $\mathcal{R} = (\Sigma, Ax, E)$ be a decomposition of an equational theory $(\Sigma, \mathcal{E})$ and $\mathcal{S}$ be a narrowing strategy. We define the set of* narrowing variants *of a term t w.r.t. $\mathcal{S}$ as $[\![t]\!]^{\mathcal{S}}_{E,Ax} = \{(t', \theta) \mid (t \leadsto^*_{\theta,E,Ax} t') \in \mathcal{S}_{\mathcal{R}}(t) \text{ and } t' = t'{\downarrow_{E,Ax}}\}$.*

Now, we can define our notion of variant–completeness.

Definition 12 (Variant Completeness and Minimality). *Let (Σ, Ax, E) be a decomposition of an equational theory $(\Sigma, \mathcal{E})$. A narrowing strategy $\mathcal{S}$ is called* $\mathcal{E}$-variant–complete *(or just variant–complete) iff for any term t $[\![t]\!]_{E,Ax} \simeq_{E,Ax} [\![t]\!]^{\mathcal{S}}_{E,Ax}$. The narrowing strategy $\mathcal{S}$ is called* $\mathcal{E}$-variant–minimal *(or just variant–minimal) iff, in addition, we have that for any term t $[\![t]\!]_{E,Ax} \approx_{Ax} [\![t]\!]^{\mathcal{S}}_{E,Ax}$ and for each pair of variants $(t_1, \theta_1), (t_2, \theta_2) \in [\![t]\!]^{\mathcal{S}}_{E,Ax}$ such that $(t_1, \theta_1) \neq_{Ax} (t_2, \theta_2)$, we have that $(t_1, \theta_1) \not\approx_{Ax} (t_2, \theta_2)$.*

This minimality property motivates the following corollary.

Corollary 2. *Let (Σ, Ax, E) be a decomposition of an equational theory $(\Sigma, \mathcal{E})$ and $\mathcal{S}$ be an $\mathcal{E}$-variant-complete narrowing strategy. For any two terms t, t' with the same top sort, the set $S = \{\theta \mid (\mathtt{tt}, \theta) \in [\![\mathtt{eq}(t, t')]\!]^{\mathcal{S}}_{\widehat{E},Ax}\}$ is a complete set of $\mathcal{E}$-unifiers for $t = t'$, where $\widehat{E}$, $\mathtt{eq}$, and $\mathtt{tt}$ are defined as in Proposition 2. If, in addition, $\mathcal{S}$ is a $\mathcal{E}$-variant-minimal narrowing strategy, then the set S is a minimal set of $\mathcal{E}$-unifiers for $t = t'$.*

In practice, the set $\mathcal{S}_{\mathcal{R}}(t)$ of narrowing sequences from a term t will be generated by an *algorithm* $\mathcal{A}_{\mathcal{S}}$. That is, $\mathcal{A}_{\mathcal{S}}$ is a computable function such that, given a pair $(\mathcal{R}, t)$, enumerates the set $\mathcal{S}_{\mathcal{R}}(t)$. If $\mathcal{E} = (\Sigma, Ax, E)$ is a decomposition of an equational theory, the strategy $\mathcal{S}_\mathcal{E}$ is variant–complete, and $[\![t]\!]_{E,Ax}$ is finite on an input term t, then $[\![t]\!]^{\mathcal{S}}_{E,Ax}$ may not be finite. Furthermore, even if $[\![t]\!]^{\mathcal{S}}_{E,Ax}$ is finite, its enumeration using the algorithm $\mathcal{A}_\mathcal{S}$ might not terminate. We are of course interested in variant–complete narrowing strategies that will *always* terminate on an input term t whenever $[\![t]\!]_{E,Ax}$ is finite, since by Corollary 2 such strategies will provide a finitary $\mathcal{E}$-unification algorithm whenever $\mathcal{E}$ has the finite variant property. This leads to the following notion of variant–termination for an algorithm $\mathcal{A}_\mathcal{S}$ restricting the class of algorithms we are interested in.

Definition 13 (Optimal Variant Termination). *Let (Σ, Ax, E) be a decomposition of an equational theory $(\Sigma, \mathcal{E})$ and $\mathcal{S}$ be an $\mathcal{E}$-variant-complete narrowing strategy. An algorithm $\mathcal{A}_\mathcal{S}$ is* variant terminating *iff $\mathcal{A}_\mathcal{S}(\mathcal{E}, t)$ terminates on input $(\mathcal{E}, t)$ iff $[\![t]\!]^{\mathcal{S}}_{E,Ax}$ is finite. An algorithm $\mathcal{A}_\mathcal{S}$ is* optimally variant terminating *iff $\mathcal{A}_\mathcal{S}$ is variant terminating and $[\![t]\!]^{\mathcal{S}}_{E,Ax}$ is variant–minimal for every term t.*

By abuse of language, we say that a narrowing strategy $\mathcal{S}$ is variant terminating (resp. optimally variant terminating) whenever $\mathcal{A}_\mathcal{S}$ is. The term "optimally variant terminating" is justified as follows.

Corollary 3. *Let $\mathcal{E} = (\Sigma, Ax, E)$ be a decomposition of an equational theory $(\Sigma, \mathcal{E})$. Let $\mathcal{S}$ be a $\mathcal{E}$-variant–complete narrowing strategy and $\mathcal{S}'$ be an optimally variant terminating narrowing strategy. Then, for each term t such that $\mathcal{S}_\mathcal{E}(t)$ is finite, then $\mathcal{S}'_\mathcal{E}(t)$ is also finite.*

4.2 Basic Narrowing Modulo Is Neither Variant–Complete Nor Optimally Variant–Terminating

In this section we show that basic narrowing modulo AC is not variant–complete. Furthermore, we show that even basic narrowing without axioms is not optimally variant–terminating, thus motivating that there is room for improvement even in the free case. We extend the standard definition of basic narrowing given in [14] to the modulo case.

Definition 14 (Basic Narrowing modulo Ax). *Let (Σ, Ax, R) be an order-sorted rewrite theory. Given a term $t \in \mathcal{T}_\Sigma(\mathcal{X})$, a substitution ρ, and a set W of variables such that $Var(t) \subseteq W$ and $Var(\rho) \subseteq W$, a basic narrowing modulo Ax step for $\langle t, \rho\rangle$ is defined by $\langle t, \rho\rangle \overset{b}{\leadsto}_{p,\theta,R,Ax} \langle t', \rho'\rangle$ if there is $p \in Pos_\Sigma(t)$, a rule $l \to r \in R$ properly renamed s.t. $Var(l) \cap W = \emptyset$, and $\theta \in CSU^{W'}_{Ax}(t|_p\rho = l)$ for $W' = W \cup Var(l)$ such that $t' = t[r]_p$, and $\rho' = \rho\theta$.*

Basic narrowing modulo AC is incomplete w.r.t. innermost rewriting modulo AC despite the free case [20], i.e., there are innermost rewriting sequences modulo AC that are not lifted to basic narrowing modulo Ax. And, therefore, basic narrowing modulo AC is not variant–complete.

Example 3. The narrowing sequence shown in Example 1 is not a basic narrowing sequence modulo AC, as after the first step it results in $\langle X, \rho_1\rangle$ and no further basic narrowing modulo AC step is possible:

$$\langle X_1 + X_2, id\rangle \overset{b}{\leadsto}_{\Lambda,\rho_1,E,Ax} \langle X, \rho_1\rangle$$
using $\rho_1 = \{X_1 \mapsto a + X', X_2 \mapsto a + X'', X \mapsto X' + X''\}$ and rule (3)

Therefore, basic narrowing modulo AC is *not variant-complete*, since the pair $(0, \sigma)$ is a variant of t. The (full or unrestricted) narrowing sequence associated to the unification problem $X_1 + X_2 \stackrel{?}{=} 0$ in the extended equational theory $\widehat{E}$ defined in Proposition 2 is:

$\mathtt{eq}(\underline{X_1 + X_2}, 0) \leadsto_{\rho_1,\widehat{E},Ax} \mathtt{eq}(X' + X'', 0)$
using $\rho_1 = \{X_1 \mapsto a + X', X_2 \mapsto a + X''\}$ and rule (3)

$\mathtt{eq}(\underline{X' + X''}, 0) \leadsto_{\rho_2,\widehat{E},Ax} \mathtt{eq}(0, 0)$ using $\rho_2 = \{X' \mapsto b, X'' \mapsto b\}$ and rule (2)

$\underline{\mathtt{eq}(0, 0)} \leadsto_{id,\widehat{E},Ax} \mathtt{tt}$ using rule $\mathtt{eq}(X, X) \to \mathtt{tt}$

Furthermore, if we add a new equation $0+0+X = 0+X$ basic narrowing modulo AC does not terminate though the number of variants does not change at all, due to the following always available narrowing step $0 + X_2 \leadsto_{\theta_1,E,Ax} 0 + X_2'$ using $\theta_1 = \{X_2 \mapsto 0 + X_2', X \mapsto X_2'\}$.

Moreover, basic narrowing in the free case is not optimally variant–terminating, as shown by the following example.

Example 4. Consider the rewrite theory $\mathcal{R} = (\Sigma, \emptyset, E)$ where E is the set of convergent rules $E = \{f(x) \to x, f(f(x)) \to f(x)\}$ and Σ contains only the unary symbol f and a constant a. The term $t = f(x)$ has only one variant: $[\![f(x)]\!]_{E,Ax} = \{(x, id)\}$. Indeed, the theory has the finite variant property (see [8]). Basic narrowing performs the following two narrowing steps:

(i) $\langle f(x), id\rangle \overset{b}{\leadsto}_{\{x \mapsto x'\},E} \langle x', \{x \mapsto x'\}\rangle$ and
(ii) $\langle f(x), id\rangle \overset{b}{\leadsto}_{\{x \mapsto f(x')\},E} \langle f(x'), \{x \mapsto f(x')\}\rangle$.

However, the second narrowing step leads to the following non-terminating basic narrowing sequence:

$$\begin{aligned}\langle f(x), id\rangle &\overset{b}{\leadsto}_{\{x \mapsto f(x')\},E} \langle f(x'), \{x \mapsto f(x')\}\rangle\\ &\overset{b}{\leadsto}_{\{x' \mapsto f(x'')\},E} \langle f(x''), \{x \mapsto f(f(x'')), x' \mapsto f(x'')\}\rangle\\ &\ldots\end{aligned}$$

and basic narrowing is unable to terminate and provide the finite number of variants associated to the term t.

In the following section we provide a narrowing strategy to compute the variants of a term that is variant–complete, variant–minimal, and optimally variant–terminating.

4.3 An Optimally Variant–Terminating, and Variant–Minimal Narrowing Strategy for Finite Variant Decompositions

For a finite variant decomposition, we achieve optimal variant termination by simply keeping track of all the variants generated so far, since we know that there is a finite set of more general variants and sooner or later narrowing will generate all the most general variants. We have developed in [7] a way of detecting such repetitions.

Definition 15 (Transition System). *[7] A* transition system *is written* $\mathcal{A} = (A, \to)$*, where* A *is a set of states, and* $\to$ *is a transition relation between states, i.e.,* $\to \subseteq A \times A$*. We write* $\mathcal{A} = (A, \to, I)$ *when* $I \subseteq A$ *is a set of initial states.*

Intuitively, we define a global strategy that keeps track of previously computed variants and discards narrowing steps that compute a previously met variant.

Definition 16 (Folding Reachable Transition Subsystem [7]). *Given a transition system* $\mathcal{A} = (A, \to, I)$ *and a relation* $G \subseteq A \times A$*, the reachable subsystem from* I *in* $\mathcal{A}$ *with folding* G *is written* $\mathcal{R}each^G_{\mathcal{A}}(I) = (Reach^G_{\to}(I), \to^G, I)$*, where* $Reach^G_{\to}(I) = \bigcup_{n\in\mathbb{N}} Frontier^G_{\to}(I)_n$ *and*

$$
\begin{array}{ll}
Frontier^G_{\to}(I)_0 = I, & \\
Frontier^G_{\to}(I)_{n+1} & = \{y \in A \mid (\exists z \in Frontier^G_{\to}(I)_n : z \to y) \wedge \\
& \qquad (\nexists k \leq n, w \in Frontier^G_{\to}(I)_k : y\ G\ w)\},
\end{array}
$$

$$
\to^G = \bigcup_{n\in\mathbb{N}} \to^G_{n+1},
$$

$$
x \to^G_{n+1} y \begin{cases} \text{if } x \in Frontier^G_{\to}(I)_n, y \in Frontier^G_{\to}(I)_{n+1}, \\ \quad x \to y; \\ \text{if } x \in Frontier^G_{\to}(I)_n, y \notin Frontier^G_{\to}(I)_{n+1}, \\ \quad \exists k \leq n : y \in Frontier^G_{\to}(I)_k, \exists w : (x \to w \wedge w\ G\ y) \end{cases}
$$

Note that, the more general relation G, the greater the chances of $\mathcal{R}each^G_{\mathcal{A}}(I)$ being a finite transition system. In [7], we study different relations G such as $\sqsubseteq_{Ax}$ or $\approx_{Ax}$ and its properties. For computing the variants, G is just the preorder $\sqsubseteq_{E,Ax}$ between variants. Given a decomposition (Σ, Ax, E) of an equational theory $(\Sigma, \mathcal{E})$ and a narrowing strategy $\mathcal{S}_{\mathcal{E}}$, we extend $\mathcal{S}_{\mathcal{E}}$ to variants as follows: given a term t and a substitution ρ s.t. $t\rho =_{Ax} t$, $\mathcal{S}_{\mathcal{E}}((t, \rho)) = \{(t, \rho) \leadsto^k_{\sigma,E,Ax} (t', \rho\sigma) \mid (t \leadsto^k_{\sigma,E,Ax} t') \in \mathcal{S}_{\mathcal{E}}(t)\}$. Given a narrowing strategy $\mathcal{S}_{\mathcal{R}}$, we write $\mathcal{S}^1_{\mathcal{R}}$ to denote narrowing derivations produced by $\mathcal{S}_{\mathcal{R}}$ of length exactly 1.

Definition 17 (Folding Narrowing Strategy). *Let* (Σ, Ax, E) *be a decomposition of an equational theory* $(\Sigma, \mathcal{E})$ *and* $\mathcal{S}_{\mathcal{E}}$ *a narrowing strategy. Let* t *be a term. Let us consider the transition system* $(\mathcal{T}_{\Sigma}(\mathcal{X}) \times \mathcal{S}ubst(\Sigma, \mathcal{X}), \mathcal{S}^1_{\mathcal{E}}, I)$ *for variants with the one-step version of the strategy* $\mathcal{S}_{\mathcal{E}}$ *and the initial state* $I = (t, id)$*. The folding* $\mathcal{S}_{\mathcal{E}}$*–narrowing strategy, denoted by* $\mathcal{S}^{\circlearrowleft}_{\mathcal{E}}(t)$*, is defined as*

$$
\begin{array}{l}
\mathcal{S}^{\circlearrowleft}_{\mathcal{E}}(t) = \{t \leadsto^k_{\sigma,E,Ax} t' \mid ((t, id) \leadsto^k_{\sigma,E,Ax} (t', \sigma)) \in \mathcal{S}_{\mathcal{E}}(t)\ \wedge \\
\qquad\qquad\qquad (t', \sigma) \in Frontier^{\sqsubseteq_{E,Ax}}_{\mathcal{S}^1_{\mathcal{E}}}(I)_k\}
\end{array}
$$

We write $Full^{\circlearrowleft}_{\mathcal{R}}$ for the folding version of the full narrowing strategy.

The following example shows that basic narrowing may be non-terminating in cases when variant narrowing does terminate.

Example 5. Considering Example 4 and using the $Full_{\mathcal{R}}^{\circlearrowleft}$ strategy we only get step (i). Step (ii) is subsumed as $(f(x'), \{x \mapsto f(x')\}) \sqsubseteq_{E,\emptyset} (x', \{x \mapsto x'\})$. So even though basic narrowing does not terminate in this case, $Full_{\mathcal{R}}^{\circlearrowleft}$ does.

The following example shows what steps can be done by $Full_{\mathcal{R}}^{\circlearrowleft}$ and termination of it on the given example.

Example 6. Using the theory from Example 2, for $t = X \oplus Y$ we get the following $Full_{\mathcal{R}}^{\circlearrowleft}$ steps. Note that we only need to consider steps with normalized substitutions as otherwise the resulting variant would be subsumed by the variant reachable using the normalized form of the same substitution.

(i) $(t, id) \leadsto_{\phi_1} (Z, \phi_1)$, with $\phi_1 = \{X \mapsto 0, Y \mapsto Z\}$,
(ii) $(t, id) \leadsto_{\phi_2} (Z, \phi_2)$, with $\phi_2 = \{X \mapsto Z, Y \mapsto 0\}$,
(iii) $(t, id) \leadsto_{\phi_3} (Z, \phi_3)$, with $\phi_3 = \{X \mapsto Z \oplus U, Y \mapsto U\}$,
(iv) $(t, id) \leadsto_{\phi_4} (Z, \phi_4)$, with $\phi_4 = \{X \mapsto U, Y \mapsto Z \oplus U\}$,
(v) $(t, id) \leadsto_{\phi_5} (0, \phi_5)$, with $\phi_5 = \{X \mapsto U, Y \mapsto U\}$,
(vi) $(t, id) \leadsto_{\phi_6} (Z_1 \oplus Z_2, \phi_6)$, with $\phi_6 = \{X \mapsto U \oplus Z_1, Y \mapsto U \oplus Z_2\}$.

There are no further steps possible from (i)-(v) as any instantiation of Z for which a narrowing step is possible would mean that the substitution is not normalized, and 0 is a normal form without variables. For the result of (vi), $(Z_1 \oplus Z_2, \phi_6)$, we are back at the beginning and can repeat all of the steps possible for (t, id), but all of the results are subsumed by the same step we already have from (t, id). So, $Full_{\mathcal{R}}^{\circlearrowleft}$ terminates for t.

Note that by the use of the folding definition we get only the shortest paths to each possible term (depending on the substitution), since the longer paths are simply subsumed by shorter ones using $\sqsubseteq_{E,Ax}$. Any folding narrowing strategy is sound as it is a further restriction of the narrowing strategy. We prove that any folding narrowing strategy is variant–complete provided the given narrowing strategy is complete according to Theorem 2.

Theorem 3 (Variant Completeness of Folding Narrowing). *Let (Σ, Ax, E) be a decomposition of an equational theory $(\Sigma, \mathcal{E})$. Let t_1 be a term and θ be an E,Ax-normalized substitution. Let $\mathcal{S}_{\mathcal{E}}$ be a complete narrowing strategy. If $t_1\theta \to^!_{E,Ax} t_2$ then there exists a term t'_2 and two E,Ax-normalized substitutions θ' and ρ s.t. $(t_1 \leadsto^*_{\theta',E,Ax} t'_2) \in \mathcal{S}_{\mathcal{E}}^{\circlearrowleft}(t)$, $\theta|_{Var(t_1)} =_{Ax} (\theta'\rho)|_{Var(t_1)}$, and $t_2 =_{Ax} t'_2\rho$.*

The following corollary establishes that folding full-narrowing is an optimally variant–terminating, and variant–minimal narrowing strategy for finite variant decompositions.

Corollary 4. *Let (Σ, Ax, E) be a decomposition of an equational theory $(\Sigma, \mathcal{E})$. The folding full–narrowing $Full_{\mathcal{E}}^{\circlearrowleft}$ is* variant–complete *and* variant–minimal, *i.e., for any term t, $[\![t]\!]_{E,Ax} \approx_{Ax} [\![t]\!]^{Full_{\mathcal{E}}^{\circlearrowleft}}_{E,Ax}$. Moreover, if (Σ, Ax, E) is a finite variant decomposition of $(\Sigma, \mathcal{E})$, then $Full_{\mathcal{E}}^{\circlearrowleft}$ is also optimally variant–terminating.*

5 Conclusions

To the best our knowledge, the general problem of finding effective strategies for narrowing modulo axioms that avoid the hopeless inefficiency of full narrowing and the incompleteness in general of basic narrowing for the modulo case, has remained unsolved up to now. We have presented *folding variant narrowing* as an effective strategy that, by computing exactly and only a mimimal complete set of variants for a term t, is optimally variant terminating, and complete both for unification purposes and for computing variants. Besides yielding in particular a new finitary unification algorithm for FVP equational theories that improves upon the variant algorithm presented in [9], and does not require anymore prior checking of FVP as described in [8], by being applicable to *any* equational theory modulo under minimal assumptions of confluence, termination, and coherence, many more applications than just cryptographic protocol analysis modulo algebraic properties in the style of the Maude-NPA [6] are opened up. In fact, several such applications, to termination methods modulo axioms [5], and to the most recent Maude CRC and ChC tools modulo axioms (see `http://maude.lcc.uma.es/CRChC/`), are already exploiting the general power of folding variant narrowing.

As always, however, much work remains ahead, particularly in the two closely-related areas of refining and optimizing the folding variant narrowing strategy, and of developing an efficient implementation. There is already an existing implementation in Maude of variant narrowing under the FVP assumption that has been shown effective in formally analyzing a good number of cryptographic protocols modulo a variety of algebraic theories describing their cryptographic infrastructure (see [6] and references there). We expect that a good part of the infrastructure of the current FVP variant narrowing strategy will be easily extensible to an optimized form of the folding variant narrowing strategy; but this will require substantial new work in design, implementation, and experimentation.

References

1. Alpuente, M., Escobar, S., Iborra, J.: Modular termination of basic narrowing. In: Voronkov, A. (ed.) RTA 2008. LNCS, vol. 5117, pp. 1–16. Springer, Heidelberg (2008)
2. Alpuente, M., Falaschi, M., Vidal, G.: Partial Evaluation of Functional Logic Programs. ACM Transactions on Programming Languages and Systems 20(4), 768–844 (1998)
3. Anantharaman, S., Narendran, P., Rusinowitch, M.: Unification modulo *cui* plus distributivity axioms. J. Autom. Reasoning 33(1), 1–28 (2004)
4. Comon-Lundh, H., Delaune, S.: The finite variant property: How to get rid of some algebraic properties. In: Giesl, J. (ed.) RTA 2005. LNCS, vol. 3467, pp. 294–307. Springer, Heidelberg (2005)
5. Durán, F., Lucas, S., Meseguer, J.: Termination modulo combinations of equational theories. In: Ghilardi, S. (ed.) FroCoS 2009. LNCS, vol. 5749, pp. 246–262. Springer, Heidelberg (2009)

6. Escobar, S., Meadows, C., Meseguer, J.: Maude-npa: Cryptographic protocol analysis modulo equational properties. In: FOSAD 2007. LNCS, vol. 5705, pp. 1–50. Springer, Heidelberg (2009)
7. Escobar, S., Meseguer, J.: Symbolic model checking of infinite-state systems using narrowing. In: Baader, F. (ed.) RTA 2007. LNCS, vol. 4533, pp. 153–168. Springer, Heidelberg (2007)
8. Escobar, S., Meseguer, J., Sasse, R.: Effectively checking the finite variant property. In: Voronkov, A. (ed.) RTA 2008. LNCS, vol. 5117, pp. 79–93. Springer, Heidelberg (2008)
9. Escobar, S., Meseguer, J., Sasse, R.: Variant narrowing and equational unification. Electr. Notes Theor. Comput. Sci. 238(3), 103–119 (2009)
10. Escobar, S., Meseguer, J., Thati, P.: Natural narrowing for general term rewriting systems. In: Giesl, J. (ed.) RTA 2005. LNCS, vol. 3467, pp. 279–293. Springer, Heidelberg (2005)
11. Giesl, J., Kapur, D.: Dependency pairs for equational rewriting. In: Middeldorp, A. (ed.) RTA 2001. LNCS, vol. 2051, pp. 93–108. Springer, Heidelberg (2001)
12. Goguen, J.A., Meseguer, J.: Equality, types, modules, and (why not?) generics for logic programming. J. Log. Program. 1(2), 179–210 (1984)
13. Hanus, M.: The Integration of Functions into Logic Programming: From Theory to Practice. Journal of Logic Programming 19&20, 583–628 (1994)
14. Hölldobler, S.: Foundations of Equational Logic Programming. In: Hölldobler, S. (ed.) Foundations of Equational Logic Programming. LNCS (LNAI), vol. 353. Springer, Heidelberg (1989)
15. Hullot, J.-M.: Canonical forms and unification. In: Bibel, W., Kowalski, R.A. (eds.) CADE 1980. LNCS, vol. 87, pp. 318–334. Springer, Heidelberg (1980)
16. Jouannaud, J.-P., Kirchner, C., Kirchner, H.: Incremental construction of unification algorithms in equational theories. In: Díaz, J. (ed.) ICALP 1983. LNCS, vol. 154, pp. 361–373. Springer, Heidelberg (1983)
17. Jouannaud, J.-P., Kirchner, H.: Completion of a set of rules modulo a set of equations. SIAM J. Comput. 15(4), 1155–1194 (1986)
18. Meseguer, J.: Conditional rewriting logic as a united model of concurrency. Theor. Comput. Sci. 96(1), 73–155 (1992)
19. Meseguer, J., Thati, P.: Symbolic reachability analysis using narrowing and its application to verification of cryptographic protocols. Higher-Order and Symbolic Computation 20(1-2), 123–160 (2007)
20. Middeldorp, A., Hamoen, E.: Completeness results for basic narrowing. Journal of Applicable Algebra in Engineering, Communication, and Computing 5, 213–253 (1994)
21. Peterson, G.E., Stickel, M.E.: Complete sets of reductions for some equational theories. J. ACM 28(2), 233–264 (1981)
22. Ryan, P.Y.A., Schneider, S.A.: An attack on a recursive authentication protocol. a cautionary tale. Inf. Process. Lett. 65(1), 7–10 (1998)
23. TeReSe (ed.): Term Rewriting Systems. Cambridge University Press, Cambridge (2003)
24. Viola, E.: E-unifiability via narrowing. In: Restivo, A., Rocca, S.R.D., Roversi, L. (eds.) ICTCS 2001. LNCS, vol. 2202, pp. 426–438. Springer, Heidelberg (2001)
25. Viry, P.: Equational rules for rewriting logic. Theor. Comput. Sci. 285(2), 487–517 (2002)

A Church-Rosser Checker Tool for Conditional Order-Sorted Equational Maude Specifications

Francisco Durán[1] and José Meseguer[2]

[1] Universidad de Málaga, Spain
[2] University of Illinois at Urbana-Champaign, IL, USA

Abstract. The Church-Rosser property, together with termination, is essential for an equational specification to have good executability conditions, and also for having a complete agreement between the specification's initial algebra, mathematical semantics, and its operational semantics by rewriting. Checking this property for expressive specifications that are order-sorted, conditional with possibly extra variables in their condition, and whose equations can be applied modulo different combinations of associativity, commutativity and identity axioms is challenging. In particular, the resulting *conditional critical pairs* that cannot be joined have often an intuitively unsatisfiable condition or seem intuitively joinable, so that sophisticated tool support is needed to eliminate them. Another challenge is the presence of different combinations of associativity, commutativity and identity axioms, including the very challenging case of associativity without commutativity for which no finitary unification algorithms exist. In this paper we present the foundations and illustrate the design and use of a completely new version of the Maude Church-Rosser Checker tool that addresses all the abovementioned challenges and can deal effectively with complex conditional specifications modulo axioms.

1 Introduction

The goal of *executable* equational specification languages is to make *computable* the abstract data types specified in them by initial algebra semantics. In practice this is accomplished by using specifications that are Church-Rosser (or at least ground Church-Rosser) and terminating, so that the equations can be used from left to right as simplification rules; the result of evaluating an expression is then the canonical form that stands as a unique representative for the equivalence class of terms equal to the original term according to the equations. This approach is fully general; indeed, a well-known result of Bergstra and Tucker [5] shows that *any* computable algebraic data type can be specified by means of a finite set of ground-Church-Rosser and terminating equations, perhaps with the help of some auxiliary functions added to the original signature. For order-sorted specifications, being Church-Rosser and terminating means not only confluence—so that a unique normal form will be reached—but also a *descent* property ensuring that the normal form will have the least possible sort among those of all other equivalent terms.

P.C. Ölveczky (Ed.): WRLA 2010, LNCS 6381, pp. 69–85, 2010.

Therefore, for computational purposes it becomes very important to know whether a given specification is indeed (ground-)Church-Rosser and terminating. A nontrivial question is how to best support this with adequate tools. One can prove the operational termination of his/her (possibly conditional) Maude equational specification by using the MTT tool [14]. A thornier issue is what to do for establishing the ground-Church-Rosser property for a terminating specification. The problem is that a specification with an initial algebra semantics can be ground-Church-Rosser even though some of its critical pairs may not be joinable. That is, the specification can often be ground-Church-Rosser without being Church-Rosser for arbitrary terms with variables. In such a situation, blindly applying a completion procedure that is trying to establish the Church-Rosser property for arbitrary terms may be both quite hopeless—the procedure may diverge or get stuck because of unorientable rules, and even with success may return a specification that is quite different from the original one—and even unnecessary, if the specification was already ground-Church-Rosser. As we further explain in Section 3, several methods that do not alter the mathematical semantics of the original specification may allow us to either prove that the specification is ground Church-Rosser, or to transform it into an equivalent one that is Church-Rosser; typically with minimal changes.

Here, we present CRC, a Church-Rosser checker to check whether a (possibly conditional) order-sorted equational specification modulo equational axioms satisfies the Church-Rosser property. Our Church-Rosser checker tool is particularly well-suited for checking specifications with an initial algebra semantics that have already been proved terminating and now need to be checked to be Church-Rosser, or at least ground-Church-Rosser. Of course, the CRC tool can also be used to check the Church-Rosser property of conditional order-sorted specifications that do not have an initial algebra semantics, such as, for example, those specified in Maude functional theories [9]. Since, for the reasons mentioned above, user interaction will typically be quite essential, completion is not attempted. Instead, if the specification cannot be shown to be Church-Rosser by the tool, proof obligations are generated and are given back to the user as a guide in the attempt to establish the ground-Church-Rosser property. Since this property is in fact inductive, in some cases the Maude inductive theorem prover can be enlisted to prove some of these proof obligations. In other cases, the user may have to modify the original specification by carefully considering the information conveyed by the proof obligations. We give in Section 3 some methodological guidelines for the use of the tool, and illustrate the use of the tool with some examples (additional examples can be found in [17]).

The present CRC tool accepts order-sorted conditional specifications, where each of the operation symbols has either no equational attributes, or any combination of associativity/commutativity/identity. To deal with the various combinations of associativity, commutativity, and identity axioms we make use of different techniques now available. Maude 2.4 supports unification modulo commutativity and modulo associativity and commutativity [10]. Identity axioms and associativity without commutativity are handled using the variant-based theory transformations presented in [15]. As pointed out in [15], the transformation cannot be used in general for the associativity without commutativity case because it does not have the finite variant property. However, the alternative semi-algorithm given there can be used in many practical situations in which the lefthand sides do have a finite set of variants. We refer the reader to [15] for further details, but the idea is that if for each operator in a module we cannot narrow on any equation's lefthand side using one of the two possible orientations of the associativity equation, then the only variant of the term is the term itself, and we can handle it just by adding the corresponding associativity equation. All this means that in practice we can often handle specifications whose operators can have *any* combination of associativity and/or commutativity and/or identity axioms. See Section 3.2 for an example.

Furthermore, it is assumed that such specifications do not contain any built-in function, do not use the `owise` attribute,[1] and that they have already been proved (operationally) terminating. The tool attempts to establish the ground-Church-Rosser property *modulo* the equational axioms specified for each of the operators by checking a sufficient condition. Therefore, the tool's output consists of a set of critical pairs and a set of membership assertions that must be shown, respectively, ground-joinable, and ground-rewritable to a term with the required sort.

The CRC tool has been implemented as an extension of Full Maude [16,13], as other tools in the Maude formal environment [11,20], and can be used on any Full Maude module satisfying the above restrictions, including structured modules, parameterized modules, etc.

The rest of the paper is structured as follows. Section 2 introduces the notion of Church-Rosser conditional order-sorted specifications modulo axioms. Section 3 presents some guidelines on how to use the tool and illustrates them with some examples. Section 4 concludes and presents some future work. Proofs of technical results are not included here for space reasons. They can be found in [19].

2 Church-Rosser (Conditional) Order-Sorted Specifications Modulo Axioms

In this section we introduce the notion of Church-Rosser order-sorted specification [23] and some standard notation on conditional rewriting (see, e.g., [29] for further details). We assume specifications of the form $\mathcal{R} = (\Sigma, A, R)$ where Σ is an A-preregular order-sorted signature, A is a set of equational axioms that are both regular and linear,[2] and R is an A-coherent set of (possibly conditional) rewrite

[1] In Maude, the `owise` attribute can be used to specify otherwise equations, i.e., equations that will be applied only if no other equation for that symbol can be applied.

[2] An equational axiom $u = v$ is regular if $\mathcal{V}ar(u) = \mathcal{V}ar(v)$, and linear if there are no repeated variables in either u or v.

rules. Let us start by introducing the notions of A-preregularity and A-coherence.

An order-sorted signature $(\Sigma, S, \leq)$ consists of a poset of sorts $(S, \leq)$ and an $S^* \times S$-indexed family of sets $\Sigma = \{\Sigma_{s_1 \ldots s_n, s}\}_{(s_1 \ldots s_n, s) \in S^* \times S}$ of function symbols. Given an S-sorted set $\mathcal{X} = \{\mathcal{X}_s \mid s \in S\}$ of *disjoint* sets of variables, the set $\mathcal{T}(\Sigma, \mathcal{X})_s$ denotes the Σ-algebra of Σ-terms of sort s with variables in $\mathcal{X}$. We denote $[t]_A$ the A-equivalence class of t.

We call an order-sorted signature A-*preregular* if for each term w the set of sorts $\{s \in S \mid \exists w' \in [w]_A \text{ s.t. } w' \in \mathcal{T}(\Sigma, \mathcal{X})_s\}$ has a least upper bound, denoted $ls[w]_A$, which can be effectively computed.[3]

We denote by $\mathcal{P}(t)$ the set of positions of a term t, and by $t|_p$ *the subterm of* t *at position* p (with $p \in \mathcal{P}(t)$). A term t with its subterm $t|_p$ replaced by the term t' is denoted by $t[t']_p$.

Given a set of equational axioms A, a substitution σ is an A-*unifier* of t and t' if $t\sigma =_A t'\sigma$, and it is an A-*match* from t to t' if $t' =_A t\sigma$.

Given a rewrite theory $\mathcal{R}$ as above, we define the relation $\rightarrow_{R/A}$, either by the inference system of rewriting logic (see [8]), or by the usual inductive description: $\rightarrow_{R/A} = \bigcup_n \rightarrow_{R/A,n}$, where $\rightarrow_{R/A,0} = \emptyset$, and for each $n \in \mathbb{N}$, we have $\rightarrow_{R/A,n+1} = \rightarrow_{R/A,n} \cup \{(u, v) \mid u =_A l\sigma \rightarrow r\sigma =_A v \wedge l \rightarrow r \text{ if } \bigwedge_i u_i \rightarrow v_i \in R \wedge \forall i, u_i\sigma \rightarrow^*_{R/A,n} v_i\sigma\}$. In general, of course, given terms t and t' with sorts in the same connected component, the problem of whether $t \rightarrow_{R/A} t'$ holds is undecidable. For this reason, a much simpler relation $\rightarrow_{R,A}$ is defined, which becomes decidable if an A-matching algorithm exists. We define (see [30]) $\rightarrow_{R,A} = \bigcup_n \rightarrow_{R,A,n}$ where $\rightarrow_{R,A,0} = \emptyset$, and for each $n \in \mathbb{N}$ and any terms u, v with sorts in the same connected component the relation $u \rightarrow_{R,A,n+1} v$ holds if either $u \rightarrow_{R,A,n} v$, or there is a position p in u, a rule $l \rightarrow r$ if $\bigwedge_i u_i \rightarrow v_i$ in R, and a substitution σ such that $u|_p =_A l\sigma$, $v = u[r\sigma]_p$, and $\forall i, u_i\sigma \rightarrow^*_{R,A,n} w_i$ with $w_i =_A v_i\sigma$.

Of course, $\rightarrow_{R,A} \subseteq \rightarrow_{R/A}$, but the question is whether any $\rightarrow_{R/A}$-step can be simulated by a $\rightarrow_{R,A}$-step. We say that $\mathcal{R}$ satisfies this A-*completeness* property if for any u, v with sorts in the same connected component we have:

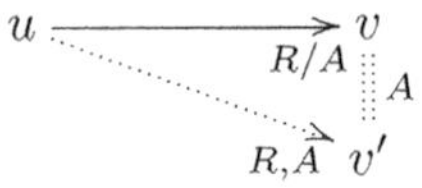

where here and in what follows dotted lines indicate existential quantification.

It is easy to check that A-completeness is equivalent to the following (strong) A-coherence property:

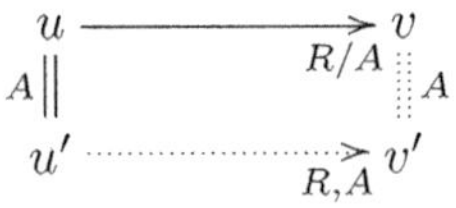

[3] The Maude system automatically checks the A-preregularity of a signature Σ for A any combination of associativity/commutativity/identity (see [9, Section 22.2.5]).

If a theory $\mathcal{R}$ is not coherent, we can try to make it so by completing the set of rules R to a set of rules $\widetilde{R}$ by a Knuth-Bendix-like completion procedure (see, e.g., [25,33,21]). For theories A that are combinations of associativity, commutativity, and identity axioms, we can make any specification A-coherent by using a very simple procedure (see, e.g., [17]).

The problem, then, is to check whether our specification $\mathcal{R}$, satisfying the above requirements, has the Church-Rosser property. As said above, for order-sorted specifications, being Church-Rosser and terminating means not only confluence, but also a descent property. After giving some auxiliary definitions, we introduce the notion of Church-Rosser conditional order-sorted specifications, and describe the sufficient conditions used by our tool to attempt checking the Church-Rosser property.

2.1 The Confluence Property

We say that a term t *A-overlaps* another term with distinct variables t' if there is a nonvariable subterm $t'|_p$ of t' for some position $p \in \mathcal{P}(t')$ such that the terms t and $t'|_p$ can be A-unified.

Definition 1. *Given an order-sorted equational specification $\mathcal{R} = (\Sigma, A, R)$, with Σ A-preregular and R A-coherent, and given conditional rewrite rules $l \to r$ if C and $l' \to r'$ if C' in R such that $(\mathcal{V}ar(l) \cup \mathcal{V}ar(r) \cup \mathcal{V}ar(C)) \cap (\mathcal{V}ar(l') \cup \mathcal{V}ar(r') \cup \mathcal{V}ar(C')) = \emptyset$ and $l|_p\sigma =_A l'\sigma$, for some nonvariable position $p \in \mathcal{P}(l)$ and A-unifier σ of $l|_p$ and l', then the triple*

$$C\sigma \wedge C'\sigma \Rightarrow l\sigma[r'\sigma]_p = r\sigma$$

is called a (conditional) critical pair.

In the uses we will make of the above definition we will always assume that the unification and the comparison for equality have been performed *modulo A*. Note also that the critical pairs accumulate the substitution instances of the conditions in the two rules, as in [7].

Given a rewrite theory $\mathcal{R} = (\Sigma, A, R)$, a critical pair $C \Rightarrow u = v$ is more general than another critical pair $C' \Rightarrow u' = v'$ if there exists a substitution σ such that $u\sigma =_A u'$, $v\sigma =_A v'$, and $C\sigma =_A C'$, where $C\sigma =_A C'$, with $C = \bigwedge_{i=1..n} u_i \to v_i$ and $C' = \bigwedge_{i=1..m} u'_i \to v'_i$, iff $n = m$ and $u_i\sigma =_A u'_i$ and $v_i\sigma =_A v'_i$ for every $i \in [1..n]$.

Then, given a specification $\mathcal{R}$, let MCP($\mathcal{R}$) denote the set of most general critical pairs between rules in $\mathcal{R}$ that, after simplifying both sides of the critical pair using the equational rules in $\mathcal{R}$, are not identical critical pairs modulo A of the form $C \Rightarrow t = t$. Under the assumption that the order-sorted equational specification $\mathcal{R}$ is operationally terminating, then, if MCP($\mathcal{R}$) $= \emptyset$, we are guaranteed that the specification $\mathcal{R}$ is *confluent* modulo A—in the obvious sense that if t can be rewritten modulo A to u and v using the rules in $\mathcal{R}$, then u and v can be rewritten modulo A to some w up to A-equality—and therefore, each term t has a unique canonical form modulo A $t{\downarrow_{\mathcal{R}}}$. Note that, due to the presence of conditional equations, we can have MCP($\mathcal{R}$) $\neq \emptyset$ with $\mathcal{R}$ still confluent, but establishing that fact may require additional reasoning. More importantly for our purposes, even in the unconditional case we can have MCP($\mathcal{R}$) $\neq \emptyset$ with $\mathcal{R}$ *ground*-confluent, that is, confluent for all ground terms. Therefore, assuming termination, MCP($\mathcal{R}$) $= \emptyset$ will ensure the confluence and, a fortiori, the ground-confluence of $\mathcal{R}$, but this is only a sufficient condition.

2.2 Context-Joinability and Unfeasible Conditional Critical Pairs

From those conditional critical pairs which are not joinable, our tool can currently discard those that are either *context-joinable* or *unfeasible*, based on a result by Avenhaus and Loría-Sáenz [2], which we generalize here to the order-sorted modulo A case. Let us first introduce some notation.

A rule $l \rightarrow u_{n+1}$ *if* $\bigwedge_{i=1..n} u_i \rightarrow v_i$ is said to be *deterministic* if $\forall j \in [1..n+1], \mathcal{V}ar(u_j) \subseteq \mathcal{V}ar(l) \cup \bigcup_{k<j} \mathcal{V}ar(v_k)$. A conditional rewrite theory is *deterministic* if each of its rules is deterministic. Given a rewrite theory $\mathcal{R}$, a term t is called *strongly irreducible* with respect to R modulo A (or *strongly* R, A-*irreducible*) if $t\sigma$ is a normal form for every normalized substitution σ. A rewrite theory $\mathcal{R}$ is called *strongly deterministic* if for every rule $l \rightarrow r$ *if* $\bigwedge_{i=1..n} u_i \rightarrow v_i$ in R each v_i is strongly R, A-irreducible.

An admissible conditional order-sorted Maude functional specification can be transformed into an equivalent deterministic rewrite theory by a very simple procedure, in which equations are turned into rewrite rules and equational conditions (ordinary and matching equations) are turned into rewrites (see [17] for a detailed algorithm).

We denote by $\rhd$ the proper subterm relation. Then, given an order $\succ$, we denote by $\succ_{st} = (\succ \cup \rhd)^+$ the smallest ordering that contains $\succ$ and $\rhd$. A partial ordering $\succ$ on $\mathcal{T}_\Sigma(\mathcal{X})$ is *well founded* if there is no infinite sequence $t_0 \succ t_1 \succ \ldots$. A partial ordering $\succ$ is *compatible with substitutions* if $u \succ u'$ implies $u\sigma \succ u'\sigma$ for any substitution σ. A partial ordering $\succ$ is *compatible with the term structure* if $u \succ u'$ implies $t[u]_p \succ t[u']_p$ for any term t and position p in t. A partial ordering $\succ$ is *compatible with the axioms* A if $v =_A u \succ u' =_A v'$ implies $v \succ v'$ for all terms u, u', v, and v' in $\mathcal{T}_\Sigma(\mathcal{X})$. A partial ordering $\succ$ is *A-compatible* if it is compatible with substitutions, compatible with the term structure, and compatible with the axioms A. Then, a *reduction ordering* is a partial ordering that is well founded and A-compatible.

A deterministic rewrite theory $\mathcal{R}$ is *quasi-reductive* w.r.t. a reduction ordering $\succ$ on $\mathcal{T}_\Sigma(\mathcal{X})$ if for every substitution σ, every rule $l \rightarrow u_{n+1}$ *if* $\bigwedge_{i=1...n} u_i \rightarrow v_i$ in R, and every $i \in [1..n]$, $u_j\sigma \succeq v_j\sigma$ for every $j \in [1..i]$ implies $l\sigma \succ_{st} u_{i+1}\sigma$.

Let a *context* $C = \{u_1 \rightarrow v_1, \ldots, u_n \rightarrow v_n\}$ be a set of oriented equations. We denote by $\overline{C}$ the result of replacing each variable x in C by a new constant $\overline{x}$. And given a term t, $\overline{t}$ results from replacing each variable $x \in \mathcal{V}ar(C)$ by the new constant $\overline{x}$.

Definition 2. *Let $\mathcal{R} = (\Sigma, A, R)$ be a deterministic rewrite theory that is quasi-reductive w.r.t. an A-compatible well-founded relation $\succ$, and let $C \Rightarrow s = t$ be a* *critical pair resulting from* $l_i \to r_i$ *if* C_i *for* $i = 1, 2$, *and* $\sigma \in Unif_A(l_1|_p, l_2)$. *We call* $C \Rightarrow s = t$ unfeasible *if there is some* $u \to v$ *in* C *such that* $\overline{u} \to_{R \cup \overline{C}, A} \overline{w_1}$, $\overline{u} \to_{R \cup \overline{C}, A} \overline{w_2}$, *and* $Unif_A(w_1, w_2) = \emptyset$ *and* w_1 *and* w_2 *are strongly irreducible with* R *modulo* A. *We call* $C \Rightarrow s = t$ context-joinable *if* $\overline{s}\downarrow_{R \cup \overline{C}}\overline{t}$.

Theorem 1. *Let* $\mathcal{R} = (\Sigma, A, R)$ *be a strongly deterministic rewrite theory that is quasi-reductive w.r.t. an A-compatible well-founded relation* $\succ$. *If every critical pair* $C \Rightarrow s = t$ *of* $\mathcal{R}$ *is either unfeasible or context-joinable, then* $\mathcal{R}$ *is confluent.*

Once all critical pairs are computed, the tool proceeds as follows. It first checks whether each conditional critical pair $C \Rightarrow s = t$ is *context joinable*:

(i) Variables in $C \Rightarrow s = t$ are added as new constants $\overline{X}$.
(ii) New *ground* rewrite rules $\overline{C}$ plus an equality operator *eq* with rules $eq(x, x) \to tt$ are added to the rules R. Call this theory $\hat{\mathcal{R}}_{\overline{C}}$.
(iii) In $\hat{\mathcal{R}}_{\overline{C}}$, we search $eq(\overline{s}, \overline{t}) \to^+ tt$ up to some predetermined depth (using the `search` command).

If the search is successful, then the conditional critical pair is context joinable. Otherwise, we then check whether $C \Rightarrow s = t$ is unfeasible as follows: For each condition $u_i \to v_i$, we perform in $\hat{\mathcal{R}}_{\overline{C}}$ the search $\overline{u_i} \to^! x : [k]$. Let $\overline{w_1} \ldots \overline{w_m}$ be the terms one obtains. If $m = 1$, then discard this term u_i and look for the next condition $u_{i+1} \to v_{i+1}$. *Otherwise*, try to find *two* different terms w_j, w_k such that (a) $Unif_A(w_j, w_k) = \emptyset$, and (b) w_j and w_k are *strongly irreducible* with $\mathcal{R}$ modulo A. If we succeed in finding a condition $u_i \to v_i$ for which associated w_j, w_k satisfy (a) and (b), then the conditional critical pair $C \Rightarrow s = t$ is *unfeasible.*[4]

2.3 The Descent Property

For an order-sorted specification it is not enough to be confluent for being Church-Rosser. The canonical form should also provide the most complete information possible about the sort of a term. This intuition is captured by our notion of Church-Rosser specifications.

Definition 3. *We call a confluent and terminating conditional order-sorted rewrite theory* $\mathcal{R} = (\Sigma, A, R)$ Church-Rosser *modulo A iff it additionally satisfies the following* descent *property: for each term t we have* $ls[t]_A \geq ls[t\downarrow_{\mathcal{R}}]_A$. *Similarly, we call a ground-confluent and terminating conditional order-sorted rewrite theory* $\mathcal{R} = (\Sigma, A, R)$ ground-Church-Rosser *modulo A iff for each ground term t we have* $ls[t]_A \geq ls[t\downarrow_{\mathcal{R}}]_A$.

Note that these notions are more general and flexible than the requirement of confluence and *sort-decreasingness* [27,22]. The issue is how to find checkable conditions for descent that, in addition to the computation of critical pairs, will ensure the Church-Rosser property. This leads us into the topic of specializations.

[4] Several optimizations, not currently available in the CRC tool are described in [17].

Given an order-sorted signature $(\Sigma, S, \leq)$, a sorted set of variables X can be viewed as a pair $(\hat{X}, \mu)$ where $\hat{X}$ is a set of variable names and μ is a sort assignment $\mu : \hat{X} \to S$. Thus, a *sort assignment* μ for X is a function mapping the names of the variables in $\hat{X}$ to their sorts. The ordering $\leq$ on S is extended to sort assignments by

$$\mu \leq \mu' \Leftrightarrow \forall x \in \hat{X}, \mu(x) \leq \mu'(x).$$

We then say that such a μ' *specializes* to μ, via the substitution

$$\rho : (x : \mu(x)) \leftarrow (x : \mu'(x))$$

called a *specialization* of $X = (\hat{X}, \mu')$ into $\rho(X) = (\hat{X}, \mu)$. Note that if the set of sorts is finite, or if each sort has only a finite number of sorts below it, then a finite sorted set of variables has a finite number of specializations.

The notion of specialization can be extended to axioms and rewrite rules. A specialization of an equation $(\forall X, l = r)$ is another equation $(\forall \rho(X), \rho(l) = \rho(r))$ where ρ is a specialization of X. A specialization of a rule $(\forall X, l \to r \text{ } \mathit{if} \text{ } C)$ is a rule $(\forall \rho(X), \rho(l) \to \rho(r) \text{ } \mathit{if} \text{ } \rho(C))$ where ρ is a specialization of X.

Thus, being *A-sort-decreasing* means that, for each rewrite rule $l \to r$ and for each specialization substitution ν, we have $ls[r\nu]_A \leq ls[l\nu]_A$. The checkable conditions that we have to add to the critical pairs to test for the descent property are called membership assertions.

Definition 4. *Let $\mathcal{R}$ be an order-sorted specification whose signature satisfies the assumptions already mentioned. Then, the set of* (conditional) membership assertions *for a conditional rule $t \to t'$ if C is defined as*

$$\{ \; t'\theta : ls[t\theta]_A \text{ } \mathit{if} \text{ } C\,\theta \;\mid\; \theta \text{ is a specialization of } \mathcal{V}ar(t) \text{ and } ls[t'\theta{\downarrow_{\mathcal{R}}}]_A \not\leq ls[t\theta]_A \; \}$$

A membership assertion $t : s$ *if* C is more general than another membership assertion $t' : s'$ *if* C' if there exists a substitution σ such that $t\sigma =_A t'$, $s \leq s'$, and $C\,\sigma =_A C'$.

Example 1. Given a specification of natural numbers and integers with the typical operations and definitions, and in particular a `square` operation defined as

```
op square : Int -> Nat .
eq square(I:Int) = I:Int * I:Int .
```

this equation gives rise to a membership assertion, because the least sort of the term `square(I:Int)` is `Nat`, but it is `Int` for the term in the righthand side. The proof obligation generated by the tool is

```
mb I:Int * I:Int : Nat .
```

This membership assertion must be proved inductively. That is, we have to treat it as the proof obligation that has to be satisfied in order to be able to assert that the specification is ground-decreasing. In this case, we have to prove that (for `I` and `J` variables of sort `Nat`) we have $\texttt{INT} \vdash_{\text{ind}} (\forall \texttt{I})(\exists \texttt{J})\ \texttt{I * I} \to^{*} \texttt{J}$, where `INT` here denotes the *rewrite theory* obtained from the original equational theory by turning each equation into a rewrite tule. This can be done using the constructor-based methods for proofs of ground reachability described in [32].

2.4 The Result of the Check

Let MMA($\mathcal{R}$) denote the set of most general membership assertions of all of the equations in the specification $\mathcal{R}$. Then, given a specification $\mathcal{R}$, the tool returns a tuple $\langle$ MCP($\mathcal{R}$), MMA($\mathcal{R}$) $\rangle$. A fundamental result underlying our tool is that the absence of critical pairs and of membership assertions in such an output is a sufficient condition for an operationally terminating specification $\mathcal{R}$ to be *Church-Rosser*.[5] In fact, for terminating unconditional specifications this check is a necessary and sufficient condition; however, for conditional specifications, the check is only a sufficient condition, because if the specification has conditional equations we can have unsatisfiable conditions in the critical pairs or in the membership assertions; that is, we can have $\langle$ MCP($\mathcal{R}$), MMA($\mathcal{R}$) $\rangle \neq \langle \emptyset, \emptyset \rangle$ with $\mathcal{R}$ still Church-Rosser. Furthermore, even if we assume that the specification is unconditional, since for specifications with an initial algebra semantics we only need to check that $\mathcal{R}$ is ground-Church-Rosser, we may sometimes have specifications that satisfy this property, but for which the tool returns a nonempty set of critical pairs or of membership assertions as proof obligations.

Of course, in other cases it may in fact be a matter of some error in the user's specification that the tool uncovers. In any case, the user has complete control on how to modify his/her specification, using the proof obligations in the output of the tool as a guide. In fact, as we explain below, several possibilities exist.

3 How to Use the Church-Rosser Checker

This section illustrates with examples the use of the Church-Rosser checker tool, and suggests some methods that—using the feedback provided by the tool—can help the user establish that his/her specification is ground-Church-Rosser.

We assume a context of use quite different from the usual context for *completing* an arbitrary equational theory. In our case we assume that the user has developed an *executable specification* of his/her intended system with an initial algebra semantics, and that this specification has already been *tested* with examples, so that the user is in fact confident that the specification is *ground-Church-Rosser*, and wants only to check this property with the tool.

[5] A detailed proof of this result will be presented elsewhere. In essence, it is a generalization of the result by Avenhaus and Loría-Sáenz [2, Theorem 4.1] to the order-sorted and modulo A case. For related results in membership equational logic see [7].

The tool can only *guarantee* success when the user's specification is unconditional and Church-Rosser and, furthermore, any associativity axiom in A for an operator has a corresponding commutativity axiom. In all other cases, the fact that the tool does not generate any proof obligations is only a *sufficient* condition, so that even when the CRC returns a collection of critical pairs and of membership assertions as proof obligations, the specification may be *ground* Church-Rosser, or for a conditional specification it may even be Church-Rosser.

An important methodological question is what to do, or not do, with these proof obligations. What should *not* be done is to let an automatic completion process add new equations to the user's specification in a mindless way. In some cases this is even impossible, because the critical pair in question cannot be oriented. In many cases it will certainly lead to a nonterminating process. In any case, it will modify the user's specification in ways that can make it difficult for the user to recognize the final result, if any, as intuitively equivalent to the original specification.

The feedback of the tool should instead be used as a guide for *careful analysis* of one's specification. As many of the examples we have studied indicate, by analyzing the critical pairs returned, the user can understand why they could not be joined. It may be a mistake that must be corrected. More often, however, it is not a matter of a mistake, but of a rule that is either *too general*—so that its very generality makes joining an associated critical pair impossible, because no more equations can apply to it—or *amenable to an equivalent formulation* that is unproblematic—for example, by reordering the parentheses for an operator that is ground-associative—or both. In any case, it is the user himself/herself who must study where the problem comes from, and how to fix it by correcting or modifying the specification. Interaction with the tool then provides a way of modifying the original specification and ascertaining whether the new version passes the test or is a good step towards that goal.

If the user's attempts to correct or modify the specification do not yet achieve a complete success, so that some proof obligations are left, *inductive* methods to discharge the remaining proof obligations may be used. Indeed, since the user's specification typically has an *initial* algebra semantics and the most common property of interest is checking that it is *ground* Church-Rosser, the proof obligations returned by the tool are *inductive* proof obligations. There are essentially two basic lines of approach, which may even be combined:

- The user may conjecture that adding a new equation $t = t'$ (or set of equations) to its specification T will make it Church-Rosser. If he can prove termination with the added equation(s) and the CRC does not generate any proof obligations for the extended specification, all is well. The only remaining issue is wether the new equation(s) have changed the module's initial algebra semantics. This can be checked by using a tool such as the Maude ITP (which does not require an equational specification to be Church-Rosser in order to perform sound inductive proofs) to verify that $T \vdash_{\text{ind}} t = t'$. A variant of this method when $t = t'$ is an associativity, or commutativity, or identity axiom, is to add it to T *not* as a simplification rule, but as an axiom.

- The other alternative is to reason inductively about the *ground joinablity* of the critical pairs, and also about the *inductive descent property* of the memberships, returned by the tool. The key point in both cases is that we should reason inductively *not* with the equational theory T (a critical pair is by construction an equational theorem in T), but with the rewrite theory $\overrightarrow{T}$ obtained by turning the equations of T into rewrite rules. An approach to inductive descent proofs with $\overrightarrow{T}$ has already been sketched in Section 2.3. For proving ground joinability, several proof methods, e.g., [31,3,26,28,4,1,6], can be used. In particular, for order-sorted specifications, constructor-based methods such as those described in [32] can be used.

A related unresolved methodological issue is what to do with *conditional* critical pairs, or conditional membership assertions, whose conditions are *unsatisfiable*. We currently discard critical pairs which the tool can show are *unfeasible* or *context-joinable*, but all remaining unjoinable pairs are left to the user. Perhaps a modular/hierarchical approach could be used, in conjunction with the inductive proof methods described above, to establish the unsatisfiability of such conditions and then discard the corresponding proof obligations.

We give in the following sections examples illustrating the use of the tool. The examples have been chosen trying to highlight those features not simultaneously supported by previous similar tools, namely, order-sortedness, conditional equations, and rewriting modulo axioms.

3.1 Hereditarily Finite Sets

The following module `HF-SETS` specifies hereditarily finite sets, that is, sets that are finite and, furthermore, their elements, the elements of those elements, and so on recursively, are all finite sets. It was developed by Ralf Sasse and José Meseguer and is inspired by the generalized sets module in Maude's prelude [9, Section 9.12.5]. It declares sorts `Set` and `Magma`, with `Set` a subsort of `Magma`. Terms of sort `Set` are generated by constructors `{}`, the empty set, and `{_}`, which makes a set out of a term of sort `Magma`. Magmas have an associative-commutative operator `_,_`. The commutative operator `_~_` checks whether two sets are equivalent. The membership relation $\in$ holding between two sets is here generalized by a predicate `_in_` holding between two magmas, and the containment relation $\subseteq$ is here modeled by a predicate `_<=_` holding between two sets.

```
(fmod HF-SETS is
   protecting BOOL-OPS .
   sorts Magma Set .
   subsort Set < Magma .
   op _`,_ : Magma Magma -> Magma [ctor assoc comm] .
   op `{_`} : Magma -> Set [ctor] .
   op `{`} : -> Set [ctor] .

   vars M M' N : Magma .                    vars S S' : Set .

   eq [01]: M, M, M' = M, M' .              eq [02]: M, M = M .

   op _in_ : Magma Magma -> Bool .
```

```
  eq [03]: M in {} = false .
  eq [04]: {} in {{M}} = false .
  eq [05]: {} in {{}} = true .
  eq [06]: {} in {{}, M} = true .
  eq [07]: {} in {{M}, N} = {} in {N} .
  eq [08]: S in {S'} = S ~ S' .
 ceq [09]: S in {S' , M} = true if S ~ S' = true .
 ceq [10]: S in {S', M} = S in {M} if S ~ S' = false .
 ceq [11]: S in S', N = true if S in S' = true .
 ceq [12]: S in S', N = S in N if S in S' = false .
 ceq [13]: S, M in M' = M in M' if S in M' = true .
 ceq [14]: S, M in M' = false if S in M' = false .

  op _<=_ : Set Set -> Bool .
  eq [15]: {} <= S = true .
  eq [16]: {M} <= S = M in S .

  op _~_ : Set Set -> Bool .
  eq [17]: S ~ S' = (S <= S') and (S' <= S) .
endfm)
```

Notice the labeling of the equations. The critical pairs returned by the tool will use them to provide information about the equations they come from. Notice also the importation of the predefined module `BOOL-OPS`, where the sort `Bool` is defined with constants `true` and `false`, and boolean operations `_and_`, `_or_`, `_xor_`, `not_`, and `_implies_`. The operators `_and_`, `_or_`, and `_xor_` are declared associative and commutative.

The Church-Rosser check gives the following result:

```
Maude> (check Church-Rosser HF-SETS .)
Church-Rosser checking of HF-SETS
Checking solution:
The following critical pairs cannot be joined:
  ccp for 07 and 09
    S':Set <= {} = true if {} ~ S':Set = true .
  ccp for 07 and 10
    S':Set <= {} = false if {} ~ S':Set = false .
  ccp for 02 and 09
    S:Set <= S':Set and S':Set <= S:Set = true
    if S:Set ~ S':Set = true .
  ccp for 09 and 10
    true = S:Set <= #2:Set and #2:Set <= S:Set
    if S:Set ~ S':Set = false /\ S:Set ~ #2:Set = true .
  ccp for 10 and 09
    S:Set <= S':Set and S':Set <= S:Set = true
    if S:Set ~ S':Set = true /\ S:Set ~ #2:Set = false .
  ccp for 10 and 10
    S:Set <= S':Set and S':Set <= S:Set
    = S:Set <= #2:Set and #2:Set <= S:Set
    if S:Set ~ S':Set = false /\ S:Set ~ #2:Set = false .
The specification is sort-decreasing.
```

The tool generates 3725 critical pairs. Most of them are joinable, and therefore discarded. From the remaining 27 critical pairs, all of which are conditional, 21 are discarded because they can be proved either context-joinable or unfeasible. Let us take a look at some of these.

Let us consider the following context-joinable critical pair:

```
ccp for 16 and 11
  S:Set in (S':Set , N:Magma) = true
  if S:Set in S':Set = true .
```

If we add the condition of this critical pair as an equation with its variables `S` and `S'` turned into constants, of sort `Set`, `#S` and `#S'`, then, the terms `true` and `#S in (##1, #S')`, with `##1` a new constant of sort `Magma`, can be joined.

The following critical pair is discarded because it is unfeasible.

```
ccp for 11 and 12
  true = S:Set in N:Magma
  if S:Set in S':Set = false /\ S:Set in S':Set = true .
```

To prove unfeasibility we focus on the conditions. With the rules

```
#S in #S' = false
#S in #S' = true
```

the term `#S in #S'` can be rewritten both to `false` and `true`. Since they do not unify and are strongly irreducible, we conclude that the critical pair is unfeasible.

Most other critical pairs are discarded for similar reasons. The only ones left are those finally returned by the tool. These critical pairs are neither context-joinable nor unfeasible. However, we can introduce new equations, that should be inductively deducible from the specification, or replace the ones we have by alternative equations, in order to eliminate such critical pairs.

Let us start with the first critical pair in the CRC output. We may argue that if the set `S'` is such that the condition is satisfied, then the term `S' <= {}` should be reducible to `true`, and try to add equations to allow this rewrite. But more easily, we may observe that the critical pair comes from equations `07` and `09` at the top, because `09` is more general than necessary. Since a set is either of the form `{}` or `{M}`, and the `{}` case is covered by equations `06` and `07`, we can eliminate this critical pair by replacing equation `09` with

```
ceq [09']: {M} in {S, M'} = true if {M} ~ S = true .
```

A new execution of the check shows that the critical pair for equations `07` and `10` is no longer given. The critical pair for equations `07` and `10` suggests a similar change for equation `10`:

```
ceq [10']: {M} in {S, M'} = {M} in {M'} if {M} ~ S = false .
```

It is not enough, however. With these new two equations, the tool gives us now four conditional critical pairs. Given these critical pairs, we realize that equations `09'` and `10'` are still problematic. The simplest change is to replace these two equations by one unconditional equation covering the two cases:

```
eq [09-10]: {M} in {S, M'} = {M} ~ S or {M} in {M'} .
```

Replacing `09'` and `10'` by `09-10` now succeeds:

```
Maude> (check Church-Rosser HF-SETS-3 .)
Church-Rosser checking of HF-SETS-3
Checking solution:
All critical pairs have been joined.
The specification is locally-confluent.
The specification is sort-decreasing.
```

Therefore, once proved operationally terminating, we can conclude that the module `HF-SETS-3` is confluent.

3.2 Lists and Sets

Let us consider now the following specification of lists and sets.

```
(fmod LIST&SET is
   sorts MBool Nat List Set .
   subsorts Nat < List Set .
   ops true false : -> MBool .
   ops _and_ _or_ : MBool MBool -> MBool [assoc comm] .
   op 0 : -> Nat .
   op s_ : Nat -> Nat .
   op _;_ : List List -> List [assoc] .
   op null : -> Set .
   op __ : Set Set -> Set [assoc comm id: null] .
   op _in_ : Nat Set -> MBool .
   op _==_ : List List -> MBool [comm] .
   op list2set : List -> Set .

   var B : MBool .                vars N M : Nat .
   vars L L' : List .             var S : Set .

   eq [01]: N N = N .
   eq [02]: true and B = B .
   eq [03]: false and B = false .
   eq [04]: true or B = true .
   eq [05]: false or B = B .
   eq [06]: 0 == s N = false .
   eq [07]: s N == s M = N == M .
   eq [08]: N ; L == M = false .
   eq [09]: N ; L == M ; L' = (N == M) and L == L' .
   eq [10]: L == L = true .
   eq [11]: list2set(N) = N .
   eq [12]: list2set(N ; L) = N list2set(L) .
   eq [13]: N in null = false .
   eq [14]: N in M S = (N == M) or N in S .
endfm)
```

It has four sorts: MBool, Nat, List, and Set, with Nat included in both List and Set as a subsort. The terms of each sort are, respectively, Booleans, natural numbers (in Peano notation), lists of natural numbers, and finite sets of natural numbers. The rewrite rules in this module then define various functions such as _and_ and _or_, a function list2set associating to each list its corresponding set, the set membership predicate _in_, and an equality predicate _==_ on lists. Furthermore, the idempotency of set union is specified by the first equation. The operators _and_ and _or_ have been declared associative and commutative, the list concatenation operator _;_ has been declared associative, the set union operator __ has been declared associative, commutative and with null as its identity, and the _==_ equality predicate has been declared commutative using the comm keyword. This module therefore illustrates how our tool can deal in principle with arbitrary combinations of associativity and/or commutativity and/or identity axioms, even though it may not succeed in some cases when some operators are associative but not commutative.

The tool gives us the following result.

```
Maude> (check Church-Rosser .)
Church-Rosser checking of LIST&SET
Checking solution:
The following critical pairs cannot be joined:
  cp for 01 and 14
    N:Nat == M:Nat = (N:Nat == M:Nat) or N:Nat == M:Nat .
  cp for 01 and 14
```

```
      (N:Nat == M:Nat) or N:Nat in #5:Set
    = (N:Nat == M:Nat) or (N:Nat == M:Nat) or N:Nat in #5:Set .
The specification is sort-decreasing.
```

These critical pairs are completely harmless. They can in fact be removed by introducing an idempotency equation for the `_or_` operator.

```
(fmod LIST&SET-2 is pr LIST&SET .
   var B : MBool .
   eq [15]: B or B = B .
endfm)
```

The tool now tells us that the specification is locally confluent and sort-decreasing, and since it is terminating (see [15]), we can conclude that it is Church-Rosser.

As explained in Section 1, to handle this specification, we apply several transformations on the original module to remove identity attributes and associativity attributes that do not come with commutativity ones. We refer the interested reader to [17] for details on the use of this transformation in the CRC, and to [15] for a detailed description of the variant transformation used.

4 Conclusions and Future Work

We have presented the foundations, design, and use of the Maude CRC 3 tool, showing how it can deal effectively with complex equational specifications that are order-sorted, conditional (possibly with extra variables in their condition), and whose equations can be applied modulo different combinations of associativity, commutativity and identity axioms and are specified in Maude as functional modules or theories. Our tool attempts to prove such specifications Church-Rosser or ground Church-Rosser under the assumption of their operational termination. Besides the much greater generality of our tool when compared with its earlier versions or with other similar tools, two very useful new features are: (i) the capacity to discharge unjoinable critical pairs by proving them to be either unfeasible or context-joinable; and (ii) the capacity to deal in principle with any combination of associativity and/or commutativity and/or identity axioms. The CRC 3 tool together with its documentation is available at `http://maude.lcc.uma.es/CRChC`.

As future work, we would like to remove the current restrictions of the CRC tool, and to provide new methods to handle conditional critical pairs (or conditional membership assertions) whose conditions are unsatisfiable. The integration of the different tools in the Maude formal environment [11], namely the ITP [12], MTT [14], CRC, ChC [18], and SCC [24] tools, in a real formal tool environment is another important pending goal. It will make it easy for the CRC to interact with other tools that can be used to discharge some of its generated proof obligations.

Acknowledgements. F. Durán was supported by Spanish Research Projects TIN2008-03107 and P07-TIC-03184. J. Meseguer was partially supported by NSF Grants CCF-0905584, CNS-07-16038, CNS-09-04749, and CNS-08-34709.

References

1. Avenhaus, J., Hillenbrand, T., Löchner, B.: On using ground joinable equations in equational theorem proving. Journal of Symbolic Computation 36(1-2), 217–233 (2003)
2. Avenhaus, J., Loría-Sáenz, C.: On conditional rewrite systems with extra variables and deterministic logic programs. In: Pfenning, F. (ed.) LPAR 1994. LNCS, vol. 822, pp. 215–229. Springer, Heidelberg (1994)
3. Bachmair, L., Dershowitz, N., Plaisted, D.A.: Completion without failure. In: Kaci, A.H., Nivat, M. (eds.) Resolution of Equations in Algebraic Structures. Rewriting Techniques, vol. 2, pp. 1–30. Academic Press, New York (1989)
4. Becker, K.: Proving ground confluence and inductive validity in constructor based equational specifications. In: Gaudel, M.-C., Jouannaud, J.-P. (eds.) CAAP 1993, FASE 1993, and TAPSOFT 1993. LNCS, vol. 668, pp. 46–60. Springer, Heidelberg (1993)
5. Bergstra, J., Tucker, J.: Characterization of computable data types by means of a finite equational specification method. In: de Bakker, J.W., van Leeuwen, J. (eds.) Seventh Colloquium on Automata, Languages and Programming. LNCS, vol. 81, pp. 76–90. Springer, Heidelberg (1980)
6. Bouhoula, A.: Simultaneous checking of completeness and ground confluence for algebraic specifications. ACM Transactions on Computational Logic 10(3) (2009)
7. Bouhoula, A., Jouannaud, J.-P., Meseguer, J.: Specification and proof in membership equational logic. Theoretical Computer Science 236(1), 35–132 (2000)
8. Bruni, R., Meseguer, J.: Semantic foundations for generalized rewrite theories. Theoretical Computer Science 351(1), 286–414 (2006)
9. Clavel, M., Durán, F., Eker, S., Lincoln, P., Martí-Oliet, N., Meseguer, J., Talcott, C. (eds.): All About Maude - A High-Performance Logical Framework: How to Specify, Program, and Verify Systems in Rewriting Logic. LNCS, vol. 4350. Springer, Heidelberg (2007)
10. Clavel, M., Durán, F., Eker, S., Lincoln, P., Martí-Oliet, N., Meseguer, J., Talcott, C.: Maude 2.4 manual (November 2008), `http://maude.cs.uiuc.edu`
11. Clavel, M., Durán, F., Hendrix, J., Lucas, S., Meseguer, J., Ölveczky, P.: The Maude formal tool environment. In: Mossakowski, T., Montanari, U., Haveraaen, M. (eds.) CALCO 2007. LNCS, vol. 4624, pp. 173–178. Springer, Heidelberg (2007)
12. Clavel, M., Palomino, M., Riesco, A.: Introducing the ITP tool: a tutorial. Journal of Universal Computer Science 12(11), 1618–1650 (2006)
13. Durán, F.: A Reflective Module Algebra with Applications to the Maude Language. PhD thesis, Universidad de Málaga, Spain (June 1999), `http://maude.csl.sri.com/papers`
14. Durán, F., Lucas, S., Meseguer, J.: MTT: The Maude termination tool (system description). In: Armando, A., Baumgartner, P., Dowek, G. (eds.) IJCAR 2008. LNCS (LNAI), vol. 5195, pp. 313–319. Springer, Heidelberg (2008)
15. Durán, F., Lucas, S., Meseguer, J.: Termination modulo combinations of equational theories. In: Ghilardi, S., Sebastiani, R. (eds.) FroCoS 2009. LNCS, vol. 5749, pp. 246–262. Springer, Heidelberg (2009)
16. Durán, F., Meseguer, J.: Maude's module algebra. Science of Computer Programming 66(2), 125–153 (2007)
17. Durán, F., Meseguer, J.: CRC 3: A Church-Rosser checker tool for conditional order-sorted equational Maude specifications (2009), `http://maude.lcc.uma.es/CRChC`

18. Durán, F., Meseguer, J.: A Maude coherence checker tool for conditional order-sorted rewrite theories. In: Olveczky, P.C. (ed.) WRLA 2010. LNCS, vol. 6381, pp. 86–103. Springer, Heidelberg (2010)
19. Durán, F., Meseguer, J.: A Church-Rosser Checker Tool for Conditional Order-Sorted Equational Maude Specifications. In: Ölveczky, P.C. (ed.) 8th Intl. Workshop on Rewriting Logic and its Applications (2010)
20. Durán, F., Ölveczky, P.C.: A guide to extending Full Maude illustrated with the implementation of Real-Time Maude. In: Roşu, G. (ed.) Proceedings 7th International Workshop on Rewriting Logic and its Applications (WRLA 2008). Electronic Notes in Theoretical Computer Science. Elsevier, Amsterdam (2008)
21. Giesl, J., Kapur, D.: Dependency pairs for equational rewriting. In: Middeldorp, A. (ed.) RTA 2001. LNCS, vol. 2051, pp. 93–108. Springer, Heidelberg (2001)
22. Gnaedig, I., Kirchner, C., Kirchner, H.: Equational completion in order-sorted algebras. Theoretical Computer Science 72, 169–202 (1990)
23. Goguen, J., Meseguer, J.: Order-sorted algebra I: Equational deduction for multiple inheritance, overloading, exceptions and partial operations. Theoretical Computer Science 105, 217–273 (1992)
24. Hendrix, J., Meseguer, J., Ohsaki, H.: A sufficient completeness checker for linear order-sorted specifications modulo axioms. In: Furbach, U., Shankar, N. (eds.) IJCAR 2006. LNCS (LNAI), vol. 4130, pp. 151–155. Springer, Heidelberg (2006)
25. Jouannaud, J.-P., Kirchner, H.: Completion of a set of rules modulo a set of equations. SIAM Journal of Computing 15(4), 1155–1194 (1986)
26. Kapur, D., Narendran, P., Otto, F.: On ground-confluence of term rewriting systems. Information and Computation 86(1), 14–31 (1990)
27. Kirchner, C., Kirchner, H., Meseguer, J.: Operational semantics of OBJ3. In: Lepistö, T., Salomaa, A. (eds.) ICALP 1988. LNCS, vol. 317, pp. 287–301. Springer, Heidelberg (1988)
28. Martin, U., Nipkow, T.: Ordered rewriting and confluence. In: Stickel, M.E. (ed.) CADE 1990. LNCS, vol. 449, pp. 366–380. Springer, Heidelberg (1990)
29. Ohlebusch, E.: Advanced Topics in Term Rewriting. Springer, Heidelberg (2002)
30. Peterson, G., Stickel, M.: Complete sets of reductions for some equational theories. Journal of ACM 28(2), 233–264 (1981)
31. Plaisted, D.: Semantic confluence tests and completion methods. Information and Control 65, 182–215 (1985)
32. Rocha, C., Meseguer, J.: Constructors, sufficient completeness, deadlock states of rewrite theories. Technical Report 2010-05-1, CS Dept., University of Illinois at Urbana-Champaign (May 2010), `http://ideals.illinois.edu`
33. Viry, P.: Equational rules for rewriting logic. Theoretical Computer Science 285(2), 487–517 (2002)

A Maude Coherence Checker Tool for Conditional Order-Sorted Rewrite Theories

Francisco Durán[1] and José Meseguer[2]

[1] Universidad de Málaga, Spain
[2] University of Illinois at Urbana-Champaign, IL, USA

Abstract. For a rewrite theory to be executable, its equations E should be (ground) confluent and terminating modulo the given axioms A, and their rules should be (ground) coherent with E modulo A. The correctness of many important formal verification tasks, including search, LTL model checking, and the development of abstractions, crucially depends on the theory being ground coherent. Furthermore, many specifications of interest are typed, have equations E and rules R that are both conditional, have axioms A involving various combinations of associativity, commutativity and identity, and may contain frozenness restrictions. This makes it essential to extend the known coherence checking methods from the untyped, unconditional, and AC or free case, to this much more general setting. We present the mathematical foundations of the Maude ChC 3 tool, which provide such a generalization to support coherence and ground coherence checking for order-sorted rewrite theories under these general assumptions. We also explain and illustrate the use of the ChC 3 tool with a nontrivial example.

1 Introduction

Traditionally, a rewrite system is a set of directed equations used to compute a value by repeatedly replacing subterms of a given formula with equal terms until a (typically unique) simplest possible form is obtained. This interpretation of a rewrite system gives an equational semantics to it, and a way of executing functional programs by rewriting. But rewriting is also useful for specifying non-equational relations, such as transitions between states. Rewriting logic [21] suggests keeping all rules with an equational interpretation as a distinguished set E of equations, and considering the remaining rules R as defining state transition steps over equivalence classes modulo E.

A rewriting logic signature is an equational specification. But, rewriting logic is parameterized by the choice of its underlying equational logic. For example, for Maude [3], the underlying equational logic is membership equational logic, so that signatures are of the form (Ω, E), where $\Omega = (K, \Sigma, S)$ is a membership equational logic signature and E is a set of (conditional) membership axioms and equations. Such a signature (Ω, E) makes explicit the set of equations in order to emphasize that rewriting will operate on congruence classes of terms *modulo* E.

P.C. Ölveczky (Ed.): WRLA 2010, LNCS 6381, pp. 86–103, 2010.

Thus, a rewrite theory has both rules and equations, so that rewriting is performed modulo such equations. However, this does not mean that an implementation of rewriting logic must have an E-matching algorithm for each equational theory E that a user might specify, which is impossible, since matching modulo an arbitrary theory is undecidable. What, e.g., Maude instead requires for rewrite theories in system modules is that:

- The equations are divided into a set A of structural axioms, for which matching algorithms exist and a set E of equations that are (ground) Church-Rosser and terminating modulo A. For some equations E, termination modulo A can be checked using the Maude Termination Tool (MTT) [9,5] and the Church-Rosser property can be checked using a Church-Rosser checker as the one presented in [15,14].
- The rules R in the module are (ground) *coherent* [22,25] with the equations E modulo A. This means that appropriate critical pairs can be filled in between rules and equations, allowing us to intermix rewriting with rules and rewriting with equations without losing completeness of rule computations by failing to perform a rewrite that would have been possible before an equational deduction step was taken. In this way, we get the effect of rewriting modulo $E \cup A$ with just a matching algorithm for A. In particular, a simple strategy available in these circumstances is to always reduce to canonical form using E before applying any rule in R. This is precisely the strategy adopted by the Maude interpreter.

Therefore, it is very important to know whether a given Church-Rosser and terminating specification is indeed ground-coherent. For this purpose, the coherence checking methods proposed by Viry [25] must be substantially generalized because: (i) they are restricted to the AC or free cases; (ii) assume that both the equations and the rules are unconditional; (iii) always require the very restrictive condition that the right-hand and left-hand sides of any equation are both linear; and (iv) are untyped. Instead, what we need to handle more expressive specifications are *generalized rewrite theories* $\mathcal{R} = (\Sigma, E \cup A, R, \phi)$ [2] such that: (i) have an initial model semantics; (ii) the equations E and the rules R can both be *conditional*; (iii) Σ is typed (here we assume Σ order-sorted); (iv) the set A of axioms may involve associativity and/or commutativity and/or identity axioms; and (v) rewriting with rules is restricted by frozenness information ϕ.

At first sight, checking coherence under these more general conditions may appear to be an even more challenging task than in the simpler situations contemplated by Viry in [25]. However, as we show in this paper, some of these more general conditions can make it *much easier* to check coherence. In particular:

(1) frozenness can eliminate many critical pairs and greatly reduce the linearity requirements on variables of equations;
(2) order-sorted type structure can: (i) eliminate many critical pairs, (ii) further relax linearity conditions on variables of equations, and (iii) eliminate many problematic non-overlap situations between equations and rules;

(3) the initial model semantics substantially relaxes the coherence requirement into a *ground coherence* one where: (i) unjoinable critical pairs can be shown ground joinable if some equational inductive proof obligations can be discharged; and (ii) by checking sufficient completeness of the equations with respect to a constructor subsignature, defined function symbols can safely be assumed to be *frozen*, which by (1) can further reduce the number of critical pairs that need to be considered and the linearity requirements on equations.

A further point to emphasize is that the present ChC tool can in principle deal with *any combination* of associativity, commutativity, and identity axioms, including the thorny cases of associativity without commutativity for which no finitary unification algorithms exist. Although in general computing critical pairs for the associativity without commutativity cases may not be possible, in many practical cases our tool can show that the relevant left-hand sides have a finite set of *variants* [6,18] when associativity is used as a rule. This then allows the application of a theory transformation described in [10] thanks to which associativity without commutativity axioms need not be used when computing critical pairs.

Our coherence checker tool (ChC) [13] is particularly well-suited for checking Maude specifications with an initial model semantics whose equations E have already been proved Church-Rosser and terminating modulo A, and now we need to check that its rules R are ground-coherent with E modulo A. Our methods can also be used to check the coherence property of conditional order-sorted specifications that do not have an initial model semantics, such as, for example, those specified in Maude system theories [4]. Since, for the reasons mentioned above, user interaction will typically be quite essential, coherence completion is not attempted. Instead, if the specification cannot be shown to be coherent or ground-coherent by the tool, proof obligations are generated and are given back to the user as a guide in the attempt to establish the ground-coherence property. Since this property is in fact inductive, in some cases the Maude inductive theorem prover can be enlisted to prove some of these proof obligations. In other cases, the user may in fact have to modify the original specification by carefully considering the information conveyed by the proof obligations. We give in Section 3 some methodological guidelines for the use of the tool, and illustrate the use of the tool with some examples.

The present ChC tool only accepts order-sorted conditional specifications, where each of the operation symbols has either no equational attributes, or any combination of associativity/commutativity/identity.[1] Furthermore, it is assumed that such specifications do not contain any built-in function, do not use the `owise` attribute, and that they have already been proved Church-Rosser and terminating. The tool attempts to establish the ground-coherence property

[1] The associativity without commutativity case is handled using a semi-algorithm proposed in [10], which works in many practical situations but not always. We refer the reader to [10] for further details.

modulo the equational axioms specified for each of the operators by checking a sufficient condition. Therefore, the tool's output consists of a set of critical pairs that the tool has not been able to join and must be shown ground-joinable.

As other tools in the Maude formal environment [5], the ChC tool has been implemented as an extension of Full Maude [12,7]. Details on how to extend Full Maude in different forms can be found in, e.g., [17,12,7,8]. Following these techniques, the ChC has been integrated within the Full Maude environment, to allow checking of modules defined in Full Maude and to get a much more convenient user interface. Of course, it would have been possible to define an interface for the tool without integrating it with Full Maude. Since all the infrastructure built for Full Maude can be used by itself, just by selecting functions from that infrastructure in the needed modules, any of the two possibilities can give rise to an interface in a very short time. However, by integrating the specifications of Full Maude and of the ChC we not only have such a needed infrastructure, but in addition we can, for example, check the coherence property of any module in Full Maude's database. We can therefore use the tool on any module accepted by Full Maude, including structured modules, parameterized modules, etc. We still have, of course, the restrictions mentioned above.

The rest of the paper is structured as follows. Section 2 introduces the notion of coherent order-sorted specification modulo axioms. Section 3 presents some directions on how to use the tool and illustrates it with an example. Section 4 concludes and presents some future work. Proofs of technical results are not included here for space reasons. They can be found in [16]. We assume that the reader is familiar with basic rewriting terminology and notations. Although we have tried to make the paper self contained, we refer the interested reader to [23] for additional details on rewriting techniques.

2 Coherent Order-Sorted Specifications Modulo Axioms

This section presents the theoretical foundations of the ChC.

2.1 Conditional Rewriting Modulo Linear and Regular Axioms A

Given an order-sorted rewrite theory $\mathcal{R} = (\Sigma, A, R)$, where A is a collection of unconditional equational axioms of the form $u = v$ that are *linear* (no repeated variables in either u or v), and *regular* ($vars(u) = vars(v)$), we define the relation $\rightarrow_{R/A}$, either by the inference system of rewriting logic (see [2]), or by the usual inductive description: $\rightarrow_{R/A} = \bigcup_n \rightarrow_{R/A,n}$, where $\rightarrow_{R/A,0} = \emptyset$, and for each $n \in \mathbb{N}$, we have $\rightarrow_{R/A,n+1} = \rightarrow_{R/A,n} \cup \{(u,v) \mid u =_A l\sigma \rightarrow r\sigma =_A v \wedge l \rightarrow r \text{ if } \bigwedge_i u_i \rightarrow v_i \in R \wedge \forall i, u_i\sigma \rightarrow^*_{R/A,n} v_i\sigma\}$. In general, of course, given terms t and t' with sorts in the same connected component, the problem of whether $t \rightarrow_{R/A} t'$ holds is undecidable.

Even if there is an effective A-matching algorithm, the relation $u \rightarrow_{R/A} v$ still remains undecidable in general, since to see if $u \rightarrow_{R/A} v$ involves searching through the possibly infinite equivalence classes $[u]_A$ and $[v]_A$ to see whether an A-match is found for a subterm of some $u' \in [u]_A$ and the result of rewriting u' belongs to the equivalence class $[v]_A$. For this reason, a much simpler relation $\rightarrow_{R,A}$ is defined, which becomes decidable if an A-matching algorithm exists. We define (see [24]) $\rightarrow_{R,A} = \bigcup_n \rightarrow_{R,A,n}$ where $\rightarrow_{R,A,0} = \emptyset$, and for each $n \in \mathbb{N}$ and any terms u, v with sorts in the same connected component the relation $u \rightarrow_{R,A,n+1} v$ holds if either $u \rightarrow_{R,A,n} v$, or there is a position p in u, a rule $l \rightarrow r$ if $\bigwedge_i u_i \rightarrow v_i$ in R, and a substitution σ such that $u|_p =_A l\sigma$, $v = u[r\sigma]_p$, and $\forall i, u_i\sigma \rightarrow^*_{R,A,n} w_i$ with $w_i =_A v_i\sigma$. Of course, $\rightarrow_{R,A} \subseteq \rightarrow_{R/A}$. The important question is the *completeness* question: can any $\rightarrow_{R/A}$-step be simulated by a $\rightarrow_{R,A}$-step? We say that $\mathcal{R}$ satisfies the A-*completeness* property if for any u, v with sorts in the same connected component we have:

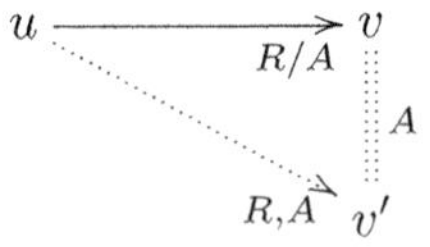

where here and in what follows dotted lines indicate existential quantification.

It is easy to check that A-completeness is equivalent to the following (strong) A-coherence[2] (or just *coherence* when A is understood) property:

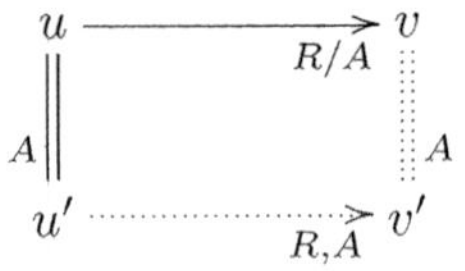

If a theory $\mathcal{R}$ is not coherent, we can try to make it so by completing the set of rules R to a set of rules $\widetilde{R}$ by a Knuth-Bendix-like completion procedure that computes critical pairs between equations in A and rules in R (see, e.g., [20,25] for the *strong* coherence completion that we use here, and [19] for the equivalent notion of extension completion). For theories A that are combinations of associativity, commutativity, left identity, and right identity axioms, the coherence completion procedure always terminates and has a very simple description (see [24], and for a more informal explanation [4, Section 4.8]).

We say that $\mathcal{R} = (\Sigma, A, R)$ is A-*confluent*, resp. A-*terminating*, if the relation $\rightarrow_{R/A}$ is confluent, resp. terminating. If $\mathcal{R}$ is A-coherent, then A-confluence is equivalent to asserting that, for any $t \rightarrow^*_{R,A} u$, $t \rightarrow^*_{R,A} v$, we have:

[2] Note that the assumption of A being regular and linear is essential for one $\rightarrow_{R/A}$-step to exactly correspond to one $\rightarrow_{R,A}$-step. For this reason, some authors (e.g., [20,25]) call conditions as the one above *strong coherence*, and consider also weaker notions of coherence.

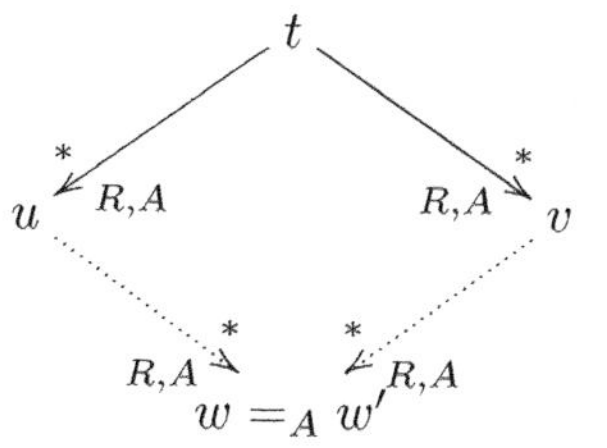

and A-termination is equivalent to the termination of the $\rightarrow_{R,A}$ relation. In what follows, given a rewrite theory $\mathcal{R} = (\Sigma, A, R)$, saying that $\mathcal{R}$ is A-coherent is equivalent to saying that the rules R are A-coherent.

The fact that we are performing *order-sorted* rewriting makes one more requirement necessary. When A-matching a subterm $t|_p$ against a rule's left-hand side to obtain a matching substitution σ, we need to check that σ is well-sorted, that is, that if a variable x has sort s, then the term $x\sigma$ has also sort s. This may however fail to be the case even though there is a term $w \in [x\sigma]_A$ which does have sort s. We call an order-sorted signature *A-preregular* if the set of sorts $\{s \in S \mid \forall w \in \mathcal{T}_\Sigma(\mathcal{X}), \exists w' \in [w]_A \text{ s.t. } w' \in \mathcal{T}_\Sigma(\mathcal{X})_s\}$ has a least upper bound, denoted $ls[w]_A$ which can be effectively computed.[3] Then we can check the well-sortedness of the substitution σ not based on $x\sigma$ above, but, implicitly, on all the terms in $[w]_A$.

Yet another property required for the good behavior of confluent and terminating rewrite theories modulo A is their being *A-sort-decreasing*. This means that $\mathcal{R}$ is A-preregular, and for each term t we have $ls[t]_A \geq ls[t{\downarrow_R}]_A$.

From this, the following lemma follows.

Lemma 1. *For R A-coherent rules, if $t \rightarrow_{R,A} t'$, then*

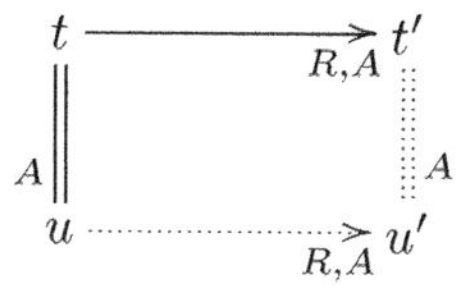

As mentioned above, for $\rightarrow_{R,A}$ to be decidable we need an A-matching algorithm. Therefore, we will consider the set of equations to be a union $E \cup A$ with A a set of axioms for which there exists a matching algorithm (as associativity, commutativity, and identity), and E the remaining equations.

2.2 Coherence of Conditional Rewrite Theories

A rule $l \rightarrow u_{n+1}$ *if* $\bigwedge_{i=1..n} u_i \rightarrow v_i$ is said to be *deterministic* if $\forall j \in [1..n]$, $\mathcal{V}ar(u_j) \subseteq \mathcal{V}ar(l) \cup \bigcup_{k<j} \mathcal{V}ar(v_k)$. A conditional rewrite theory is *deterministic*

[3] The Maude system automatically checks the A-preregularity of a signature Σ for A any combination of associativity, commutativity, left identity, and right identity axioms (see [4, Chapter 22.2.5]).

if each of its rules is deterministic. Given a rewrite theory $\mathcal{R}$, a term t is called *strongly irreducible* with respect to R modulo A (or *strongly* R, A*-irreducible*) if $t\sigma$ is a normal form for every normalized substitution σ. A rewrite theory $\mathcal{R}$ is called *strongly deterministic* if for every rule $l \to r$ *if* $\bigwedge_{i=1..n} u_i \to v_i$ in R each v_i is strongly R, A-irreducible.

We assume an order-sorted rewrite theory of the form $\mathcal{R} = (\Sigma, E \cup A, R, \phi)$, where:

(1) ϕ is the frozenness information [2].
(2) $(\Sigma, E \cup A)$ is an order-sorted equational theory with possibly conditional equations, which can be converted into a strongly deterministic rewrite theory that is *operationally terminating* modulo A. Furthermore, the equations E are *confluent* modulo A. Also, the axioms in A are a collection of regular and linear unconditional equational axioms and are all at the *kind level*, i.e., each connected component in the poset $(S, \leq)$ of sorts has a top sort, and the variables in the axioms A all have such top sorts.
(3) R is a collection of rewrite rules $l \to r$ *if* C, where C is an *equational condition*, which again can be turned into a deterministic rewrite rule of the form $l \to r$ *if* $u_1 \to_E v_1 \wedge \ldots \wedge u_n \to_E v_n$ with the $v_1, \ldots, v_n$ strongly E, A-irreducible.
(4) Both the equations E and the rules R are A*-coherent*. Therefore, the relations $\to_{R/A}$ (resp. $\to_{E/A}$) and $\to_{R,A}$ (resp. $\to_{E,A}$) essentially coincide.

Definition 1. *A rewrite theory* $\mathcal{R} = (\Sigma, E \cup A, R, \phi)$ *satisfying (1)-(4) above is called* coherent *(resp.* ground coherent*) iff for each* Σ*-term* t *(resp. ground* Σ*-term* t*) such that* $t \to_{E,A} u$*, and* $t \to_{R,A} v$ *we have*

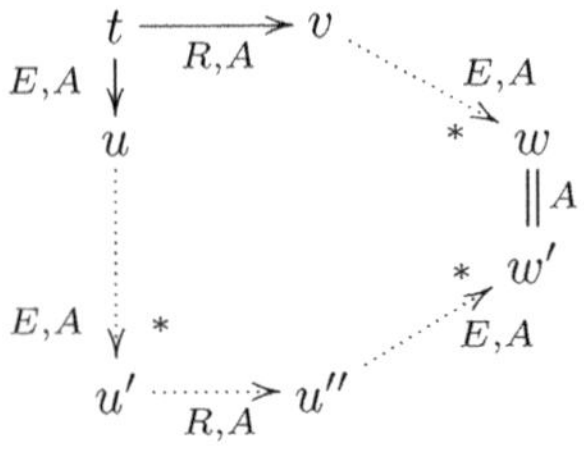

(C)

Likewise, $\mathcal{R}$ *is called* locally coherent *(resp.* ground locally coherent*) iff for each* Σ*-term* t *(resp. ground* Σ*-term* t*) such that* $t \to_{E,A} u$*, and* $t \to_{R,A} v$ *we have*

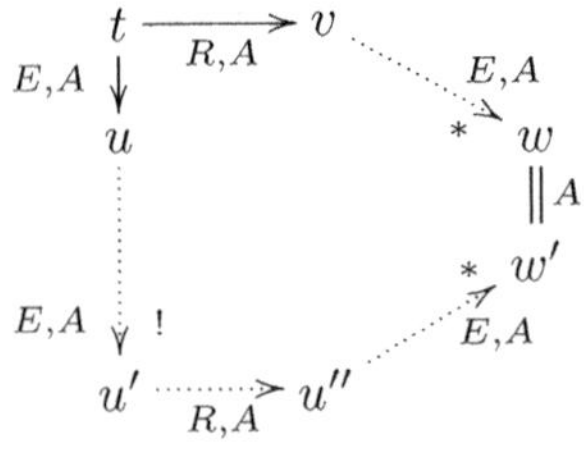

(LC)

where dotted arrows are existentially quantified, and $s \to^{!}_{E,A} t$ *iff* $s \to^{*}_{E,A} t$ *and* t *is* E, A*-irreducible.*

Theorem 1. *$\mathcal{R}$ is coherent (resp. ground coherent) iff $\mathcal{R}$ is locally coherent (resp. locally ground coherent).*

Since for all terms t, t is coherent iff t is locally coherent, we can approach the verification of coherence for $\mathcal{R}$ as follows: We can reason by cases on the situations $u \;{}_{E,A}\!\swarrow\; t \;\searrow_{R,A}\; v$ depending on whether they are or not *overlap* situations. For this we need the notion of a conditional critical pair, and the notion of conditional critical pair joinability.

Definition 2. *Given conditional rewrite rules with disjoint variables $l \to r$ if C in R and $l' \to r'$ if C' in E, their set of conditional critical pairs modulo A is defined as usual: either we find a non-variable position p in l such that $\alpha \in Unif_A(l|_p, l')$ and then we form the conditional critical pair*

$$\begin{array}{lllll} \alpha(C) \wedge \alpha(C') & \Rightarrow & \alpha(l[l']_p) \underset{A}{=\!=} \alpha(l) \xrightarrow[R]{} \alpha(r) & & \\ & & \downarrow E & & (I) \\ & & \alpha(l[r']_p) & & \end{array}$$

or we have a non-variable and nonfrozen *position p' in l' such that $\alpha \in Unif_A(l'|_{p'}, l)$ and we form the conditional critical pair:*

$$\begin{array}{lllll} \alpha(C) \wedge \alpha(C') & \Rightarrow & \alpha(l') \underset{A}{=\!=} \alpha(l'[l]_{p'}) \xrightarrow[R]{} \alpha(l'[r]_{p'}) & & \\ & & \downarrow E & & (II) \\ & & \alpha(r') & & \end{array}$$

We typically write these critical pairs as $\alpha(C) \wedge \alpha(C') \Rightarrow \alpha(l[r']_p) \to \alpha(r)$ and $\alpha(C) \wedge \alpha(C') \Rightarrow \alpha(r') \to \alpha(l'[r]_{p'})$.

We say that a critical pair of type (I) is *joinable* iff for any substitution τ such that $E \cup A \vdash \tau\alpha(C) \wedge \tau\alpha(C')$ we then have

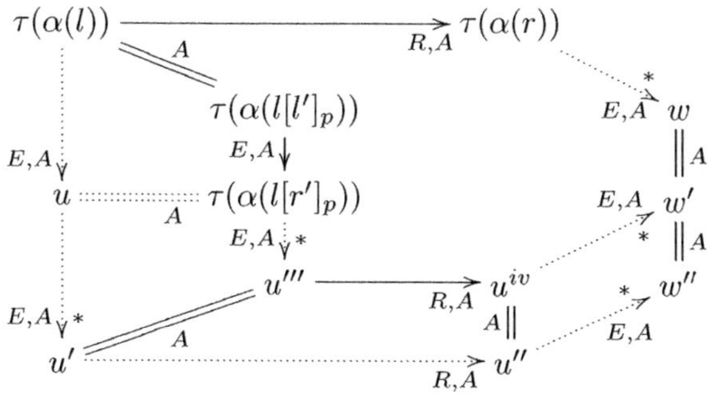

Of course, by $(C) \Leftrightarrow (LC)$ *it is enough to make this check* with $u''' = u'''\downarrow_{E,A}$.

Similarly, we say that a critical pair of type (II) is *joinable* iff for any substitution τ such that $E \cup A \vdash \tau\alpha(C) \wedge \tau\alpha(C')$ we then have

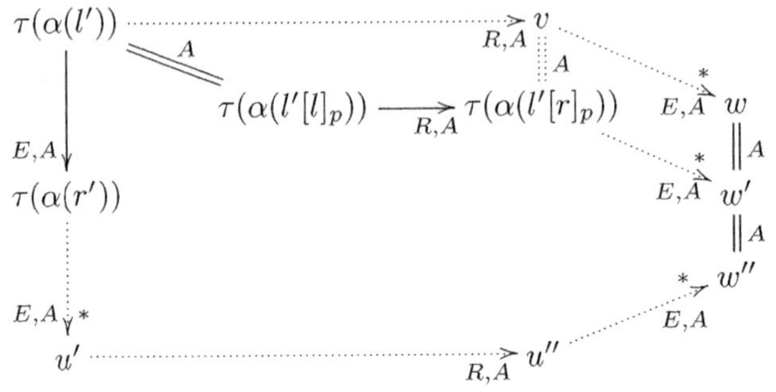

where, again, by $(C) \Leftrightarrow (LC)$ it is enough to perform the check with $u' = u'\downarrow_{E,A}$.

Of course, joinability of all conditional critical pairs is a *necessary* condition for coherence. The challenge now is to find a set of *sufficient conditions* for coherence that includes the joinability of conditional critical pairs.

Specifically, non-overlapping situations between equations and rules require additional conditions. In the case of *coherence* checking, we need to worry about non-overlapping of R under E, that is, for $l' \rightarrow_E r'$ *if* C' in E and $l \rightarrow_R r$ *if* C in R we need to worry about situations of the form:

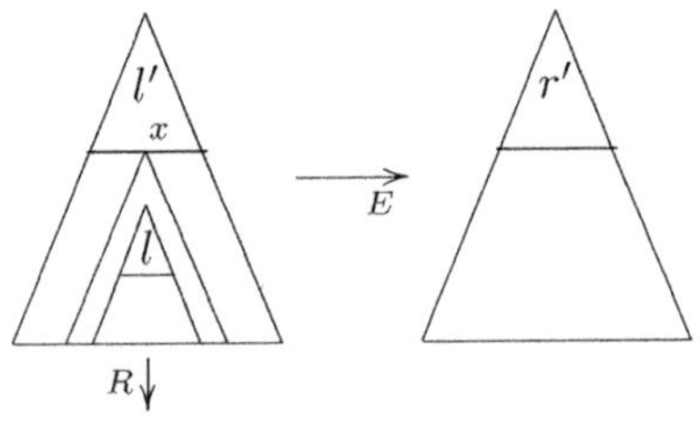

This situation can be problematic in two related ways: (1) when $l' \rightarrow_E r'$ is unconditional but not linear, or (2) when $l' \rightarrow_E r'$ *if* C' is conditional. The problem with (1) is well-understood since [25]. The problem with (2) was also mentioned by Viry in [25]; it has to do with the fact that the satisfiability of the condition C' in an equation $l' \rightarrow_E r'$ *if* C' *depends* on the substitution θ (may hold or not depending on the given θ). But since R *rewrites* the substitution θ, we do *not know if C' will hold anymore after a one-step rewrite with the rule* $l \rightarrow_R r$ *if* C.[4]

Theorem 2. *Given $\mathcal{R}$ as above, then if:*

(i) all conditional critical pairs are joinable and

[4] Note that we can view cases of unconditional $l \rightarrow r$ with l *non-linear* as special cases of (2), since we can *linearize* l, and give an explicit equality condition instead. E.g., `x + x = x` becomes `x + y = x if x = y`.

(*ii*) *for any equation* $l' \to r'$ *if* C' *in* E, *for each* $x \in \mathcal{V}ar(l')$ *such that* x *is non-frozen in* l', *then* either
   (*a*) x *is such that* $x \notin vars(C')$, x *is also non-frozen in* r', *and* x *is linear in both* l' *and* r', *or*
   (*b*) *the sort* s *of* x *is such that* no rewriting with $\to_{R,A}$ is possible for terms of such sort s,

then $\mathcal{R}$ *is coherent.*

Condition (*ii*)-(*b*) of Theorem 2 requires a fixpoint calculation. An algorithm that checks that situations where a non-frozen variable x in a left-hand side of an equation fails to satisfy (*ii*)-(*a*) or (*ii*)-(*b*) is impossible is provided in [13].

2.3 Context-Joinability and Unfeasibility of Conditional Critical Pairs

From those conditional critical pairs which cannot be joined, the tool can currently automatically discard those that are *context-joinable* or *unfeasible*, based on a result by Avenhaus and Loría-Sáenz [1], which we generalize here to the order-sorted case and modulo A. Let us first introduce some notation.

Let a *context* $C = \{u_1 \to_E v_1, \ldots, u_n \to_E v_n\}$ be a set of oriented equations. We denote by $\overline{C}$ the result of replacing each variable x by a new constant $\overline{x}$, and by $\overline{X}$ the set of such new constants. Given a term t, $\overline{t}$ results from replacing each variable $x \in \mathcal{V}ar(C)$ by the constant $\overline{x}$.

We denote by $\rhd$ the proper subterm relation. Then, given an order $\succ$, we denote by $\succ_{st} = (\succ \cup \rhd)^+$ the smallest ordering that contains $\succ$ and $\rhd$. A partial ordering $\succ$ on $\mathcal{T}_\Sigma(\mathcal{X})$ is *well founded* if there is no infinite sequence $t_0 \succ t_1 \succ \ldots$. A partial ordering $\succ$ is *compatible with substitutions* if $u \succ u'$ implies $u\sigma \succ u'\sigma$ for any substitution σ. A partial ordering $\succ$ is *compatible with the term structure* if $u \succ u'$ implies $t[u]_p \succ t[u']_p$ for any term t and position p in t. A partial ordering $\succ$ is *compatible with the axioms* A if $v =_A u \succ u' =_A v'$ implies $v \succ v'$ for all terms u, u', v, and v' in $\mathcal{T}_\Sigma(\mathcal{X})$. A partial ordering $\succ$ is *A-compatible* if it is compatible with substitutions, compatible with the term structure, and compatible with the axioms A. Then, a *reduction ordering* is a partial ordering that is well founded and A-compatible.

A deterministic rewrite theory $\mathcal{R}$ is *quasi-reductive* w.r.t. a reduction ordering $\succ$ on $\mathcal{T}_\Sigma(\mathcal{X})$ if for every substitution σ, every rule $l \to u_{n+1}$ *if* $u_1 \to v_1 \wedge \ldots \wedge u_n \to v_n$ in R , and every $i \in [1..n]$, $u_j\sigma \succeq v_j\sigma$ for every $j \in [1..i]$ implies $l\sigma \succ_{st} u_{i+1}\sigma$.

Definition 3. *Let* E *be an order-sorted deterministic term rewrite systems that is quasi-reductive modulo* A *w.r.t. an* A*-compatible order* $\succ$, *and let* $C \Rightarrow s \to t$ *be a conditional critical pair resulting from* $l \to r$ `if` C_1 *in* R *and* $l' \to r'$ `if` C_2 *in* E, *and* $\sigma \in Unif_A(l|_p, l')$ *(resp.* $\sigma \in Unif_A(l'|_q, l)$*). We call* $C \Rightarrow s \to t$ unfeasible *if there are terms* t_0, t_1, t_2 *such that* $\sigma(l) \succ_{st} t_0$ *(resp.* $\sigma(l') \succ_{st} t_0$*),* $\overline{t_0} \to^*_{E\cup\overline{C},A} t_1$, $\overline{t_0} \to^*_{E\cup\overline{C},A} t_2$, *and* t_1, t_2 *are not unifyable and strongly* $E \cup \overline{C}$, A*-irreducible.*

A Maude order-sorted conditional specification can be converted into an order-sorted deterministic rewrite theory with a simple procedure (see, e.g., [13]). Maude checks that the conditional equational specifications entered are deterministic (c.f. [4]), and we assume it is operationally terminating, and therefore there exists a well-founded A-compatible order $\succ_{st}$ such that we can use the results in [1] and their extension to the Maude case [15], to discard those conditional critical pairs generated that are unfeasible.

Definition 4. *Given a rewrite theory* $\mathcal{R} = (\Sigma, E \cup A, R)$, *a non-joinable conditional critical pair* $C \Rightarrow u \rightarrow v$ *(coming from a conditional critical pair* $C \Rightarrow u \;{}_{E,A}\!\swarrow t \searrow_{R,A}\; v$*) is* context-joinable *if and only if in the extended rewrite theory* $\mathcal{R}_C = (\Sigma \cup \overline{X}, E \cup \overline{C} \cup A, R)$ *we have:*

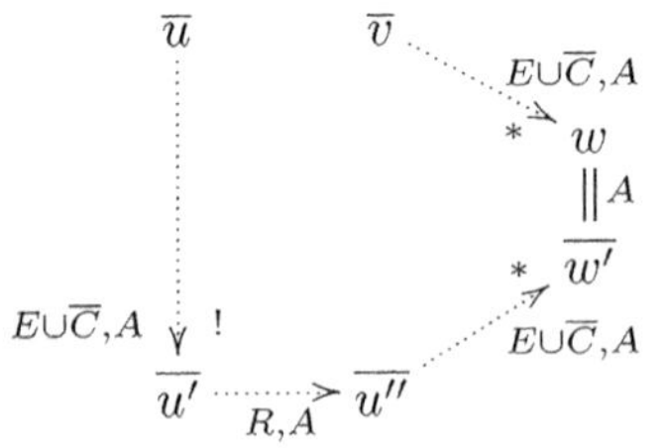

Lemma 2. *If the conditional critical pair* $C \Rightarrow u \rightarrow v$ *is context joinable, then for all substitutions* σ *such that* σC *holds we have*

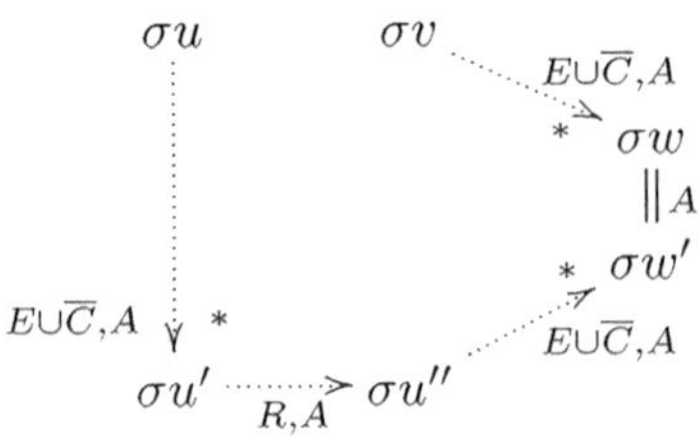

and therefore, the coherence property holds for the conditional critical pair $C \Rightarrow u \;{}_{E,A}\!\swarrow t \searrow_{R,A}\; v$.

2.4 The Ground Coherence Case

Assume that Σ has a sub-signature of constructors Ω that has been verified to be *sufficiently complete* with respect to the equations E modulo A. Then, we can view each $f \in \Sigma$ with a different syntactic form from Ω as a *frozen* operator, since any ground term in E, A-canonical form will *not* contain the symbol f. This automatically excludes all problematic non-overlaps with R below E except for:

(i) constructor equations, and

(ii) equations $f(t_1, \ldots, t_n) \to r$ *if* C in E with $f \in \Sigma - \Omega$, and with f having the identity, left identity, or right identity attributes, and such that the left-hand side of the equation resulting from the variant-based transformation to remove the identity attributes has a *non-frozen variable* (see [10] for details on the variant-based transformation).

Therefore, for ground coherence under the assumption of frozenness of defined symbols, we only have to check condition (ii) in Theorem 2 on equations of types (i) and (ii) above.

Furthermore, for a conditional critical pair $\alpha(C) \wedge \alpha(C') \Rightarrow u \to v$ which we have not been able to show context-joinable, we can use constructor-based inductive methods to try to prove its *inductive one-step reachability*, that is, that $\mathcal{R} \vdash_{ind} \alpha(C) \wedge \alpha(C') \Rightarrow u \to v$. We can illustrate such inductive proof methods with a simple example of a rewrite theory operating on cells containing numbers modulo 4, where the rules double the cell's contents each time.

```
(mod DOUBLE is
   sorts Nat/4 State .
   op 0 : -> Nat/4 [ctor] .
   op s : Nat/4 -> Nat/4 [ctor] .
   op '[_'] : Nat/4 -> State [ctor] .
   op _+_ : Nat/4 Nat/4 -> Nat/4 .

   vars N M : Nat/4 .

   eq s(s(s(s(N)))) = N .
   eq N + 0 = N .
   eq N + s(M) = s(N + M) .

   rl [double-0]: [0] => [0] .
   rl [double-s]: [s(N)] => [s(N) + s(N)] .
 endm)
```

The equations in this theory can be proved terminating using Maude's MTT, Church-Rosser using Maude's CRC, and sufficiently complete using Maude's SCC. The ChC gives a nontrivial critical pair:

```
Maude> (check ground coherence .)

Coherence checking of DOUBLE
Coherence checking solution:
The following critical pairs cannot be rewritten:
  cp [#1:Nat/4]
    => [#1:Nat/4 + #1:Nat/4].
```

However, a simple constructor-based induction on `N` proves that

$$\texttt{DOUBLE} \vdash_{ind} \texttt{[N]} \to \texttt{[N + N]}.$$

Indeed, using rule `double-0` we can reduce the base case to checking

$$\texttt{DOUBLE} \vdash_{ind} \texttt{[0]} = \texttt{[0 + 0]},$$

which can be trivially discharged; and the inductive step can be proved by using rule `double-s` and discharging the trivial equality

$$\texttt{DOUBLE} \vdash_{ind} \texttt{[s(N) + s(N)]} = \texttt{[s(N) + s(N)]}.$$

In general, the inductive equational goals generated this way may not be so trivial as in this example, and may require additional steps of inductive equational reasoning. Also, it may be the case that more than one rule can be applied to rewrite the constructor-instantiated lefthand side of a critical pair, so that we get a disjunctive proof obligation. For example, if we had added to the DOUBLE module a rule to reset any cell contents to 1, namely,

```
rl [reset]: [N] => [s(0)] .
```

then, in the induction step of the inductive proof for the same critical pair [N] $\rightarrow$ [N + N], we would now get the (still equally trivial in this case) disjunctive goal

$$\texttt{DOUBLE} \vdash_{ind} \texttt{[s(N) + s(N)]} = \texttt{[s(N) + s(N)]} \vee \texttt{[s(0)]} = \texttt{[s(N) + s(N)]}.$$

3 How to Use the Maude Coherence Checker

This section illustrates the use of the Maude ChC tool, and suggests some methods that—using the feedback provided by the tool—can help the user establish that his/her specification is ground-coherent.

We assume a context of use in which the user has already developed an *executable specification* of his/her intended system with an initial model semantics, and that this specification has already been checked to have confluent and terminating equations and to have been *tested* with examples, so that the user is in fact confident that the specification is *ground-coherent*, and wants only to check this property with the tool.

The ChC tool not generating any proof obligations is only a *sufficient* condition: in some cases of interest the specification may be *ground* coherent, but not coherent; or may be coherent but the ChC cannot check this automatically because of conditional rules or equations. Then, a collection of critical pairs will be returned by the tool as proof obligations. The ChC does *not* attempt an automatic completion process to add new rules to the user's specification. In many cases this could easily lead to a nonterminating process. Even if such a completion were to terminate, it could easily modify and enlarge the user's specification in undesirable ways. Instead, the feedback of the ChC tool should be used as a guide for *careful analysis* about one's specification. By analyzing the critical pairs returned, the user can understand why they could not be joined. In any case, it is the user himself/herself who must study where the coherence problems come from, and how to fix them by modifying the specification. Interaction with the tool then provides a way of modifying the original specification and ascertaining whether the new version passes the test or is a good step towards that goal.

We present in the following section a simple example that illustrates the use of the tool for different combinations of the associativity, commutativity, and identity axioms. The interested reader can find in [14] additional examples in which conditional equations and rules are used, cases in which conditional critical pairs are discarded using inductive proofs, and so on.

3.1 An Unordered Communication Channel

Consider a communication channel in which messages can get out of order. There is a sender and a receiver. The sender is sending a sequence of data items, for example numbers. The receiver is supposed to get the sequence in the exact same order in which they were in the sender's sequence. To achieve this in-order communication in spite of the unordered nature of the channel, the sender sends each data item in a message together with a sequence number; and the receiver sends back an `ack` message indicating that has received the item. The Full Maude specification of the protocol is as follows:

```
(mod UNORDERED-CHANNEL is
  sorts Nat NatList Msg Conf State .
  subsort Msg < Conf .
  op 0 : -> Nat [ctor] .
  op s : Nat -> Nat [ctor] .
  op nil : -> NatList [ctor] .
  op _;_ : Nat NatList -> NatList [ctor] .     *** list constructor
  op _@_ : NatList NatList -> NatList .        *** list append
  op '[_',_'] : Nat Nat -> Msg [ctor] .
  op ack : Nat -> Msg [ctor] .
  op null : -> Conf [ctor] .
  op __ : Conf Conf -> Conf [ctor assoc comm id: null] .
  op '{_',_|_|_',_'} : NatList Nat Conf NatList Nat -> State [ctor] .

  vars N M J K : Nat .          var C : Conf .
  vars L P Q : NatList .

  eq nil @ L = L .              eq (N ; L) @ P = N ; (L @ P) .

  rl [snd]: {N ; L, M | C | P, K} => {N ; L, M | [N, M] C | P, K} .
  rl [rec]: {L, M | [N, J] C | P, J}
    => {L, M | ack(J) C | P @ (N ; nil), s(J)} .
  rl [rec-ack]: {N ; L, J | ack(J) C | P, M} => {L, s(J) | C | P, M} .
endm)
```

The contents of the unordered channel is modeled as a *multiset* of messages of sort `Conf`. The entire system state, involving the sender, the channel, and the receiver is a 5-tuple of sort `State`, where the components are:

- a buffer for the sender containing the current list of items to be sent,
- a counter for the sender keeping track of the sequence number for items to be sent,
- the contents of the unordered channel,
- a buffer for the receiver storing the sequence of items already received, and
- a counter for the receiver keeping track of the sequence number for items received.

One essential property of this protocol is of course that it achieves *in-order communication* in spite of the unordered communication medium. We can specify this in-order communication property as an *invariant* in Maude. We will assume that all initial states are of the form:

```
{n1 ; ... ; nk ; nil, 0 | null | nil, 0}
```

That is, the sender's buffer contains a list of numbers `n1 ; ... ; nk ; nil` and has the counter set to `0`, the channel is empty, and the receiver's buffer is also empty. Also, the receiver's counter is initially set to `0`.

In specifying the invariant, the auxiliary notion of a list prefix may be useful. Given lists L and L' we say that L is a *prefix* of L' iff either: (1) $L = L'$, or (2) there is a nonempty list L'' such that L @ $L'' = L'$.

```
(mod UNORDERED-CHANNEL-INVARIANT is inc UNORDERED-CHANNEL .
  sort Truth .
  ops tt ff : -> Truth [ctor] .
  op _~_ : Nat Nat -> Truth [comm] . *** equality predicate
  op _and_ : Truth Truth -> Truth [assoc comm id: tt] .

  vars M N K P : Nat .          var C : Conf .
  vars L L' L'' : NatList .     var B : Truth .

  eq 0 ~ 0 = tt .
  eq 0 ~ s(N) = ff .
  eq s(N) ~ s(M) = N ~ N .
  eq ff and ff = ff .

  op prefix : NatList State -> Truth .
  eq [I1]: prefix(M ; L, {L', N | C | K ; L'', P})
    = (M ~ K) and prefix(L, {L', N | C | L'', P}) .
  eq [I3]: prefix(L, {L, N | C | nil, K}) = tt .
  eq [I4]: prefix(nil, {L', N | C | M ; L'', K}) = ff .
endm)
```

The equational part of the specification can be checked terminating and Church-Rosser using the MTT [9] and the CRC [14]. And the rules can be shown to be ground coherent with the equations by using the ChC tool.

```
Maude> (check ground coherence .)

Coherence checking of UNORDERED-CHANNEL
Coherence checking solution:
All critical pairs have been rewritten and all equations are non-
  ↪constructor.
The specification is ground coherent.
```

The problem with this simple example is that one cannot verify the invariant using the `search` command in Maude, because, due to the snd rule, the number of messages that can be present in the channel is unbounded, so that there is an infinite number of reachable states. One should therefore use an *abstraction*.

```
(mod UNORDERED-CHANNEL-ABSTRACTION is
  pr UNORDERED-CHANNEL-INVARIANT .
  vars M N P K : Nat .
  vars L L' L'' : NatList .
  var C : Conf .

  eq [A1]: {L, M | [N, P] [N, P] C | L', K}
    = {L, M | [N, P] C | L', K} .
endm)
```

There are of course several key properties that such an abstraction should satisfy:

(1) the set of states reachable from any initial state should be finite,
(2) the equational theory should be confluent and terminating,
(3) the rules should be coherent with the equations, and
(4) the abstraction should preserve the invariant.

Properties (1), (2) and (4) can easily be checked. For (3) we can use the ChC.

```
Maude> (check ground coherence .)

Coherence checking of UNORDERED-CHANNEL-ABSTRACTION
Coherence checking solution:
The following critical pairs cannot be rewritten:
  cp for A1 and rec
    {L:NatList, M:Nat | #3:Conf [N:Nat, J:Nat] | P:NatList, J:Nat}
    => {L:NatList, M:Nat | #3:Conf ack(J:Nat) [N:Nat, J:Nat]
                         | P:NatList ; N:Nat, s(J:Nat)}.
  cp for A1 and rec
    {L:NatList, M:Nat | [N:Nat, J:Nat] | P:NatList, J:Nat}
    => {L:NatList, M:Nat | ack(J:Nat) [N:Nat, J:Nat]
                         | P:NatList ; N:Nat, s(J:Nat)}.
```

These critical pairs indicate that a rule is missing. We can add the rule:

```
(mod UNORDERED-CHANNEL-ABSTRACTION-2 is
  inc UNORDERED-CHANNEL-ABSTRACTION .
  vars M N K : Nat . vars L L' : NatList . var C : Conf .

  rl [rec2]: {L, M | [N, K] C | L', K}
    => {L, M | [N, K] ack(K) C | L' ; N, s(K)} .
endm)
```

After checking properties (1), (2) and (4) above, we can check also he coherence of the specification.

```
Maude> (check ground coherence .)

Coherence checking of UNORDERED-CHANNEL-ABSTRACTION-2
Coherence checking solution:
All critical pairs have been rewritten, and no rule can be applied
below non-frozen and non-linear variables of equations.
```

4 Conclusions and Future Work

We have presented the theoretical foundations and design of the Maude Coherence Checker. This tool addresses an important need of rewriting logic specifications, namely, checking coherence and ground coherence for very general order-sorted rewrite theories whose equations and rules can be conditional and can be applied modulo various combinations of associativity and/or commutativity and/or identity axioms, and whose operators may have frozenness restrictions. As we have shown, some of these more general requirements, plus the initial model semantics of rewrite theories, can make it in fact *easier* to check coherence and ground coherence than in the much more restrictive untyped, unconditional, and unfrozen case considered by Viry [25]. The tool, together with its documentation, is available at `http://maude.lcc.uma.es/CRChC`.

More work remains ahead. An important issue is that of formal tool integration. The ChC and the CRC are already integrated within a single tool; but as we have explained, the checking of ground coherence can generate inductive equational goals that should be discharged by the Maude ITP. Therefore, a closer integration between the ChC and the ITP would be highly desirable.

Acknowledgements. F. Durán was supported by Spanish Research Projects TIN2008-03107 and P07-TIC-03184. J. Meseguer was partially supported by NSF Grants CCF-0905584, CNS-07-16038, CNS-09-04749, and CNS-08-34709.

References

1. Avenhaus, J., Loría-Sáenz, C.: On conditional rewrite systems with extra variables and deterministic logic programs. In: Pfenning, F. (ed.) LPAR 1994. LNCS, vol. 822, pp. 215–229. Springer, Heidelberg (1994)
2. Bruni, R., Meseguer, J.: Semantic foundations for generalized rewrite theories. Theoretical Computer Science 351(1), 286–414 (2006)
3. Clavel, M., Durán, F., Eker, S., Lincoln, P., Martí-Oliet, N., Meseguer, J., Quesada, J.: Maude: Specification and programming in rewriting logic. Theoretical Computer Science 285, 187–243 (2002)
4. Clavel, M., Durán, F., Eker, S., Lincoln, P., Martí-Oliet, N., Meseguer, J., Talcott, C.: All About Maude - A High-Performance Logical Framework. LNCS, vol. 4350. Springer, Heidelberg (2007)
5. Clavel, M., Durán, F., Hendrix, J., Lucas, S., Meseguer, J., Ölveczky, P.: The Maude formal tool environment. In: Mossakowski, T., Montanari, U., Haveraaen, M. (eds.) CALCO 2007. LNCS, vol. 4624, pp. 173–178. Springer, Heidelberg (2007)
6. Comon-Lundh, H., Delaune, S.: The finite variant property: How to get rid of some algebraic properties. In: Giesl, J. (ed.) RTA 2005. LNCS, vol. 3467, pp. 294–307. Springer, Heidelberg (2005)
7. Durán, F.: A Reflective Module Algebra with Applications to the Maude Language. PhD thesis, U. de Málaga, Spain (June 1999), `http://maude.csl.sri.com/papers`
8. Durán, F.: The extensibility of Maude's module algebra. In: Rus, T. (ed.) AMAST 2000. LNCS, vol. 1816, pp. 422–437. Springer, Heidelberg (2000)
9. Durán, F., Lucas, S., Meseguer, J.: MTT: The Maude termination tool (system description). In: Armando, A., Baumgartner, P., Dowek, G. (eds.) IJCAR 2008. LNCS (LNAI), vol. 5195, pp. 313–319. Springer, Heidelberg (2008)
10. Durán, F., Lucas, S., Meseguer, J.: Termination modulo combinations of equational theories. In: Ghilardi, S. (ed.) FroCoS 2009. LNCS, vol. 5749, pp. 246–262. Springer, Heidelberg (2009)
11. Durán, F., Meseguer, J.: A Church-Rosser checker tool for Maude equational specifications. Technical Report ITI-2000-5, Dpto. de Lenguajes y Ciencias de la Computación, U. de Málaga (October 2000), `http://maude.cs.uiuc.edu`
12. Durán, F., Meseguer, J.: Maude's module algebra. Science of Computer Programming 66(2), 125–153 (2007)
13. Durán, F., Meseguer, J.: ChC 3: A coherence checker tool for conditional order-sorted rewrite Maude specifications (2009), `http://maude.lcc.uma.es/CRChC`
14. Durán, F., Meseguer, J.: CRC 3: A Church-Rosser checker tool for conditional order-sorted equational Maude specifications (2009), `http://maude.lcc.uma.es/CRChC`
15. Durán, F., Meseguer, J.: A Church-Rosser checker tool for conditional order-sorted equational Maude specifications. In: Ölveczky, P.C. (ed.) 8th Intl. Workshop on Rewriting Logic and its Applications (2010)
16. Durán, F., Meseguer, J.: A Maude coherence checker tool for conditional order-sorted rewrite theories, long version (2010), `http://maude.lcc.uma.es/CRChC`
17. Durán, F., Ölveczky, P.C.: A guide to extending Full Maude illustrated with the implementation of Real-Time Maude. In: Roşu, G. (ed.) Proceedings 7th Intl. Workshop on Rewriting Logic and its Applications (WRLA 2008). Electronic Notes in Theoretical Computer Science. Elsevier, Amsterdam (2008)

18. Escobar, S., Meseguer, J., Sasse, R.: Variant narrowing and equational unification. In: Rosu, G. (ed.) Proc. 7th Intl. Workshop on Rewriting Logic and its Applications (WRLA 2008). Electronic Notes in Theoretical Computer Science, vol. 238, pp. 103–119. Elsevier, Amsterdam (2008)
19. Giesl, J., Kapur, D.: Dependency pairs for equational rewriting. In: Middeldorp, A. (ed.) RTA 2001. LNCS, vol. 2051, pp. 93–108. Springer, Heidelberg (2001)
20. Jouannaud, J.-P., Kirchner, H.: Completion of a set of rules modulo a set of equations. SIAM Journal of Computing 15(4), 1155–1194 (1986)
21. Meseguer, J.: Conditional rewriting logic as a unified model of concurrency. Theoretical Computer Science 96(1), 73–155 (1992)
22. Meseguer, J.: A logical theory of concurrent objects and its realization in the Maude language. In: Agha, G., Wegner, P., Yonezawa, A. (eds.) Research Directions in Concurrent Object-Oriented Programming, pp. 314–390. The MIT Press, Cambridge (1993)
23. Ohlebusch, E.: Advanced Topics in Term Rewriting. Springer, Heidelberg (2002)
24. Peterson, G., Stickel, M.: Complete sets of reductions for some equational theories. Journal of ACM 28(2), 233–264 (1981)
25. Viry, P.: Equational rules for rewriting logic. Theoretical Computer Science 285(2), 487–517 (2002)

K-Maude: A Rewriting Based Tool for Semantics of Programming Languages*

Traian Florin Șerbănuță and Grigore Roșu

University of Illinois at Urbana-Champaign

Abstract. K is a rewriting-based framework for defining programming languages. K-Maude is a tool implementing K on top of Maude. K-Maude provides an interface accepting K modules along with regular Maude modules and a collection of tools for transforming K language definitions into Maude rewrite theories for execution or analysis, or into LaTeX for documentation purposes. The current K-Maude prototype was successfully used in defining several languages and language analysis tools, both for research and for teaching purposes. This paper describes the K-Maude tool, both from a user and from an implementer perspective.

1 Introduction

There is overwhelming evidence by now that rewriting logic [4] is a powerful framework for programming language design, semantics and analysis (there are too many papers on these topics to cite; we recommend the interested reader to consult the rewriting logic semantics project [5,12] and the references there). There are two major reasons for that: (1) on the one hand, existing language definitional approaches such as structural operational semantics (with [14] or without [7] evaluation contexts, modular [6] or not) and natural semantics [3] can be faithfully captured by rewriting logic, so one can use rewriting logic and Maude [2] to define and analyze languages using these formalisms, and (2) on the other hand, rewriting logic, thanks to its generality and powerful tool support, encourages the development of new language definitional approaches.

The K framework [11,9] is a semantic definitional framework inspired from rewriting logic but specialized and optimized for programming languages. It consists of three components: a concurrent rewrite abstract machine, a language definitional technique, and a specialized notation. The aim of the concurrent rewrite abstract machine is to increase the potential for concurrency of a rewrite theory by allowing rules which overlap but do not change the overlapped subterm (e.g., two threads writing in different locations in the store) to apply concurrently; the concurrency aspect of K is beyond the scope of this paper, so we do not discuss it here. We will briefly recall the K language definitional technique in Sec. 2, but this paper is essentially related to the K specialized notation. Roșu and Șerbănuță [11] present an overview of the K framework in its full generality, while Roșu [9] discusses K in depth, relating it to other definitional frameworks.

* Supported in part by NSF grants CCF-0916893, CNS-0720512, and CCF-0448501, NASA contract NNL08AA23C, a Samsung SAIT grant, and several Microsoft gifts.

P.C. Ölveczky (Ed.): WRLA 2010, LNCS 6381, pp. 104–122, 2010.

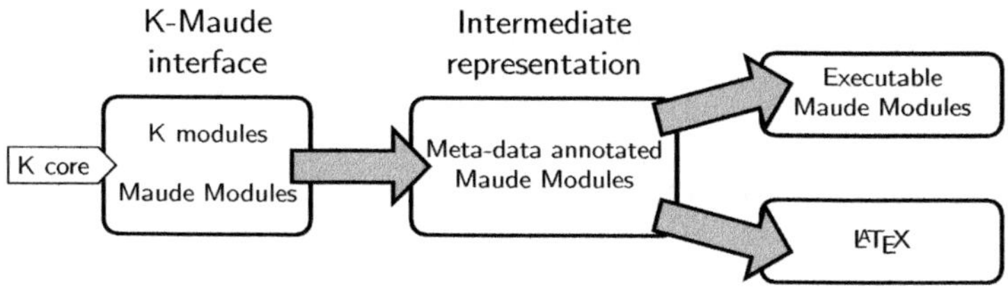

Fig. 1. K-Maude overview. Grayed arrows correspond to translating tools.

The K technique has been manually (without automated tool support) used in the context of rewriting logic and Maude for more than five years, for teaching programming language and program verification courses as well as for several research projects. Such manual uses of K in Maude turned our to be verbose and error prone, because Maude is a general purpose rewrite engine not specifically optimized for programming languages. Thus, the idea of developing a K specialized layer on top of Maude came naturally. The resulting integrated toolkit is called K-Maude and is the subject of this paper. Figure 1 shows the architecture of K-Maude. The gray arrows represent translators implemented as part of the toolkit. The K core contains the ingredients of the K technique, that are handy in most language definitions, such as ones for defining computations, configurations, environments, stores, etc. The K-Maude interface is what the user typically sees: besides usual Maude modules (K-Maude fully extends Maude), one can also include K-Maude files (with extension `.kmaude`) containing modules using the K specialized notation.

A first component of K-Maude translates K modules to Maude modules. The resulting Maude modules encode K-specific features as meta-data attributes and serve as an intermediate representation of K-Maude definitions. Since this representation is just an artifact of using Maude, we will refrain from describing it and we will identify it with the K module it stands for. This intermediate representation can be further translated to different back-ends. We provide two such translators, one to executable/analyzable Maude and one to LaTeX. The former yields actual executable language definitions in Maude which can serve as interpreters for the defined languages or as a basis for formal analysis. This paper addresses only the K-Maude interface and the translation from K-Maude definitions to executable Maude definitions; the specific analysis efforts within the K framework [8,10] are out of the scope of this paper.

The K-Maude to LaTeX translator is meant to serve for documentation purposes. Indeed, we believe that K can be used by ordinary language designers as a formal notation for rigorously specifying the semantics of their languages, the same way context-free grammars are used for formally specifying syntax, so a user-friendly LaTeX notation may be preferred.

Section 2 briefly discusses the K definitional framework. Section 3 gives a user perspective of K-Maude, both w.r.t. its built-in features and how it can be used. Section 4 describes how K-Maude is translated to Maude, so that language designers can execute and formally analyze their K language definitions using Maude, and Section 5 describes how K-Maude is translated to LaTeX, so that language designers can visualize their language definitions.

2 K: A Rewriting-Based Framework for Computations

K [11,9] is a rewriting-based language definitional framework based on intuitions from the chemical abstract machine (CHAM) [1], evaluation contexts [14] and continuations [13]. The idea underlying language semantics in K is to represent the program configuration as a "nested soup" structure, which contains the context needed for the computation, with elements of the context represented as multisets or lists each *wrapped* inside a corresponding *cell*; a cell may also contain other cells. Objects wrapped by cells generally include standard items such as environments, stores, etc, as well as items specific to the given semantics. Mathematically, cells are written using the notation $\langle \ldots \rangle_{\mathsf{env}}$; here 'env' denotes the *cell label* and '...' will represent the contents of the cell. When written in ASCII, such as in K-Maude, we prefer to use the XML-like notation ⟨env⟩ ... ⟨/env⟩. One regularly used cell, labeled by 'k', represents the current *computations structure* of sort K, or simply the *computation*, which is a $\curvearrowright$-separated list of tasks, such as $t_1 \curvearrowright t_2 \curvearrowright \ldots \curvearrowright t_n$. Another, labeled by $\top$, represents the entire configuration structure.

Figure 2 presents the definition of IMP++, a concurrent imperative language using the K framework, as written in K-Maude. The definition follows the general Maude module syntax, but, in addition to Maude syntax, it uses several K specific constructs which will be detailed in the sequel. Although the definition is written using pure ASCII in the K-Maude tool, we have replaced some of the ASCII symbols with mathematical symbols when typesetting, to improve readability. That is, we have replaced, here, and everywhere else in the paper, the following: the mapping construct '|->' by '$\mapsto$', the K arrow '~>' by '$\curvearrowright$', the operation definition keyword '->' by '$\rightarrow$', and the rewriting construct '=>' by '$\Rightarrow$'.

Although not enforced by the K-Maude tool, the IMP++ presented below is divided into three modules: syntax, configuration, and semantics. The K modules are introduced by the special keywords **kmod_is_endkm**. The left column presents the syntax of IMP++ (lines 1–31) and the default configuration of a running program (lines 33–46), while the right column presents the executable semantics of IMP++. Let us start describing IMP++ from its configuration module (lines 33–46). IMP++'s computations consist of expressions (both arithmetic and boolean), and statements (line 35). Among them, booleans and integers are distinguished as results, that is, finished computations (line 36). Execution-wise, IMP++ is an environment-based multi-threaded language (lines 39–41); The $*$ postfixed to the name of the thread cell indicates its multiplicity. All threads are grouped in a threads cell and share a common store (line 42), as well as an input and an output stream (line 44). The entire configuration is contained in a top cell $\top$.

```
kmod IMPPP−SYNTAX is
 including PL−INT + PL−ID
 sort AExp subsorts Int Id < AExp
 op _+_ : AExp AExp →AExp
   [gather(E e) prec 33 strict]
 op _/_ : AExp AExp →AExp
   [gather(E e) prec 31 strict]
 op ++_ : Id → AExp [prec 0]
 op read : → AExp
 sort BExp subsort Bool < BExp
 op _<=_ : AExp AExp →BExp
   [prec 37 seqstrict]
 op not_ : BExp → BExp
   [prec 53 strict]
 op _and_ : BExp BExp →BExp
   [gather(E e) prec 55 strict (1)]
 sort Stmt   op '{'} : → Stmt
 op _; : AExp → Stmt [prec 90 strict]
 op __ : Stmt Stmt → Stmt
   [prec 100 gather(e E)]
 op _=_; : Id AExp → Stmt
   [prec 80 gather (e E) strict (2)]
 op if_then_else_ : BExp Stmt Stmt
   → Stmt [strict (1)]
 op while_do_ : BExp Stmt → Stmt
 op print_; : AExp → Stmt [strict]
 op spawn_ : Stmt → Stmt [prec 90]
 op haltThread ; : → Stmt
 op var_; : Id → Stmt [prec 2]
 op '{_'} : Stmt → Stmt [gather(&)]
endkm

kmod IMPPP−CONFIGURATION is
 including IMPPP−SYNTAX + K
 subsort AExp BExp Stmt <K
 subsort Bool Int < KResult
 configuration
  ⟨T⟩
   ⟨threads⟩⟨thread*⟩
    ⟨k⟩.K⟨/k⟩ ⟨env⟩.Map⟨/env⟩
   ⟨/thread*⟩⟨/threads⟩
   ⟨store⟩.Map⟨/store⟩
   ⟨nextLoc⟩0⟨/nextLoc⟩
   ⟨in⟩.List⟨/in⟩ ⟨out⟩.List⟨/out⟩
  ⟨/T⟩
endkm
```

```
kmod IMPPP−SEMANTICS is
 including IMPPP−CONFIGURATION
 rule ⟨k⟩X:Id ⇒ I:Int⟨_/k⟩
      ⟨env_⟩X ↦ N:Nat⟨_/env⟩
      ⟨store_⟩N ↦ I⟨_/store⟩
 rule I1:Int + I2:Int ⇒ I1 +Int I2
 rule I1 / I2 ⇒ I1 /Int I2 if I2 =/=Bool 0
 rule ⟨k⟩++ X ⇒I +Int 1⟨_/k⟩
      ⟨env_⟩X ↦N⟨_/env⟩
      ⟨store_⟩N ↦(I ⇒ I +Int 1)⟨_/store⟩
 rule ⟨k⟩read ⇒ I⟨_/k⟩
      ⟨in⟩ListItem(I) ⇒ .⟨_/in⟩
 rule I1 <= I2 ⇒ I1 <=Int I2
 rule not T:Bool ⇒ notBool T
 rule true and B:BExp ⇒B
 rule false and _ ⇒ false
 rule {} ⇒ .    rule I ; ⇒ .
 rule S1:Stmt S2:Stmt ⇒ S1 ↷ S2
 rule ⟨k⟩X = I ; ⇒ .⟨_/k⟩
      ⟨env_⟩X ↦N⟨_/env⟩
      ⟨store_⟩N ↦(_ ⇒ I)⟨_/store⟩
 rule if true then S1 else _ ⇒ S1
 rule if false then _else S2 ⇒ S2
 rule ⟨k⟩while B do S:Stmt ⇒
  if B then S while B do S else {}⟨_/k⟩
 rule ⟨k⟩print I ; ⇒ .⟨_/k⟩
      ⟨out_⟩. ⇒ ListItem(I)⟨/out⟩
 rule ⟨k⟩spawn S ⇒ .⟨_/k⟩
      ⟨env⟩Env:Map⟨/env⟩
      (. ⇒ ⟨thread_⟩
             ⟨k⟩S⟨/k⟩
             ⟨env⟩Env⟨/env⟩
           ⟨_/thread⟩)
 rule ⟨thread_⟩⟨k⟩.K⟨/k⟩ ⟨_/thread⟩ ⇒ .
 rule ⟨k⟩haltThread ; ↷ _ ⇒ .⟨/k⟩
 rule ⟨k⟩var X ; ⇒ .⟨_/k⟩
      ⟨env⟩Env ⇒ Env[N / X]⟨/env⟩
      ⟨store_⟩. ⇒ N ↦0⟨_/store⟩
      ⟨nextLoc⟩N ⇒ N +Nat 1⟨/nextLoc⟩
 rule ⟨k⟩{S} ⇒ S ↷ env(Env)⟨_/k⟩
      ⟨env⟩Env⟨/env⟩
 op env : Map → K
 rule env(_) ↷ env(Env) ⇒ env(Env)
 rule ⟨k⟩env(Env) ⇒ .⟨_/k⟩
      ⟨env⟩ _ ⇒ Env⟨/env⟩
endkm
```

Fig. 2. Full definition of IMP++ in K-Maude

Arithmetic expressions are constructed from variables and integers with addition and division (lines 2–7, 49–53), but additionally include variable increment (lines 8, 54–56), to exhibit side effects, and external input (lines 9,57–58). Boolean expressions (lines 10–16, 59–62) are constructed from comparing arithmetic expressions, with conjunction and negation as connectives. Statements consists of standard constructs as the empty statement, the expression statement, sequential composition, assignment, conditional, and loop (lines 17–25, 63–71), to which were added the following: output (lines 26, 72–73), thread creation and dissolution (lines 27, 74–80), abrupt thread termination (lines 28, 81), as well as local variable declarations and blocks (lines 29–30, 82–91). To exhibit features of K (and K-Maude) not shown by the IMP++ definition, we will also discuss features from the CHALLENGE [11] definition, an elaborate extension of IMP++.

Due to K's own use of the '.' symbol (for generic unit), we do not use '.' as terminator in K modules, as mandatory in Maude modules; instead, we rely on reserved keywords such as **including**, **sort**[s], **subsort**[s], **op**[s], **configuration**, **context**, and **rule**, to disambiguate declarations. The syntax uses mostly standard Maude syntax and conventions for expressing the CFG as a collection of sorts (for non-terminals), and subsorts and (mixfix) operation declarations (for grammar productions). However, in addition to Maude's attributes (such as precedence and gathering), K specific attributes can be added, such as **strict**, which is used to specify that (certain) arguments of a language construct need to be evaluated first (and their effects on the global state be propagated) before giving semantics to the construct itself.

K Rewrite Rules

A K definition consists of two types of sentences: structural rules (often reversible, like equations) and computational rules (typically non-reversible).

Structural rules carry no computational meaning; instead, borrowing a concept from CHAMs, structural rules can *heat* and *cool* computations. When a computation is heated, it breaks into smaller pieces, exposing subexpressions of more complex expressions for evaluation. Cooling reverses this process, reassembling the (potentially modified) pieces into a computation with the same "shape". The following are examples of structural rules:

$$a_1 + a_2 \rightleftharpoons a_2 \curvearrowright a_1 + \square$$

$$a_1 + a_2 \rightleftharpoons a_1 \curvearrowright \square + a_2$$

$$\text{if } b \text{ then } s_1 \text{ else } s_2 \rightleftharpoons b \curvearrowright \text{if } \square \text{ then } s_1 \text{ else } s_2$$

Language syntax is completely abstract in K, in the sense that each language construct is a 'KLabel' which is applied to other computations, i.e., terms of sort K; for convenience, and also supported by the K-Maude tool, we continue to use the mix-fix notation for syntax, like above. Unlike in evaluation contexts, $\square$ is not a "hole", but rather part of a KLabel, carrying the obvious "plug" intuition; e.g., the KLabels involving $\square$ above are '$\square + \_$', '$\_ + \square$', and '$\text{if } \square \text{ then\_else\_}$'.

Many structural rules can be automatically generated by annotating constructs in the language syntax with **strict** attributes: a **strict** attribute generates the appropriate structural rules for each strict argument. If an operator is intended to be strict in only some of its arguments, then the positions of the strict arguments are listed as arguments of the **strict** attribute; for example, the first two equations above directly correspond to the attribute **strict** for addition in IMP++ (line 5), i.e., strict in all arguments, while the last one corresponds to **strict**(1) attribute used for the IMP++ conditional (line 24). One can also define evaluation contexts in K, by indicating the "hole" where evaluation should take place; for example, assuming an extension of IMP++ with pointers and a C-like dereferencing operator $*\_$ (like the CHALLENGE definition [11]), a context declaration '**context** $*$ [HOLE] $=$ _;' says that the argument of $*\_$ needs to be evaluated before the assignment can be defined.

Computational rules represent actual steps of computation. However, to account for the differences between K rules and regular rewrite rules, we chose to introduce K rules with the **rule** keyword, even when they have the form of regular rewrite rules, e.g., the rules for addition and conditional (line 52, and lines 68–69, respectively). **rule** can also be used to express structural rules, by adding the **structural** attribute to the end of the rule.

In-place rewriting. In addition to regular rewrite rules, of the form '**rule** l $\Rightarrow$ r', K allows one to also write rules using the following contextual notation:

$$C[\underset{t'_1}{\underline{t_1}}, \underset{t'_2}{\underline{t_2}}, ..., \underset{t'_n}{\underline{t_n}}]$$

which says that in (multi-)context C (that is a term with multiple, ordered holes), each pattern t_i rewrites to t'_i for each $i \in \{1, ..., n\}$. An n-hole context could formally be described as a term over the set of variables $\{\Box_i\}_{1\leq i\leq n}$ containing exactly one occurrence of each variable $\Box_i$. An instantiation of an n-context C with terms $t_1, \ldots, t_n$, written $C[t_1, \ldots, t_n]$ is obtained by applying on C the substitution yielding t_i for each $\Box_i$.

One motivation for in-place rewriting rules is that they allow for a more compact and less error-prone representation for rules matching large configurations but effecting only small changes. Another motivation is that the context C can be concurrently shared by various rules, which can apply concurrently provided that none of them changes C (C is "read only"). If one ignores this concurrency aspect, then one can translate each K contextual rule into a rewrite rule $C[t_1, t_2, ..., t_n] \to C[t'_1, t'_2, ..., t'_n]$; this is precisely what K-Maude does. In K-Maude, the mathematical in-place rewriting $\frac{l}{r}$ is injected in-place as 'l $\Rightarrow$ r'. By default, the in-place rewriting construct '$\_ \Rightarrow \_$' is greedy; parenthesis can be used for disambiguation purposes (see, e.g., lines 56, and 76–79).

Anonymous variables. Another advantage emerging from the single-term representation of rules induced by in-place rewriting is that variables occurring only once in the rule can now be "anonymized", that is, replaced by the anonymous variable symbol '_', since they are only used for matching purposes. This is especially true in the case of matching inside cells, of which we usually use/replace only one object in a rule, but we need to match the contents of the entire cell. A special notation for cells is used to help the intuition that the cells might be "open" at one end, or both. For example, in the rule for reading a variable from the store (lines 49–51), one can either use $\langle \mathrm{k}\rangle \mathrm{X} \Rightarrow \mathrm{I}\langle \_/\mathrm{k}\rangle$ or $\langle \mathrm{k}\rangle(\mathrm{X} \Rightarrow \mathrm{I}) \curvearrowright \_\langle /\mathrm{k}\rangle$ to specify that X is to be matched and replaced at the beginning of computation, and use either $\langle \mathrm{env}\_\rangle \mathrm{X} \mapsto \mathrm{N}\langle \_/\mathrm{env}\rangle$ or $\langle \mathrm{env}\rangle\_\ \mathrm{X} \mapsto \mathrm{N}\ \langle /\mathrm{env}\rangle$ to specify that $\mathrm{X} \mapsto \mathrm{N}$ is to be matched in the middle of the environment; also, one can use either $\langle \mathrm{out}\_\rangle\texttt{.} \Rightarrow \mathrm{ListItem(I)}\langle /\mathrm{out}\rangle$ or $\langle \mathrm{out}\rangle\_(\texttt{.} \Rightarrow \mathrm{ListItem(I)})\langle /\mathrm{out}\rangle$ to add an integer to the end of the output list (line 73). One could think of this notation as having the following intuition: $\langle\rangle$ is a membrane delimiter, which can carry inside attributes such as name (e.g., 'out'), visual information specifying that a membrane is closing the cell ('/'), as well as information specifying whether a part of the cell (the left, the right, or both) is subsumed by the pattern ('_'). Therefore one can read the "tag" $\langle \_/\mathrm{k}\rangle$ as: the membrane closing the k cell while subsuming the final part of the cell. By convention we will always attach '/' to the name of the cell, and '_' to the membrane wall closest to the cell contents.

As previously mentioned, the unit elements for the List, Set, Map, and even K sorts is '.'; however, since cells are not typed, whenever disambiguation is needed, one can postfix the name of the sort to the '.'; e.g., the initial values in the configuration term (lines 37–45), or the empty computation '.K' used in the rule for dissolving a thread (line 80). The preprocessor transforms them to the right constants used in the Maude representation. Also, the preprocessor allows one to avoid variable declarations by declaring variables inline in the rules, using the Maude syntax, e.g., I:Int. However, once a variable was declared inline, the other apparitions of the variable in the module need not be sorted anymore, as they would assume the already declared sort.

The notation used by the K-Maude tool is a one-dimensional ASCII rendition of the K mathematical notation [11]. For example, the K mathematical rendering of the IMP++ assignment rule (lines 65–67) is:

$$\langle \underset{\cdot}{\underline{X = I;}}\ \_\rangle_{\mathsf{k}}\ \langle \_\ X \mapsto N\ \_\rangle_{\mathsf{env}}\ \langle \_\ N \mapsto \underset{I}{\underline{\_}}\ \_\rangle_{\mathsf{store}}$$

The above rule says that if the assignment $X = I$ is the first computational task, and if X is at location N in the environment, then replace whatever is at location N in the store by I and discard the assignment. Note that in the mathematical notation the membranes wrapping the cells, e.g., $\langle \mathrm{env}\_\rangle\ \langle \_/\mathrm{env}\rangle$ are replaced by "thinner" membranes, e.g., $\langle\_\quad \_\rangle_{\mathsf{env}}$, but still maintain all relevant information.

3 K-Maude Interface

For the purpose of this paper, K can be regarded as a notational layer on top of rewriting logic, specialized and optimized for writing definitions of complex programming languages and models. Since our aim for K-Maude is to fully support rewriting logic and Maude, we implemented it as an extension of Maude. Consequently, one is free to use or not the K notation when writing language definitions. An extreme approach, which could be convenient for existing Maude users who want to gradually get exposed to K, is to only include the provided K-Maude core and then follow the K language definitional technique but use plain Maude, the same way one can give SOS or other semantic definitions using plain Maude [12]. This section presents the ingredients of both the K technique and the specific K notation used by the K-Maude tool. However, we will insist more on notation here, and refer the interested reader to the relevant material discussing in depth the corresponding concepts within the K technique [11].

3.1 K-Maude Core

The core syntax of K-Maude can be found in file 'k-prelude.maude', module `K-TECHNIQUE`. It starts by providing means to build computations as sequences of abstract syntax trees, as well as distinguishing the result computations among them. Then it defines lists, bags, sets, and maps as sorts, together with means to inject computations as elements in these data-structures. Finally, the core provides minimal support for describing configurations as "nested soups" of cells, along with two default cell names, 'k' and 'T'.

Basic K syntax [11, Sec. 5.2]. A computation is a term of a specific sort K, and is defined as a list of tasks with identity '`.`' and constructor '$\_ \curvearrowright \_$' as well as a way of building structured computations by applying labels (one per language construct) on top of lists of computations:

op $\_ \curvearrowright \_$: $K\ K \rightarrow K$ [prec 100 assoc id: .] .
op $\_(\_)$: *KLabel List*{*K*} $\rightarrow$ *K* [prec 0 gather(& &)] .

In K-Maude, the sort List{K} is built from K using '$\_ ,, \_$' as a constructor (to allow us to use the single ',' in language definitions) and '`.List{K}`' as unit (to disambiguate from the unit of computations). Finished computations are distinguished to allow for a computational treatment of strictness rules. The sort KResult is meant to describe *results*, or computations which need no further evaluation, and the sort List{KResult} is the corresponding subsort of List{K}. We additionally introduce KResultLabel and KHybridLabel, as subsorts of KLabel, together with their corresponding application constructors:

op $\_(\_)$: KResultLabel *List*{*K*} $\rightarrow$ *KResult* [ditto] .
op $\_(\_)$: KHybridLabel *List*{*KResult*} $\rightarrow$ *KResult* [ditto] .

The distinction between the two is that, while the first encapsulates the entire list of computations below into a result, the second is "hybrid", that is, it only becomes a result when all the computations it "wraps" become results.

Lists, bags, sets, and maps. K-Maude provides generic sorts List, Set, Bag and Map, constructed from their corresponding element sorts ListItem, SetItem,

BagItem and MapItem, with constructor '__' and having unit '.'. Moreover, the following injections of K into the element sorts are provided: ListItem, SetItem, BagItem, and '$\_\mapsto\_$' (to inject a pair of K's into MapItem).

Labelled Cells. The configuration is defined as a structured "soup" of cells. Therefore, we use the already defined sort Bag to hold such a collection of cells. Having a unique sort for cells makes cell nesting easy; a cell holding other cells simply needs to take a Bag as an argument. To declare a cell, one only needs to specify its label, as a 'CellLabel'.

sort *CellLabel* .

op $\langle\_\rangle\_\langle/\_\rangle$: *CellLabel K CellLabel* → *BagItem* [prec 0] .
op $\langle\_\rangle\_\langle/\_\rangle$: *CellLabel List CellLabel* → *BagItem* [prec 0] .
op $\langle\_\rangle\_\langle/\_\rangle$: *CellLabel Bag CellLabel* → *BagItem* [prec 0] .
op $\langle\_\rangle\_\langle/\_\rangle$: *CellLabel Set CellLabel* → *BagItem* [prec 0] .
op $\langle\_\rangle\_\langle/\_\rangle$: *CellLabel Map CellLabel* → *BagItem* [prec 0] .

ops k T : → *CellLabel* .

K-Maude currently allows five kinds of cells, each containing either a computation, a list, a bag, a set, or a map. The syntax for the cells is defined as that of an XML element, with an opening and a closing tag, which must match (i.e., the corresponding CellLabels must be equal).

3.2 K-Maude Specific Modules

The K-Maude modules are introduced by **kmod_is_endkm**, to distinguish them from usual Maude modules. A script, external to Maude, is used to preprocess the K-modules into a form which can be parsed by Maude, for example by a adding the terminator '.' at the end of declarations, by changing **kmod** into 'mod' and **endkm** into 'endm', and by wrapping specific K attributes in 'metadata' strings.

3.3 Language Syntax and Annotations

As already mentioned, the syntax of the language is defined as an algebraic signature, associating to each language construct a mix-fix operation, following the equivalence between CFG grammars and mix-fix algebraic signatures. For example, the BNF rule for conditional *Stmt* ::=`if` *BExp* `then` *Stmt* `else` *Stmt* translates into the operation declaration **op** if_then_else_ : BExp Stmt Stmt →Stmt.

Syntax attributes [11, Sec. 5.4]. In addition to the existing operation declaration attributes provided by Maude, K-Maude introduces several new attributes:
– **strict** specifies what arguments need to be evaluated before evaluating the language construct itself. For example, the **strict**(2) attribute of the assignment declaration in IMP++ (line 22) states that the semantic rule for assignment can assume that the second argument is evaluated;

- **seqstrict** is similar, but also states the evaluation *sequence* of the arguments;
- **hybrid** specifies that the language construct would become a value once all its arguments have been fully evaluated.

3.4 Defining the Program Configuration [11, Sec. 5.1]

Currently, the program configuration is specified by providing a term which should stand for the initial configuration, introduced by the **configuration** keyword. The cell labels present in the configuration are then inferred and declared as CellLabel constants by the preprocessing tool.

Specifying the structure of the configuration serves not only for documentation purposes, but has consequences in both the modularity and the compactness of definitions, since the semantics rules need to mention only the context required for them to apply, as detailed in the next section.

3.5 Defining Language Semantics

The K-specific semantic constructs which are supported by K-Maude are the K contexts and the K (structural or computational) rules. Contexts can be thought of as evaluation contexts, specifying the order of evaluation, while K rules provide notational shortcuts to make definitions more compact.

Context strictness. K-contexts are usually used for specifying strictness constraints which depend on a context rather than on a single construct. For example, the context strictness declaration '**context** $*$ [HOLE] = _' (see the CHALLENGE definition [11]) specifies the evaluation to an L-value of a pointer in the assignment construct, allowing the rule for pointer-assignment to assume it has a value in place of the first argument (the hole); the fact that it would also have a value instead of the second argument was specified by the strictness annotation for '_=_'. Actually, all strictness annotations are turned by the tool into context strictness ones during the compilation process, before the actual heating and cooling equations are being generated.

K rules [11, Sec. 5.3]. K rules are introduced by the **rule** keyword, and basically describe a special pattern term enriched with syntax for expressing the K-specific features described below. There are two types of K rules, structural and computational; in K-Maude we distinguish them by adding the attribute **structural** to the former. The intuition is that the structural rules prepare the program state for a computational step. Therefore, the K-Maude tool translates the former into equations and the latter into rules.

In-place rewriting. A K rule is a term which should contain at least one occurrence of the T1 $\Rightarrow$ T2 construct, which is used as a textual representation for the K visual replacement pattern: $\frac{T1}{T2}$. For example, the increment rule of IMP++ (lines 54–56) contains two non-trivial in-place replacements, while also sharing quite a bit of the context: if the construct '++ X' is found on top of the computation, and the environment contains the mapping of X to a location N, and the store maps N to an integer I, then '++ X' is locally replaced by the value associated to $I+1$, and I is locally replaced by $I+1$ in the store. The '_ ⇒ _' arrow is greedy, i.e., it will expand to the nearest enclosing boundaries. Therefore, one might sometimes need to use parenthesis to clearly fix those boundaries for parsing reasons, e.g., changing the value at a location in the store in line 56, but also semantic reasons, e.g., new thread creation in lines 76–79.

Anonymous variables. Specified by '_', anonymous variables can be used to replace all variables whose name is not needed in the match-and-replace process. For example, the IMP++ 'haltThread' rule (line 81) uses an anonymous variable to abstract the remainder of the computation, since it will be discarded by the rule.

Cell comprehension. Maude-K allows partial specification of the contents of a cell, by adding '_' as an attribute inside the membrane delimiter '⟨⟩' on the side of the cell which should be abstracted away. For example, the IMP++ rule for output (lines 72–73), saying that the integer argument of a 'print' statement is appended to the contents of the 'output' cell, abstracts away both the rest of the computation and the existing output list; these are not changed by the rule.

Context abstraction and context transformation [11, Sec. 5.5]. The main reason for specifying the structure of the configuration is that one does not need to mention the full context required for the application of a rule, but only the parts which are relevant. Within a rigid configuration structure in which the path to each cell is unambiguous, it becomes straight-forward to infer what context needs to be added to a rule to adapt it to the running configuration. A simple instance of using context abstraction is the IMP++ assignment rule (lines 65–67). The rule for assignment should be the same in any definition containing an environment and a store. Although in our definition the store is not at the same level with the computation and the environment, we can still use this rule in the specification, because it can be easily inferred which store the rule refers to.

The following rule could be used to define a rendez-vous synchronization construct (see the CHALLENGE definition [11, Sec. 6]) as follows:

rule ⟨k⟩ rv I ⇒ . ⟨_/k⟩ ⟨k⟩ rv I ⇒ . ⟨_/k⟩ .

Note that, although the two computation cells need to be in two different threads, there is no danger of confusion, since the multiplicity of the 'k' cell is one, so the only way to make sense of this rule is to have each computation in its own thread, since the multiplicity of the 'thread' cell may vary.

Default contexts. Another aspect of the context abstraction with impact on modularity is filling the context with default values on the right-hand-side of an (in-place) rewriting pattern. One such example is the IMP++ rule for thread creation (lines 74–79). Note that we have specified the thread cell as being incomplete in both sides. This is used as a notation to specify that the thread cell is incompletely specified, and thus it should be context-transformed, filling all gaps with default values. For this specific configuration, this notation was not necessary, but this allows for modular changes of configuration when adding cells having constant initial values when a thread is started, such as a function call stack or a set of locks hold by the thread—see the CHALLENGE definition [11, Sec. 6] for a complex example.

It is arguable that context abstraction could have undesirable effects on badly written specifications. However, due to its deterministic nature, we believe it to be rather useful and intuitive. Besides saving the need for providing additional context (which could get quite large and tedious to write), and thus providing brevity to specification, it also enables reusing, since now a rule specifies only the minimal needed context. Moreover, K-Maude desugars K rules to pure rewriting logic rules and equations, so one could always inspect the resulting rules to ensure no unexpected behaviors are introduced when resolving the context abstraction.

Rewrite rules. One can additionally use regular rewrite rules and equations when giving semantics to the language constructs. The K-specific syntactic conventions presented above also apply to them; e.g., one can use '_' as an anonymous variable in the left-hand-side of a rule, and even context abstraction in the right-hand-side, which is useful, for example, to set up the initial configuration when starting the execution of a program.

4 From K-Maude to Maude

This section describes the technical part of the K-Maude tool. As the semantics of the K framework itself is given using rewriting logic, it comes natural that the executable semantics of K, as given by the K-Maude tool, is given by reduction to pure Maude (executable) rewrite theories. That is, each of the K-specific features is transformed into its rewriting logic representation.

Syntax. As mentioned in Sec. 3.3, the K-Maude interface allows for the definition of syntax as an algebraic signature, using subsorting and mixfix operations to emulate CFG grammar descriptions. This allows the programs to look more natural, but also, more importantly, it improves the readiness of the semantic rules. Nevertheless, as previously mentioned, the K framework takes a fairly abstract view on syntax, that is, a tree built as labels applied to (possibly empty) lists of subtrees. To achieve that, the K-Maude tool transforms all syntax into labels. With syntax being just labels and with the distinction between value (of sort KResult) and non-value computations, the strictness attributes are easily desugared as heating only on non-value computations and cooling only on values.

Semantics. The semantic part of a K definition is gradually transformed into an executable Maude module as follows: First, the configuration term is used to resolve context abstraction. Next, cell comprehension is resolved by adding anonymous variables, which, in their turn are replaced with proper (fresh) variables of the right sort. Then, K rules are transformed into rewriting logic equations and rules, by resolving the in-place rewriting. Finally, all computation terms (including the test programs specified by the user) are transformed into ASTs.

Besides the original preprocessor, which wraps the K definitions so that they can be recognized and parsed by Maude, all syntax and semantic transformations are entirely defined within Maude, taking advantage of its reflective capabilities and of the predefined Maude modules used to represent and transform meta-terms and meta-modules. In the sequel we give more details.

4.1 From Syntax to K Syntax and K Representation

To take advantage of Maude's parsing capabilities and to keep semantics as human readable as possible, the user of K-Maude is allowed to use Maude's mix-fix multi-sorted algebraic signatures to define the syntax of the desired language or calculus. However, in K we want to keep computations to a minimal structure to facilitate easy and generic traversal functions, which are crucial for advanced reflective features such as code generation (see, e.g., the CHALLENGE definition [11, Sec.6]). To achieve this, the K-Maude tool automatically generates the labels for the abstract (running) syntax from the input (user) syntax.

Abstract syntax. The K running syntax only consists of K labels, as defined in the core syntax of computations presented above. Since the semantic rules mix the syntax with semantics-specific constructs and use them in contexts where computations are required, the user has to subsort all syntactic categories to computation sorts K and KResult, depending on whether they represent proper computations or values, respectively. The tool uses this information to generate the appropriate labels, i.e., constants of the 'KLabel' or the 'KResultLabel' sorts, for each operation symbol. For example, for the conditional construct, its corresponding K label declaration is "**op** 'if_then_else_ : $\rightarrow$ KLabel". To avoid label symbol conflicts, the K label symbols are generated by simply quoting the identifier used to declare the mixfix syntactic construct.

Handling data types. There are certain sorts, such as integers, booleans, and identifiers, which need to be handled in a special way, to be able to identify them when giving the semantics. To address that, we allow certain sorts to be identified as builtins at the user level, by introducing a new computation sort Builtins and subsorting all such sorts to it. These sorts will be injected into labels in an appropriate manner, following the subsorting chain to either KResult or K. For example, integers are injected into KResultLabel as '**op** Int_ : Int $\rightarrow$ KResultLabel', since they are subsorted to KResult, while identifiers are injected to KLabel through '**op** Id_ : Id $\rightarrow$ KLabel'.

Translating terms. The constant labels and constant injections defined above are used to completely replace the original syntax. For example, the fragment

`if` a <= 2 `then` a = 2 ; `else` {}

gets translated to:

'`if_then_else`('_<=_(*Id* a(. *List*{*K*}),,*Int* 2(. *List*{*K*})),,
'_=_;(*Id* a(. *List*{*K*}),,*Int* 2(. *List*{*K* })),,'‘{’}(. *List*{*K*}))

4.2 Strictness

Strictness annotations provided as attributes to operator declarations are translated into K context declarations, one for each position in which the operation should be strict. Then, each context is transformed into two equations: one which pulls the strict argument (represented by the hole) out of the context for evaluation, and another one which, once the argument becomes a value, plugs it back into its original context.

Strict operator attributes. For each argument position declared strict for an operation, a context declaration is generated, containing a hole. For example, the **strict**(2) declaration for the assignment operation in IMP++ (line 22) would generate the following context declaration: '**context** '_=_;(K1:K,,[HOLE])', while the **seqstrict** declaration for '_<=_' is desugared into two context declarations, '**context** '_<=_([HOLE],,K1:K)' and '**context** '_<=_(K1:KResult,,[HOLE])'. Sequential strictness in ensured by requiring the first argument of the last context above to be an evaluated computation.

Strict contexts. Although we could identify proper computations by a side condition testing that they are not of sort KResult, we prefer to introduce a new category of computations, KProper, with the intuition that KProper and KResult form a partition of the K sort. Since all computations are built by applying labels on other lists of computations, we therefore also introduce the sort KProperLabel, and change all existing label definitions such that any K label which is not a result label will be a proper label. For example, the label associated to the conditional would now have KProperLabel as its resulting sort; the same holds for the 'Id_' injection. Having KProper computations, the strict contexts are desugared as follows: two equations are generated for each context, one for pulling out the *proper* computation for evaluation and the other for plugging in the *result* computation. For the assignment operation declared strict in the second argument, the generated equations are:

eq $\langle$ k $\rangle$ K1:*K* = Kcxt:*KProper* $\curvearrowright$Rest:*K* $\langle/$ k $\rangle$
 = $\langle$ k $\rangle$ Kcxt:*KProper* $\curvearrowright$ freezer ("'_=_;(K1:*K*,,'[HOLE'']:*K*)")(
 freezeVar("K1:*K*")(K1:*K*)) $\curvearrowright$Rest:*K* $\langle/$ k $\rangle$.
eq $\langle$ k $\rangle$ Kcxt:*KResult* $\curvearrowright$ freezer ("'_=_;(K1:*K*,,'[HOLE'']:*K*)")(
 freezeVar("K1:*K*")(K1:*K*)) $\curvearrowright$Rest:*K* $\langle/$ k $\rangle$.
 = $\langle$ k $\rangle$ K1:*K* = Kcxt:*KResult* $\curvearrowright$Rest:*K* $\langle/$ k $\rangle$

These equations apply only at the top of the continuation, because they should only affect the current evaluation redex. Again, as a way to generate unique and meaningful identifiers, we have chosen to have a generic wrapper freezer which takes the printed form of an entire context, represented as a string, and returns a K label. Moreover, all the variable arguments are wrapped by a label obtained from applying the special freezeVar constructor over the string representation of the variable name. This serves not only to easily identify variables visually, but also to prevent variable contents from mixing in the case of variables of sort List{K}.

4.3 K Semantics

This section describes and exemplifies the process of translating the K semantic constructs to Maude constructs, obtaining an executable definition as a result.

Applying Context Transformers. Although K-Maude allows the specification to omit the configuration context (for modularity and compactness purposes), this context needs to be filled in by the tool as a first step towards obtaining a runnable definition. To do that, we use the tree associated to the **configuration** declaration to iteratively match the cells having the maximal level in the tree, and to wrap them (if not already wrapped) by their corresponding parent cell in the configuration tree, and then continue. Let us present how the context transformers algorithm works on the examples discussed in Sec. 3.5.

The assignment rule. For this rule, the 'k' and 'env' cells are the deepest in the configuration tree; they both are subcells of the 'thread' cell. Since the 'store' cell corresponds to a higher level in the configuration tree, the 'k' and 'env' cells are wrapped by a 'thread' cell in the first iteration of the algorithm:

$$\textbf{rule}\ \langle \mathsf{thread}\_\rangle\ \langle \mathsf{k}\rangle\ \mathrm{X} = \mathrm{I}\ ;\ \Rightarrow\ .\ \langle \_/\mathsf{k}\rangle\ \langle \mathsf{env}\_\rangle\ \mathrm{X} \mapsto \mathrm{N}\ \langle \_/\mathsf{env}\rangle\ \langle \_/\mathsf{thread}\rangle$$
$$\langle\ \mathsf{store}\ \_\rangle\ \mathrm{N} \mapsto (\_\ \Rightarrow\ \mathrm{I})\ \langle \_/\ \mathsf{store}\rangle$$

However, the 'store' cell is still higher in the configuration than the 'thread' cell, so the 'thread' cell itself needs to be wrapped by the 'threads' cell:

$$\textbf{rule}\ \langle \mathsf{threads}\_\rangle\ \langle \mathsf{thread}\_\rangle\ \langle \mathsf{k}\rangle\ \mathrm{X} = \mathrm{I}\ ;\ \Rightarrow\ .\ \langle \_/\mathsf{k}\rangle\ \langle \mathsf{env}\_\rangle\ \mathrm{X} \mapsto \mathrm{N}\ \langle \_/\mathsf{env}\rangle$$
$$\langle /\_\mathsf{thread}\rangle\ \langle /\_\mathsf{threads}\rangle\ \ \langle\ \mathsf{store}\ \_\rangle\ \mathrm{N} \mapsto (\_\ \Rightarrow\ \mathrm{I})\ \langle \_/\ \mathsf{store}\rangle$$

The levels of the cells in the new term correspond to their levels in the configuration term; therefore the algorithm concludes successfully.

The rendez-vous rule. $\textbf{rule}\ \langle \mathrm{k}\rangle\ \mathrm{rv\ I}\ \Rightarrow\ .\ \langle \_/\mathrm{k}\rangle\ \langle \mathrm{k}\rangle\ \mathrm{rv\ I}\ \Rightarrow\ .\ \langle \_/\mathrm{k}\rangle$
Although the two computations are here at the same level, their multiplicity does not correspond to the one declared in the configuration term. Therefore the context transformers will wrap each of them in their container 'thread' cell:

$$\textbf{rule}\ \langle \mathsf{thread}\_\rangle\ \langle \mathsf{k}\rangle\ \mathbf{rv}\ \mathrm{I}\ \Rightarrow\ .\ \langle \_/\mathsf{k}\rangle\ \langle \_/\mathsf{thread}\rangle$$
$$\langle \mathsf{thread}\_\rangle\ \langle \mathsf{k}\rangle\ \mathbf{rv}\ \mathrm{I}\ \Rightarrow\ .\ \langle \_/\mathsf{k}\rangle\ \langle \_/\mathsf{thread}\rangle$$

Since the thread cell has variable multiplicity, the process is complete.

Default cell values. Consider a simple 'run' construct, which given the program to be run and a list of input values creates an initial configuration for running the program with the given input. As only the 'k' cell and the 'in' cell would have non-default values in the initial configuration, we can write the rule for initiating the computation as:

$$\textbf{rule}\ \mathrm{run(P,L)}\ \Rightarrow\ \langle \mathsf{T}\_\rangle\ \langle \mathsf{k}\rangle\ \mathrm{P}\ \langle /\mathsf{k}\rangle\ \langle \mathsf{in}\rangle\ \mathrm{L}\ \langle /\mathsf{in}\rangle\ \langle \_\mathsf{T}\rangle$$

Since an incomplete cell appears in the right-hand-side, it will be replaced by the corresponding default configuration (sub)term in which the user-specified cells substitute their corresponding cell in the configuration. Moreover, a cell having multiplicity zero or more is only included only if one of its sub-cells was specified by the user. For our example, the generated rule would be:

rule run(P,L) ⇒ ⟨T⟩
  ⟨threads⟩ ⟨thread⟩ ⟨k⟩ P ⟨/k⟩ ⟨env⟩ . *Map*⟨/env⟩ ⟨/thread⟩⟨/threads⟩
  ⟨store⟩ . *Map*⟨/store⟩ ⟨nextLoc⟩0⟨/nextLoc⟩ ⟨in⟩ L ⟨/in⟩ ⟨out⟩ . *List*⟨/out⟩
⟨/T⟩

Resolving variables. Once the context transformations have been applied (taking advantage of the cell comprehension feature), the next step towards obtaining a standard rewriting theory is to resolve cell comprehension and anonymous variables by replacing them with variables of the right sort. To do that, the K definition is traversed, and each term is recursively visited. The visitor uses contextual information to infer the constructor and the variables needed to resolve cell comprehension, and then it uses the full signature to resolve the anonymous variables. For example, the assignment rule presented above will look as follows after this step:

rule ⟨threads⟩ ?1:*Bag* ⟨thread⟩ ?2:*Bag*
    ⟨k⟩ (X = I ; ⇒ .) ↷ ?3:*K* ⟨/k⟩ ⟨env⟩ ?4:*Map* X ↦N ⟨/env⟩
  ⟨/thread⟩ ⟨/threads⟩ ⟨store⟩ ?5:*Map* N ↦(?6:*Int* ⇒ I) ⟨/store⟩

Note that although set comprehension uses ellipses on both sides of the cell, we only need one variable, since the constructor is associative and commutative. The names for the replacement variables start with '?' and have appended numbers for disambiguation.

Resolving in-place rewriting. Transforming K rules into rewrite rules and equations becomes relatively simple upon the completion of the steps above. From each K rule C[l1 ⇒ r1 ,..., ln ⇒ rn], the two terms of the corresponding rewrite rule (l ⇒ r) or equation (l = r), can be inferred as being l = C[l1 ,..., ln], and r = C[r1 ,..., rn]. This inference process is defined by building the two terms l and r together while traversing the K rules. If the rule has the structural attribute, then it would be transformed into an equation; otherwise, into a rewrite rule. At the completion of this step, the assignment rule is:

rl ⟨threads⟩ ?1:*Bag* ⟨thread⟩ ?2:*Bag*
    ⟨k⟩ X = I ; ↷ ?3:*K* ⟨/k⟩ ⟨env⟩ ?4:*Map* X ↦N ⟨/env⟩
  ⟨/thread⟩ ⟨/threads⟩ ⟨store⟩ ?5:*Map* N ↦?6:*Int* ⟨/store⟩
⇒ ⟨threads⟩ ?1:*Bag* ⟨thread⟩ ?2:*Bag*
    ⟨k⟩ . ↷ ?3:*K* ⟨/k⟩ ⟨env⟩ ?4:*Map* X ↦N ⟨/env⟩
  ⟨/thread⟩ ⟨/threads⟩ ⟨store⟩ ?5:*Map* N ↦I ⟨/store⟩

Reduction to the K abstract syntax. After all previous transformation have applied, the rule is transformed to the AST form. Additionally, this step reduces the compositions of constructors with their identities (due to the use of · in rules) which were introduced at the previous step. The final running version of the assignment rule would thus be:

$$
\begin{array}{l}
\mathrm{rl}\ \langle \mathsf{threads}\rangle\ ?1{:}\mathit{Bag}\ \langle \mathsf{thread}\rangle\ ?2{:}\mathit{Bag} \\
\quad\quad \langle \mathsf{k}\rangle\ \text{'\_=\_;}(\mathit{Id}\ \mathrm{X}(.\mathit{List}\{K\}),,\mathit{Int}\ \mathrm{I}(.\mathit{List}\{K\}))\ \curvearrowright\ ?3{:}K\ \langle/\mathsf{k}\rangle \\
\quad\quad \langle \mathsf{env}\rangle\ ?4{:}\mathit{Map}\ \mathit{Id}\ \mathrm{X}(.\mathit{List}\{K\})\ \mapsto \mathit{Int}\ \mathrm{N}(.\mathit{List}\{K\})\ \langle/\mathsf{env}\rangle \\
\quad \langle/\mathsf{thread}\rangle\ \langle/\mathsf{threads}\rangle \\
\quad \langle \mathsf{store}\rangle\ ?5{:}\mathit{Map}\ \mathit{Int}\ \mathrm{N}(.\mathit{List}\{K\})\ \mapsto \mathit{Int}\ ?6{:}\mathit{Int}(.\mathit{List}\{K\})\ \langle/\mathsf{store}\rangle \\
\Rightarrow\ \langle \mathsf{threads}\rangle\ ?1{:}\mathit{Bag}\ \langle \mathsf{thread}\rangle\ ?2{:}\mathit{Bag} \\
\quad\quad \langle \mathsf{k}\rangle\ ?3{:}K\ \langle/\mathsf{k}\rangle\ \langle \mathsf{env}\rangle\ ?4{:}\mathit{Map}\ \mathit{Id}\ \mathrm{X}(.\mathit{List}\{K\})\ \mapsto \mathit{Int}\ \mathrm{N}(.\mathit{List}\{K\})\ \langle/\mathsf{env}\rangle \\
\quad\ \ \langle/\mathsf{thread}\rangle\ \langle/\mathsf{threads}\rangle \\
\quad\ \ \langle \mathsf{store}\rangle\ ?5{:}\mathit{Map}\ \mathit{Int}\ \mathrm{N}(.\mathit{List}\{K\})\ \mapsto \mathit{Int}\ \mathrm{I}(.\mathit{List}\{K\}))\ \langle/\mathsf{store}\rangle
\end{array}
$$

5 From K-Maude to LaTeX

To facilitate the visualization, understanding, and debugging of K definitions, as well as their inclusion in research papers and presentations, K-Maude allows for annotations (as special attributes) specifying how various constructs should be represented in LaTeX, and provides a tool (written in Maude, as well) which automatically generates a LaTeX document from a provided K-Maude definition. The LaTeX-specific annotation is wrapped in the **latex** attribute. For example, the following environment cell definition requires that '<=' be typeset as '$\leq$'.

op _<=_ : *AExp AExp* → *BExp* [**latex**({#1}\leq{#2})]

Typesetting styles. The LaTeX generated from K modules is fully configurable, as each specific part of a definition is enclosed in LaTeX macros. The compilation script then takes the output produced by Maude and includes a style file in the preamble, containing definitions for all the macros. Moreover, it allows for the user to provide its own style file which is loaded after the main one, and can customize part of the macros. K-Maude currently provides two such main styles, differing only in the way they typeset cells. One of them typesets rules using only the mathematical K notation, producing rules as the one at the end of Sec. 2, or the ones in [11]. The other, presented below, uses a more graphical notation for cells, and it is thus better for visualizing definitions.

Formatted output. Sort, subsort, and operation declarations are converted to their equivalent BNF notation, since this notation is prevalent in programming languages definitions. For example, the IMP++ syntax for arithmetic expressions (lines 11–22) is automatically typeset to:

$$
\begin{array}{rl}
\mathit{AExp} ::= & \mathit{Int} \mid \mathit{Id} \\
& \mid \mathit{AExp} + \mathit{AExp}\ [\mathrm{strict}] \\
& \mid \mathit{AExp}\ /\ \mathit{AExp}\ [\mathrm{strict}] \\
& \mid {+}{+}\ \mathit{Id} \\
& \mid \mathsf{read}
\end{array}
$$

K cells are represented using the `tikz` package as rectangles with rounded sides and with the cell label attached to the top. Completely specified cells have both sides rounded. Incomplete cells, on either side, have the corresponding side "ripped". For example, the Imp++ assignment rule (lines 39–41) is typeset as:

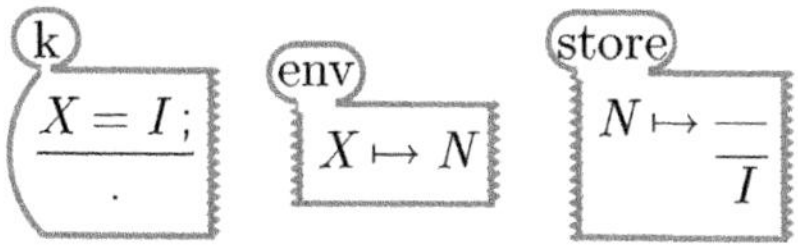

The configuration term for Imp++ (lines 38–45) is typeset to:

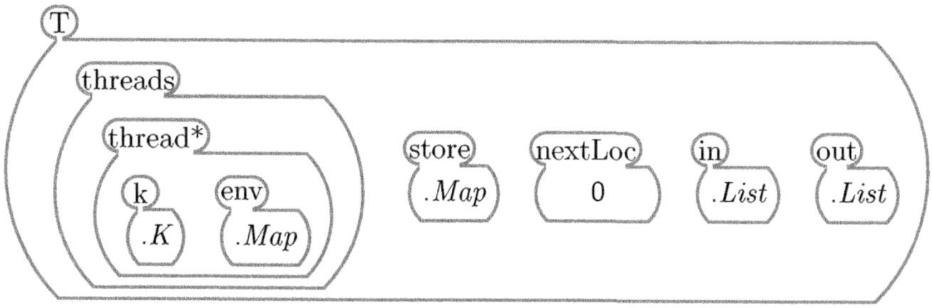

To ensure that the definition is typeset in the order it was written in, and that the cells inside a rule are typeset in the order specified by the user, we use modified versions of the `K-TECHNIQUE` module and of the Maude `META-MODULE` module for the K to LaTeX transformation. More precisely, both modules are altered by removing all commutativity attributes. This basically means that, for the purpose of this transformation, bags and sets of (meta-) rules, equations, membership axioms, operation declarations, subsorts, and sorts, are all regarded as lists.

6 Conclusions

We described K-Maude, an implementation of the K language definitional framework in Maude. The K-Maude interface comes as an extension of Maude, allowing users to define a language using K modules with specific K syntax in addition to the existing Maude modules. The K-specific modules extend the Maude module syntax with constructs aiming at simplifying the language definition task by abstracting away irrelevant details. These multi-layered abstractions allow for concise language definitions with a high potential for reuse of language features. K-Maude defines several meta-transformations which gradually translate K modules into either executable Maude modules, to obtain interpreters, or into LaTeX, to obtain formal language semantics documentation.

References

1. Berry, G., Boudol, G.: The chemical abstract machine. Theoretical Computer Science 96(1), 217–248 (1992)
2. Clavel, M., Durán, F., Eker, S., Meseguer, J., Lincoln, P., Martí-Oliet, N., Talcott, C.: All About Maude - A High-Performance Logical Framework. LNCS, vol. 4350. Springer, Heidelberg (2007)
3. Kahn, G.: Natural semantics. In: Brandenburg, F.J., Wirsing, M., Vidal-Naquet, G. (eds.) STACS 1987. LNCS, vol. 247, pp. 22–39. Springer, Heidelberg (1987)
4. Meseguer, J.: Conditional rewriting logic as a unified model of concurrency. Theoretical Computer Science 96(1), 73–155 (1992)
5. Meseguer, J., Roşu, G.: The rewriting logic semantics project. Theoretical Computer Science 373(3), 213–237 (2007)
6. Mosses, P.D.: Modular structural operational semantics. Journal of Logic and Algebraic Programming (60-61), 195–228 (2004)
7. Plotkin, G.D.: A structural approach to operational semantics. Journal of Logic and Algebraic Programming (60-61), 17–139 (2004)
8. Roşu, G., Ellison, C., Schulte, W.: Matching logic: An alternative to Hoare/Floyd logic. In: AMAST 2010. LNCS, Springer, Heidelberg (2010)
9. Roşu, G.: K: a rewrite-based framework for modular language design, semantics, analysis and implementation. Tech. Rep. UIUCDCS-R-2006-2802, University of Illinois (2006)
10. Roşu, G., Schulte, W., Şerbănuţă, T.F.: Runtime verification of C memory safety. In: Peled, D. (ed.) RV 2009. LNCS, vol. 5779, pp. 132–151. Springer, Heidelberg (2009)
11. Roşu, G., Şerbănuţă, T.F.: An overview of the K semantic framework. Journal of Logic and Algebraic Programming 79(6), 397–434 (2010)
12. Şerbănuţă, T.F., Roşu, G., Meseguer, J.: A rewriting logic approach to operational semantics. Information and Computation 207, 305–340 (2009)
13. Strachey, C., Wadsworth, C.P.: Continuations: A mathematical semantics for handling full jumps. Higher-Order and Symbolic Computation 13(1/2), 135–152 (2000)
14. Wright, A.K., Felleisen, M.: A syntactic approach to type soundness. Information and Computation 115(1), 38–94 (1994)

Collecting Semantics under Predicate Abstraction in the K Framework*

Irina Măriuca Asăvoae and Mihail Asăvoae

Faculty of Computer Science
Alexandru Ioan Cuza University, Iaşi, Romania
{mariuca.asavoae,mihail.asavoae}@info.uaic.ro

Abstract. The K framework is a specialization of rewriting logic for defining programming language semantics. This paper introduces the model checking with predicate abstraction technique into the K framework. To express this technique in K, we go to the foundations of predicate abstraction, that is abstract interpretation, and use its collecting semantics. As such, we propose a suitable description in K for collecting semantics under predicate abstraction of a simple imperative language. Next, we prove that our K specification for collecting semantics is a sound approximation of the K specification for concrete semantics. This work makes a further step towards the development of program verification methodologies in rewriting logic semantics project in general and the K framework in particular.

1 Introduction

Programs are expected to work correctly with respect to certain requirements. To ensure their desired behavior, one needs to be able to formally reason about programs and about programming languages. Existing formal approaches range from manually-constructed proofs to highly automated techniques. The latter includes model checking [2] and static analysis [15] as methods of ensuring correctness and finding certain classes of bugs.

In the context of software model checking, a program is translated, via convenient abstractions, into a state transition system. Abstraction helps to reduce the space size and therefore to improve the chances of a runnable/terminating verification process. However, the reduction in the number of states introduces additional behaviors, and leads to an over-approximation of the initial program.

Abstract interpretation [3] makes precise the fact that formal verification of the concrete program is reduced to verification of the simplified, abstract program, if the abstraction is sound. Several program reasoning methods make use of collecting semantics. Essentially, collecting semantics [3] abstracts each program point by set of states, and stores the collected information according to the property of interest.

* This work has been supported by Project POSDRU/88/1.5/S/47646 and by Contract ANCS POS-CCE 62 (DAK).

P.C. Ölveczky (Ed.): WRLA 2010, LNCS 6381, pp. 123–139, 2010.

Predicate abstraction [6] is a popular technique to build abstract models for programs which relies on defining a set of predicates over program variables. Valid executions in the abstract representation correspond to valid executions in the original program, whereas invalid runs in the abstract semantics need to be checked for feasibility in the concrete counterpart.

The K framework [17,19] proposes a rewrite logic-based approach specialized for the design and analysis of programming languages. A definition (or specification) of a programming language in K consists of a multiset of cells called program configuration, together with semantic sentences. A program configuration represents the structural support to define program executions. Semantic sentences include equations and rewrite rules, with equations controlling the abstraction degree and with rewrite rules controlling the observability degree of a K definition. The resulting specification is modular, semantics-based and executable. Therefore, K permits, in a unified framework, "evaluations" of both programs written in a defined programming language and the corresponding reasoning tools developed for the particular language.

In this paper we explore the potential of the K framework to define program reasoning methods. More specifically, we use K to define model checking with predicate abstraction. Because of the semantics-based characteristic of K specifications, we use collecting semantics as a means of delivering the work. As such, we propose a K description for the collecting semantics under predicate abstraction for a simple imperative language. In order to check the consistency of this K specification, we go along the lines of abstract interpretation standards, and prove that the defined programs' abstract executions are a sound approximation of the programs' concrete executions (the latter is also specified in K [19]). The present work is meant to be the incipient part of a larger project which aims to define K specifications for program analysis and verification.

Even though we frame our work in this paper within K for notational convenience, since K can be "desugared" to a large extent into rewriting logic (see [17,19]), the results in this paper apply very well also to rewriting logic. In fact, we fully adhere to the rewriting logic project [14,21].

Related work. There is extensive work in *software model checking*, and most of it spawned from abstract interpretation [3] and model checking [2]. If we follow the line of *predicate abstraction* that emerged in [6], we could mention only a few and important forward steps in improving the technique with counterexample-based refinement [4], localized, on-demand abstraction refinement [7], and then optimization of abstract computation using Craig interpolants [10].

Rewriting logic [12,9] *theories* allow for nondeterminism and concurrency, while a LTL Model Checker for Maude is described in [5], and used in [1] for Java programs. A methodology for equational abstraction in the context of rewrite logic, with direct application to the Maude model checker, is proposed in [13]. An alternative (to ours) *predicate abstraction* approach is introduced to model checking under rewriting logic in [16]. A comparative study of various program semantics defined in the context of rewriting logic can be found in [21].

The K framework is extensively described in [17,19], and it is used to define a series of languages, such as Scheme in [11], and a non-trivial object oriented language called KOOL in [8], as well as type systems, explicit state model checkers, and Hoare style program verifier [18]. The latest development within the K framework is K-Maude, a rewriting based tool for semantics of programing languages, introduced in [20].

Having the above brief history map, let us pinpoint a few correlations with the current work. In the K framework area, this paper contributes with incorporation of model checking under predicate abstraction as program meta-executions, showing that the K definitional style for concrete semantics of programming languages can be consistently used for program verification methods. However, since K is a rewrite-based framework, a question arises about how is our work positioned with respect to previously described abstractions in rewriting logic systems. Firstly, the equational abstraction creates the abstract state space via equivalence classes, which inherently introduces the overhead of equivalence checking in the infrastructure. In our case, the abstract state is denoted and calculated in the traditional predicate abstraction style (as predicates conjunction), the overhead being transferred to the specialized SMT-solver for the calculation of the abstract transition. Here we need to boast only the potential benefits our approach could bring into rewriting logic from state of the art abstraction based model checking techniques. Secondly, predicate abstraction support is already introduced in rewriting logic systems by [16]. There, the concrete transitional system is provided as a rewrite theory, which is injected with the abstraction predicates to produce the rewrite theory for the abstract transitional system. Then, model checking is performed on the latter theory. There is an important aspect of our approach, which does not seem to be easy to address with the technique in [16]: we are able to also obtain the inverse transformation, from abstract to concrete. This aspect is particularly important when the model checking in the abstract system fails to verify the property, and a refinement of the abstraction needs to be performed.

Outline of the paper. The structure of the paper is as follows: Section 2 introduces *the K framework* by defining concrete semantics for a simple imperative programming language, *SIMP*. Section 3 defines a *collecting semantics under predicate abstraction* for *SIMP*, which can be used to reason about program correctness. Section 4 states the formal correspondence between the concrete semantics and the collecting semantics of *SIMP*, as defined in K. Finally, in Section 5 we draw conclusions and present directions for further work.

2 Preliminaries

K is a rewrite logic-based framework for design and analysis of programming languages. A K specification consists of *configurations* and *rules*. The configurations, formed of K cells, are (potentially) labeled and nested structures that represent program states. The rules in K are divided into two classes: *computational rules*, that may be interpreted as transitions in a program execution, and *structural rules*, that modify a term to enable the application of a computational rule. The K framework allows one to define modular and executable programming language semantics.

We present the K framework by means of an example - a simple imperative language *SIMP* with simple integer arithmetic, basic boolean expressions, assignments, if statements, while statements, sequential composition, and blocks. For this purpose we rely extensively on [19].

The K syntax with annotations and semantics of *SIMP* is given in Fig. 1. The left column states the *SIMP* abstract syntax, the middle column introduces a special K notation, called strictness attribute, and the right column presents the K rules for *SIMP* language semantics. Because the abstract syntax is given in a standard way, we proceed explaining, via an example, the strictness attribute called *seqstrict* (denoted here as *sq* for space efficiency). The strictness attribute that corresponds to the addition rule $Aexp + Aexp$ is translated into the set of heating/cooling rule pairs: $a_1 + a_2 \rightleftharpoons a_1 \curvearrowright \square + a_2$ and $i_1 + a_2 \rightleftharpoons a_2 \curvearrowright i_1 + \square$. These structural rules state how an arithmetic expression with addition is evaluated sequentially: first the lefthand side term (here a_1) is reduced to an integer, and only then the righthand side term a_2 is reduced to some integer. The resulted integers are added using the internal operation of integer addition $+_{Int}$ as represented by the rule $i_1 + i_2 \rightarrow i_1 +_{Int} i_2$ in the *SIMP* semantics (right column). The assignment statement has the attribute $sq(2)$ which means that strictness attribute *seqtrict* is applied only to the second argument.

The K modeling of a program configuration is a wrapped multiset of cells written $\langle c \rangle_l$, where c is the multiset of cells and l is the cell label. Examples of labels include: top $\top$, current computation k, store, call stack, output, formal analysis results, etc. The *SIMP* program configuration is:

$$Configuration \equiv \langle \langle K \rangle_{\mathsf{k}} \langle \mathsf{Map}[\,Var \mapsto Int\,] \rangle_{\mathsf{state}} \rangle_{\top}$$

where the top cell $\langle \ldots \rangle_{\top}$ contains two other cells: the computation $\langle K \rangle_{\mathsf{k}}$ and the store $\langle \mathsf{Map}[\,Var \mapsto Int\,] \rangle_{\mathsf{state}}$. The k cell has a special meaning in K, maintaining computational contents, much as programs or fragments of programs. The computations, i.e. terms of special sort k, are nested list structures of computational tasks. Elements of such a list are separated by an associative operator "$\curvearrowright$", as in $s_1 \curvearrowright s_2$, and are processed sequentially: s_2 is computed after s_1. The "$\cdot$" is the identity of "$\curvearrowright$". The contents of state cell is an element from $\mathsf{Map}[\,Var \mapsto Int\,]$, namely a mapping from program variables to integer values (maps are easy to define algebraically and, like lists and sets, they are considered builtins in K).

The third column in Fig. 1 contains the semantic rules of *SIMP*. The K rules generalize the usual rewrite rules, namely K rules manipulate parts of the rewrite term in different ways: write, read, and don't care. This special type of rewrite rule is conveniently represented in a bidimensional form. In this notation, the lefthand side of the rewrite rule is placed above a horizontal line and the righthand side is placed below. The bidirectional notation is flexible and concise, one could underline only the parts of the term that are to be modified. Ordinary rewrite rules are a special case of K rules, when the entire term is replaced; in this case, the standard notation $left \rightarrow right$ is used.

Syntax	Annotation	Semantics
$AExp ::= Var \mid Int$		$\langle \frac{x}{\sigma[x]} \ldots\rangle_{\mathsf{k}} \ \langle \sigma \rangle_{\mathsf{state}}$
$\mid AExp + AExp \mid \ldots$	*sq*	$i_1 + i_2 \rightarrow i_1 +_{Int} i_2$
$BExp ::= AExp$ <= $AExp$	*sq*	i_1 <= $i_2 \rightarrow i_1 \leq_{Int} i_2$
$\mid AExp = AExp$	*sq*	$i_1 = i_2 \rightarrow i_1 =_{Int} i_2$
$\mid$ `not` $BExp$	*sq*	`not` $t \rightarrow \neg_{Bool}\, t$
$\mid BExp$ `and` $BExp$	*sq*(1)	*true* `and` $b \rightarrow b$ *false* `and` $b \rightarrow$ *false*
$Stmt ::=$ `skip`		`skip` $\rightarrow \cdot$
$\mid Var := AExp$	*sq*(2)	$\langle \frac{x := i}{\cdot} \ldots\rangle_{\mathsf{k}} \ \langle \ldots\, x \mapsto \frac{\_}{i} \ldots\rangle_{\mathsf{state}}$
$\mid Stmt\ ;\ Stmt$		$s_1 ; s_2 \rightharpoonup s_1 \curvearrowright s_2$
$\mid \{Stmt\}$		$\{s\} \rightharpoonup s$
$\mid$ `if` $BExp$ `then` $Stmt$ `else` $Stmt$	*sq*(1)	`if` *true* `then` s_1 `else` $s_2 \rightarrow s_1$ `if` *false* `then` s_1 `else` $s_2 \rightarrow s_2$
$\mid$ `while` $BExp$ `do` $Stmt$		$\langle \frac{\texttt{while}\ b\ \texttt{do}\ s}{\texttt{if}\ b\ \texttt{then}\{s;\ \texttt{while}\ b\ \texttt{do}\ s\}\ \texttt{else}\ \cdot} \ldots\rangle_{\mathsf{k}}$
$Pgm ::=$ `vars` $\mathsf{Set}[Var];\ Stmt$		$\langle \frac{\texttt{vars}\ xs; s}{s} \rangle_{\mathsf{k}} \ \langle \frac{\cdot}{xs \mapsto is} \rangle_{\mathsf{state}}$

Fig. 1. K syntax of *SIMP* (left) with annotations (middle) and semantics (right) with $x \in Var$, $xs \in \mathsf{Set}[Var]$, $i, i_1, i_2 \in Int$, $is \in \mathsf{Set}[Int]$, $b \in BExp$, $s, s_1, s_2 \in Stmt$

The first K rule is a *computational rule* using bidimensional notation to describe the variable lookup operation. The underlined term x in cell k means that x is a "write" term, and is to be replaced by $\sigma[x]$ from the state cell. The absence of the horizontal line under σ indicates that this is a "read" term and the state remains unchanged. The notation "$\langle x\, \_\rangle$" in cell k says that x is placed in the beginning of the term contained by this cell.

The assignment rule has the statement $x := i$ as the current computation task (the first element in cell k), with a "don't care" value "_" for x *somewhere* in the store (as shown by the notation $\langle \_\ x\ \_\rangle_{\mathsf{state}}$). The value of x is updated with i in the store and the assignment is replaced by the empty computation.

The variable lookup and assignment rules are computational rules. Recall that structural rules are used to only rearrange the term to enable the application of the computational rules. One such example is the "while" rule from the right column in Fig. 1, which unfolds one step of a while-loop statement into a conditional statement. The structural transformation is represented with a dotted line to convey the idea that this transformation is lighter-weight than in computational rules. We recall that the usual rewrite rules are special cases of K rules and the K framework proposes "$\rightarrow$" for computational rules, and "$\rightharpoonup$" for structural rules. The former notation is used for "if" rules, while the latter is used for the sequential composition.

The application of the initialization rule of a program (last rule in the right column in Fig. 1) leaves the computation cell containing the entire set of statements, and the memory cell containing an initial mapping of program variables

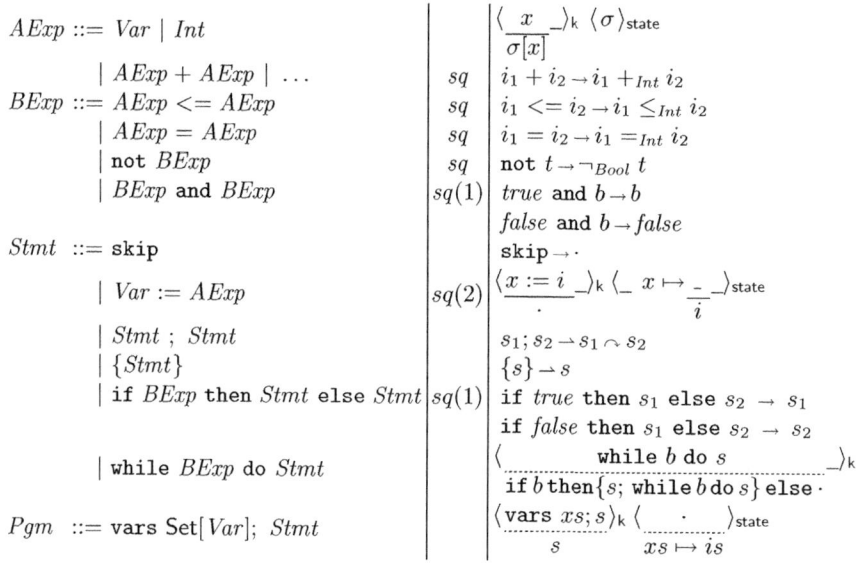

```
vars x, y, err;
x := 0;   err := x;
while (y <= 0) do {
  x := x + 1;   y := y + x;   x := -1 + x;
  if not (x = 0) then err := 1 else skip;
}
```

Fig. 2. Example of a *SIMP* program

xs into integers. The program terminates when computation is completely consumed, meaning when the computation cell is $\langle \cdot \rangle_{\mathsf{k}}$.

Example 1. A *SIMP* program *pgmX* is given in Fig. 2 as `vars x, y, err;` sX, where sX denotes the statements of the program. In a concrete execution, initialized with $\langle\, \langle\, sX \,\rangle_{\mathsf{k}}\ \langle\_\mathtt{y} \mapsto -3\_\rangle_{\mathsf{state}}\, \rangle_{\top}$, the first computational rule applied is the rule for assignments, such that the state cell becomes $\langle\_\mathtt{y} \mapsto -3, \mathtt{x} \mapsto 0\_\rangle_{\mathsf{state}}$. This execution terminates with $\langle \mathtt{x} \mapsto 0, \mathtt{y} \mapsto 0, \mathtt{err} \mapsto 0\rangle_{\mathsf{state}}$. However, if the while condition in the program is $(0 <= \mathtt{y})$ and the program is initialized with $\langle\, \langle\, sX \,\rangle_{\mathsf{k}}\ \langle\_\mathtt{y} \mapsto 3\_\rangle_{\mathsf{state}}\, \rangle_{\top}$, then the execution does not terminate.

3 Collecting Semantics under Predicate Abstraction

Collecting semantics defines the set of program executions from the property of interest point of view and has several instantiations: computation traces, transitive closure of the program transition relation, reachable states, and so on. In this work we *collect forward abstract computation traces* using predicate abstraction. We describe next the details of this setting.

First, we recall the notion of abstract computation in the predicate abstraction environment. Abstraction is a mapping from a set of concrete states to an abstract state. An abstract transition between two abstract states exists if there is at least a concrete transition between concrete states from the preimage of each abstract state, respectively. In predicate abstraction, an abstract state is represented by a predicate φ. The formal definition for φ is $\varphi ::= p \in \Pi \mid \neg p \mid \varphi \wedge \varphi \mid \mathit{true} \mid \mathit{false}$, where Π is a finite set of predicates from *AtomPreds* - all atomic predicates of interest over program variables. In other words, $\varphi \in \mathcal{L}(\Pi)$ = the lattice generated by the atomic predicates from Π. Formally, this lattice is defined as $\langle \mathcal{L}(\Pi), \sqcap, \sqcup, \bot, \top \rangle$ where $\sqcap$ stands for the logic operator $\vee$, $\sqcup$ stands for the logic operator $\wedge$, while $\bot$ and $\top$ stand for *true* and *false*, respectively (more details on this can be found in [15]). Intuitively, an abstract state φ corresponds to the set of concrete states for which the values of the program variables make the formula φ true. The correspondence between the concrete states and the predicates φ is provided by a Galois connection from the powerset of all concrete states to $\mathcal{L}(\Pi)$. (Additional details on this Galois connection are given in Section 4, where are of use.) We denote a transition in predicate abstraction by a function $post^\sharp$ standing for the abstract transition $\varphi \xrightarrow{s} post^\sharp(\varphi, s)$. The formal definition of $post^\sharp$ is in Fig. 4. We do not elaborate on it now, since it makes use of notations introduced later in this section. Of importance here is to have an understanding of the abstract computation with predicate abstraction, as this is a component of the collecting semantics.

We proceed to define in K the program meta-executions using collecting semantics under a fixed predicate abstraction. The finite set of predicates Π is given, as well as the property of interest AGϕ. (We refer meta-executions also as abstract executions.)

The *abstract configuration* in K is defined as:

$$Configuration^{\sharp} \equiv \langle\langle\langle K^{\sharp}\rangle_{\mathsf{k}^{\sharp}} \langle State^{\sharp}\rangle_{\mathsf{state}^{\sharp}} \langle \mathsf{List}[Label] \rangle_{\mathsf{path}}\rangle_{\mathsf{trace}^{\sharp}*} \langle Store^{\sharp}\rangle_{\mathsf{store}^{\sharp}} \langle \Phi \rangle_{\mathsf{inv}}\rangle_{\top^{\sharp}}$$

In order to have the intuition behind $Configuration^{\sharp}$, imagine that $PdcT$ is a parallel divide and conquer algorithm performing the traversal of a digraph. $PdcT$ traverses the digraph in a standard fashion but, when it encounters a node with more than one neighbor, it is going to clone itself on each neighboring direction. The instances of $PdcT$ communicate via a shared memory where everyone deposits its own visited nodes. When an instance of $PdcT$ encounters a node existing in the shared memory, it terminates its job, as that part of the digraph is already under the administration of another instance. $Configuration^{\sharp}$ is similar to a state in the running of $PdcT$. Namely, a trace$^{\sharp}$ cell resembles with the state of an instance of $PdcT$, and the store$^{\sharp}$ cell resembles with the state of the shared memory. Moreover, the rules for collecting semantics under predicate abstraction, from Fig. 6, are similar with the transitions between states of $PdcT$.

Next, we provide detailed description of each cell of the K configuration for collecting semantics under predicate abstraction.

The k$^{\sharp}$ cell maintains the "abstract" computation of the program to be verified. This is in essence the control flow graph of a program, or of a program fragment. More formally, we provide the definition of an abstract computation $K^{\sharp}$ as follows:

$$\langle K^{\sharp}\rangle_{\mathsf{k}^{\sharp}} \begin{cases} Label ::= \text{positive integers representing program points} \\ Var ::= \text{symbols denoting program variables} \\ Asg ::= asg(Var, AExp) \\ Cnd ::= cnd(BExp) \\ TransAsg ::= Label : Asg \\ TransCnd ::= Label : Cnd \\ Trans ::= TransAsg \mid TransCnd \\ Ks ::= Trans \mid if(Ks, Ks) \mid while(TransCnd, Ks) \mid skip \mid \mathsf{List}_{\curvearrowright}[Ks] \\ K^{\sharp} ::= Ks \curvearrowright \lfloor Label \rfloor \end{cases}$$

There is no obvious novelty in the abstract computation k$^{\sharp}$ besides adding labels to each basic statement - assignments and conditions. However, a closer look shows that we categorize the statements into basic statements (*Trans*, involved in computational rules), and composed statements (*ifs* and *whiles*, involved in structural rules). Moreover, note that after constructing the list of abstract computational tasks Ks we finalize by tailing $\lfloor Label \rfloor$ to it ($\lfloor Label \rfloor$ marks the end of

$$stmt^\sharp : Label \times Stmt \rightharpoonup LabelK^\sharp Pair \qquad k^\sharp : Stmt \rightharpoonup K^\sharp$$
$$|\_,\_| : K^\sharp \times Label \rightharpoonup LabelK^\sharp Pair \qquad \lfloor\_\rfloor : Label \rightharpoonup K^\sharp$$
$$\_ : \_ \ : Label \times (Asg \cup Cnd) \rightharpoonup Trans \qquad \_ \frown \_ : K^\sharp \times K^\sharp \rightharpoonup K^\sharp$$

$$\begin{aligned}
&stmt^\sharp(\ell_{in},\ \cdot) = |\ \cdot,\ \ell_{in}|\\
&stmt^\sharp(\ell_{in},\ \texttt{skip}; S) = stmt^\sharp(\ell_{in},\ S)\\
&stmt^\sharp(\ell_{in},\ X := A; S) = |\ell_{in} : asg(X, A) \frown K,\ \ell_{fin}|\\
&\qquad \textbf{if } |K,\ \ell_{fin}| := stmt^\sharp(\ell_{in}+1,\ S)\\
&stmt^\sharp(\ell_{in},\ S_1;\ S_2) = |K_1 \frown K_2,\ \ell_{fin}|\\
&\qquad \textbf{if } |K_1,\ \ell_{aux}| := stmt^\sharp(\ell_{in},\ S_1) \text{ and } |K_2,\ \ell_{fin}| := stmt^\sharp(\ell_{aux},\ S_2)\\
&stmt^\sharp(\ell_{in},\ \{S\}) = stmt^\sharp(\ell_{in},\ S)\\
&stmt^\sharp(\ell_{in},\ \texttt{if } B \texttt{ then } S_1 \texttt{ else } S_2) = |\ell_{in} : cnd(B) \frown if(K_1, K_2),\ \ell_{fin}|\\
&\qquad \textbf{if } |K_1,\ \ell_{aux}| := stmt^\sharp(\ell_{in}+1,\ S_1) \text{ and } |K_2,\ \ell_{fin}| := stmt^\sharp(\ell_{aux},\ S_2)\\
&stmt^\sharp(\ell_{in},\ \texttt{while } B \texttt{ do } S) = |while(\ \ell_{in} : cnd(B), K),\ \ell_{fin}|\\
&\qquad \textbf{if } |K,\ \ell_{fin}| := stmt^\sharp(\ell_{in}+1,\ S)\ )
\end{aligned}$$

$$k^\sharp(S) = K \frown \lfloor \ell_{fin} \rfloor \ \ \textbf{if } |K,\ \ell_{fin}| := stmt^\sharp(1, S)$$

Fig. 3. The rewrite-based rules for abstract computation $k^\sharp$ of a *SIMP* program

the program, and provides base case computational rules). Then, this is delivered as $K^\sharp$ - the content of the abstract computation cell $\mathsf{k}^\sharp$.

In Fig. 3 we give an explicit rewrite-based method for labeling a program, and its transformation into an abstract computation (the last rule).

Example 2. The abstract computation for the program in Fig. 2 is:

$1\!:asg(\mathtt{x}, 0)\frown 2\!:asg(\mathtt{err}, \mathtt{x})\frown while(3\!:cnd(\mathtt{y} \leq 0), 4\!:asg(\mathtt{x}, \mathtt{x}+1)\frown 5\!:asg(\mathtt{y}, \mathtt{y}+\mathtt{x})$ $\frown 6\!:asg(\mathtt{x}, -1+\mathtt{x})\frown 7\!:cnd(\neg(\mathtt{x}=0))\frown if(8\!:asg(\mathtt{err}, 1), \cdot))\frown \lfloor 9 \rfloor$

A $\mathsf{state}^\sharp$ cell is an abstract state which actually stands for a subset of states in the concrete execution. Since we use predicate abstraction with atomic predicates Π, the abstract state is a formula $\varphi \in \mathcal{L}(\Pi)$. However, here we prefer an equivalent representation which writes a formula in $\varphi \in \mathcal{L}(\Pi) - \{\top\}$ as $\bigwedge_{p\in\Pi} op(\varphi, p)$, where op is defined as:

$$op(\varphi, p) = \begin{cases} p, \text{ if } \varphi \Rightarrow p \\ \neg p, \text{ if } \varphi \Rightarrow \neg p \\ \bot_p, \text{ otherwise} \end{cases}$$

Obviously, *true* is $(\bot_p)_{p\in\Pi}$, and for example, if the set of atomic predicates Π is $\{\mathtt{x} \geq 0,\ \mathtt{x} = 0,\ \mathtt{y} = 1\}$, then the formula $\varphi := \mathtt{x} > 0$ is defined with the above representation as $\langle (\mathtt{x} \geq 0)\ \neg(\mathtt{x}=0)\ \bot_{(\mathtt{y}=1)} \rangle_{\mathsf{state}^\sharp}$. Also, we recall that this abstract state corresponds to the set of concrete states which map x to a positive integer. Hence, we use the equivalent representation of an abstract state:

$$\langle\, State^\sharp \,\rangle_{\mathsf{state}^\sharp} \quad \begin{cases} State^\sharp ::= Valid \mid False \\ Valid ::= \mathsf{Map}[\Pi \mapsto \{(), \neg(), \bot_{()}\}] \end{cases}$$

In this representation, the set of predicates Π defining the abstraction is implicitly contained in a state$^\sharp$ cell. Note that *False* stands for the top element of $\mathcal{L}(\Pi)$, and is actually the abstract state corresponding to the empty set of concrete states. Meanwhile, *Valid* stands for any element from $\mathcal{L}(\Pi) - \{\top\}$. Because it represents a nonempty set of concrete states, we say that $\Gamma \in \mathit{Valid}$ is a "valid" abstract state. Moreover, we make the implicit assumption that $\mathit{false} \in \mathit{False}$ is always differentiated from a valid abstract state Γ, as if *false* and Γ are cells with distinct labels (while state$^\sharp$ cell can contain either of them). However, for simplicity, we do not embellish the notation any further. Finally, in Fig. 4 we define $post^\sharp$, the update operator for the abstract state.

$post^\sharp_\phi : \mathit{Valid} \times (\mathit{Asg} \cup \mathit{Cnd}) \rightarrow \mathit{State}^\sharp$ is defined as follows:

$$post^\sharp_\phi(\Gamma, s) = \begin{cases} \wedge_{p \in \Pi}\ post_{(\Gamma,s)}(p), \text{ if } \wedge_{p \in \Pi}\ post_{(\Gamma,s)}(p) \Rightarrow \phi \\ \mathit{false}, \text{ otherwise} \end{cases}$$

$$\text{with } \ post_{(\Gamma, cnd(b))}(p) = \begin{cases} \mathit{false}, \text{ if } \wedge_{p \in \Pi} \Gamma(p) \wedge b \Rightarrow \mathit{false} \\ p, \text{ if } \wedge_{p \in \Pi} \Gamma(p) \wedge b \Rightarrow p \\ \neg p, \text{ if } \wedge_{p \in \Pi} \Gamma(p) \wedge b \Rightarrow \neg p \\ \bot_p, \text{ otherwise} \end{cases}$$

$$\text{and } \ post_{(\Gamma, asg(x,a))}(p) = \begin{cases} p, \text{ if } \wedge_{p \in \Pi} \Gamma(p) \wedge (x' = a) \Rightarrow p[x'/x] \\ \neg p, \text{ if } \wedge_{p \in \Pi} \Gamma(p) \wedge (x' = x) \Rightarrow \neg p[x'/x] \\ \bot_p, \text{ otherwise} \end{cases}$$

Fig. 4. The update operator $post^\sharp$ for the abstract state cell state$^\sharp$

A cell of type path is a list of labels which represents a trace of a possible abstract execution. Note that we refer to this as *trace* because many details are cut out from the abstract execution. Instead, we keep as representative the program points where the abstract execution took place, in their order of appearance.

We finalize the description of the trace$^\sharp$ cell with the observation that trace$^\sharp$ models a *forward abstract computation trace.* Namely, the cells k$^\sharp$ and state$^\sharp$ capture the *abstract computation*, the cell path stands for *trace*, while *forward* comes from $post^\sharp$, the abstract state update operator. Note that trace$^\sharp *$ in $\mathit{Configuration}^\sharp$ indicates the existence of many trace$^\sharp$ cells, and this encapsulates the *collecting* attribute of the semantics.

The content of a store$^\sharp$ cell, denoted as Σ, is a set of pairs $(\!|\ell, \Gamma|\!)$ of labels from the abstract computation and elements from $\mathcal{L}(\Pi) - \{\top\}$, formally defined as $\mathit{Store}^\sharp ::= \mathsf{Set}[\mathsf{Pair}(\mathit{Label},\ \mathit{Valid})]$. The abstract store update is defined as $\Sigma[\ell \rightarrow \Gamma] = \Sigma \cup \{(\!|\ell, \Gamma|\!)\}$. A more standard definition for the abstract store would involve a mapping, such as $\mathit{Store}^\sharp ::= \mathsf{Map}[\mathit{Label} \mapsto \mathsf{Set}[\mathit{Valid}]]$. Note that the two representations are equivalent. However, we prefer the former one in order to suggest the *collecting* nature of the current semantics.

An inv cell maintains the formula to be validated. In this work we restrict this formula to invariants AGϕ, where $\phi ::= p \in \mathit{AtomPreds} \mid \neg p \mid \phi \wedge \phi \mid \mathit{false} \mid \mathit{true}$, and A, G are the CTL operators "always" and "general". On short, AGϕ is translated as "formula ϕ is satisfied in any (abstract) state, on any computational path". Φ is defined similarly with $State^\sharp$ (Π is replaced by the set of atomic predicates from ϕ). Usually, Π includes the atomic predicates from ϕ.

$$Initialization^\sharp \equiv pgm^{\phi}_{\Pi} \rightharpoonup \langle\langle\langle k^\sharp(s)\rangle_{\mathsf{k}^\sharp}\langle \sqcap\{\varphi\in\mathcal{L}(\Pi) | \phi \Rightarrow \varphi\}\rangle_{\mathsf{state}^\sharp}\langle\cdot\rangle_{\mathsf{path}}\rangle_{\mathsf{trace}^\sharp}\langle\cdot\rangle_{\mathsf{store}^\sharp}\langle\phi\rangle_{\mathsf{inv}}\rangle_{\mathsf{T}^\sharp}$$

$$Termination^\sharp \equiv \begin{cases} \langle \_ \ \langle\langle\cdot\rangle_{\mathsf{k}^\sharp} \ \langle\cdot\rangle_{\mathsf{state}^\sharp} \ \langle P\rangle_{\mathsf{path}} \ \rangle_{\mathsf{trace}^\sharp} \ \_\rangle_{\mathsf{T}^\sharp} \rightharpoonup \langle P\rangle_{\mathsf{CE}} \\ \langle\langle \_ \rangle_{\mathsf{store}^\sharp} \ \langle\phi\rangle_{\mathsf{inv}} \ \rangle_{\mathsf{T}^\sharp} \rightharpoonup \langle\cdot\rangle_{\mathsf{CE}} \end{cases}$$

Fig. 5. Initialization and termination for K abstract executions of *SIMP*

Fig. 5 provides the K structural rules for initialization and termination of the abstract executions, where pgm^{ϕ}_{Π} is a shorthand for the input cell containing the program $pgm = \texttt{vars}\ xs;\ s$, the abstraction predicates set Π, and the formula to verify ϕ. The initialization of an execution in collecting semantics for the program *pgm* has one $\mathsf{trace}^\sharp$ cell containing the abstract computation of the program, an initial abstract state corresponding to the best over-approximation of the property ϕ in the lattice $\mathcal{L}(\Pi)$, an empty path, an empty abstract store, and the property to be verified upon the program. (Note that we consider "·" as the unit element for any cell.) The choice of the initial $\mathsf{state}^\sharp$ cell, $\sqcap\{\varphi \in \Pi | \ \phi \Rightarrow \varphi\}$, is the abstract representation of all concrete states σ_0, where ϕ is true (i.e. $\{\sigma_0 \mid \sigma_0 \models \phi\}$). As a matter of fact, we could as well generalize the initial abstract state cell $\mathsf{state}^\sharp$ to contain any element from the lattice $\mathcal{L}(\Pi)$. The termination of an execution in collecting semantics is expected to provide a path representing a potential counterexample to the validity of the property ϕ for *pgm*. In the case when there is no counterexample, the property is valid in the abstract model, and also in the program. Otherwise, no conclusion could be derived with respect to the validity of property ϕ for the given program.

The semantic rules for an execution with collecting semantics under predicate abstraction are described in Fig. 6. Note that in these rules the cells are considered to appear in the inner most environment wrapping them, according with the *locality principle* [19]. Next, we explain in details each of these rules.

The first two rules, (R1-2), deal with the case when the abstract computation reaches the final label of the program either with a valid abstract state Γ or with *false*. If the abstract state is valid then its containing $\mathsf{trace}^\sharp$ cell is voided (because there is no abstract computation left for it, and along the current abstract trace only valid states were encountered, meaning that the property ϕ is satisfied in any abstract state along this trace). If the abstract state is *false* then an error is found just before the end of the program. Whenever an error is found, meaning an abstract state where ϕ is not valid, we end the abstract execution from that particular $\mathsf{trace}^\sharp$ cell and keep its representation in the path cell as a witness to the potential discovery of a bug (so called counterexample). This happens in the rules annotated with ⊠. (Note that □ annotation stands for "good" termination.)

The rules (R3-4) present two other base cases, when the statement labeled ℓ is at the top of the abstract computation (i.e. $\langle \ell : \_ \ \_\rangle_{\mathsf{k}^\sharp}$). The rule (R3) covers the case when a particular program point is reached again, with the same

(R1)$_{\square}$:
$$\langle\langle\lfloor\ell\rfloor\rangle_{\mathsf{k}\sharp}\ \langle\Gamma\rangle_{\mathsf{state}\sharp}\ \langle\_\rangle_{\mathsf{path}}\rangle_{\mathsf{trace}\sharp} \rightharpoonup \cdot$$

(R2)$_{\boxtimes}$:
$$\langle\frac{\lfloor\ell\rfloor}{\cdot}\rangle_{\mathsf{k}\sharp}\ \langle\frac{\mathit{false}}{\cdot}\rangle_{\mathsf{state}\sharp}\ \langle\_\ \frac{\cdot}{\ell}\rangle_{\mathsf{path}}$$

(R3)$_{\square\checkmark}$:
$$\frac{\langle\langle\ell:\_\ \_\rangle_{\mathsf{k}\sharp}\ \langle\Gamma\rangle_{\mathsf{state}\sharp}\ \langle\_\rangle_{\mathsf{path}}\rangle_{\mathsf{trace}\sharp}}{\cdot}\ \langle\_\ (\!|\ell,\Gamma|\!)\ \_\rangle_{\mathsf{store}\sharp}$$

(R4)$_{\boxtimes}$:
$$\langle\frac{\ell:\_\curvearrowright\_}{\cdot}\rangle_{\mathsf{k}\sharp}\ \langle\frac{\mathit{false}}{\cdot}\rangle_{\mathsf{state}\sharp}\ \langle\_\ \frac{\cdot}{\ell}\rangle_{\mathsf{path}}$$

(R5)$_{\rightarrowtail}$:
$$\langle\frac{\ell:\mathit{asg}(x,a)}{\cdot}\ \_\rangle_{\mathsf{k}\sharp}\ \langle\frac{\Gamma}{\mathit{post}^{\sharp}_{\phi}(\Gamma,\mathit{asg}(x,a))}\rangle_{\mathsf{state}\sharp}\ \langle\_\ \frac{\cdot}{\ell}\rangle_{\mathsf{path}}\ \langle\frac{\Sigma}{\Sigma[\ell\rightarrowtail\Gamma]}\rangle_{\mathsf{store}\sharp}\ \langle\phi\rangle_{\mathsf{inv}}$$

if $(\!|\ell,\Gamma|\!)\notin\Sigma$

(R6)$_{\rightarrowtail}$:
$$\langle\frac{\ell:\mathit{cnd}(b)\curvearrowright \mathit{if}(K_1,\ K_2)}{\mathit{skip}\curvearrowright K_1}\curvearrowright K\rangle_{\mathsf{k}\sharp}\ \langle\frac{\Gamma}{\mathit{post}^{\sharp}_{\phi}(\Gamma,\mathit{cnd}(b))}\rangle_{\mathsf{state}\sharp}\ \langle\frac{P}{P,\ell}\rangle_{\mathsf{path}}$$
$$\frac{\cdot}{\langle\langle \mathit{skip}\curvearrowright K_2\curvearrowright K\rangle_{\mathsf{k}\sharp}\ \langle \mathit{post}^{\sharp}_{\phi}(\Gamma,\mathit{cnd}(\neg b))\rangle_{\mathsf{state}\sharp}\ \langle P,\ \ell\rangle_{\mathsf{path}}\rangle_{\mathsf{trace}\sharp}}\ \langle\frac{\Sigma}{\Sigma[\ell\rightarrowtail\Gamma]}\rangle_{\mathsf{store}\sharp}\ \langle\phi\rangle_{\mathsf{inv}}$$
if $(\!|\ell,\Gamma|\!)\notin\Sigma$

(R7)$_{\curlyvee}$:
$$\langle\langle \mathit{skip}\curvearrowright\_\rangle_{\mathsf{k}\sharp}\ \langle\mathit{false}\rangle_{\mathsf{state}\sharp}\ \langle\_\rangle_{\mathsf{path}}\rangle_{\mathsf{trace}\sharp} \rightharpoonup \cdot$$

(R8)$_{\curlyvee}$:
$$\langle\frac{\mathit{skip}}{\cdot}\ \_\rangle_{\mathsf{k}\sharp}\ \langle\Gamma\rangle_{\mathsf{state}\sharp}$$

(R9)$_{\multimap}$:
$$\langle\frac{\mathit{while}(\ell:\mathit{cnd}(b),\ K)}{\ell:\mathit{cnd}(b)\curvearrowright \mathit{if}(K\curvearrowright \mathit{while}(\ell:\mathit{cnd}(b),K),\ \cdot)}\ \_\rangle_{\mathsf{k}\sharp}$$

Fig. 6. K rules for collecting semantics under predicate abstraction of *SIMP*

abstract state Γ. This is expressed by the fact that the abstract store, $\mathsf{store}^\sharp$ cell, contains the pair $(\!|\ell, \Gamma|\!)$. We can void the current $\mathsf{trace}^\sharp$ cell, because this particular abstract trace will not increment the $\mathsf{store}^\sharp$ cell any further. However, if a particular program point is reached with the *false* abstract state, as in (R4), we maintain the path as a counterexample.

Rule $(\text{R5})_{\rightarrow}$ performs an abstract execution of an assignment statement encountered at the top of the abstract computation, $\mathsf{k}^\sharp$ cell. This means that the abstract state is updated by the abstract postcondition $post^\sharp$, while the current abstract state is used to update $\mathsf{store}^\sharp$, by adding the pair $(\!|\ell, \Gamma|\!)$ to it. Note that this addition is made only if the pair is not already in the abstract store, according to the definition of the $\mathsf{store}^\sharp$ update.

The rules $(\text{R7-8})_{\curlyvee}$, containing *skip* at the top of the abstract computation, are both following a branching rule $(\text{R6})_{\rightarrow}$. When the abstract execution encounters a branching condition, denoted by $\ell : cnd(b) \curvearrowright if(K_1, K_2)$, the rule $(\text{R6})_{\rightarrow}$ spawns another abstract trace $newt^\sharp$. In this way, the current abstract trace maintains the "then" branch, with the boolean condition b, while $newt^\sharp$ maintains the "else" branch, with the boolean condition $\neg b$. However, it might be the case that not both branches are possible executions (e.g. if the boolean condition is *false*, then only the "else" branch is feasible). In order to filter these cases, when spawning the two traces, we also add a *skip* flag at the top of the abstract computation. The structural rules $(\text{R7-8})_{\curlyvee}$ filter the *skip* flag: if the abstract state obtained by adding the conditional evaluates to *false*, then we remove this $\mathsf{trace}^\sharp$ cell, otherwise we continue the execution removing the *skip* flag.

The last rule, $(\text{R9})_{\multimap}$ unfolds the while statement once. Note that the last three rules, (R7-9), are structural rules that transform the abstract computation. Also, we emphasize the R's annotations provide additional rules' classification.

Example 3. For the program in Fig. 2 and the property AG($\mathtt{err}=0$) the abstract execution with the predicate abstraction given by $\Pi = \{\mathtt{err} = 0\}$ terminates with $\langle 1, 2, 3\rangle_{\mathsf{CE}}$, while if $\Pi = \{\mathtt{err}=0, \mathtt{x}=0\}$ the abstract execution terminates with $\langle 1, 2, 3, 4, 5, 6, 7, 8, 3\rangle_{\mathsf{CE}}$. However, with the predicate abstraction given by $\Pi = \{\mathtt{err}=0, \mathtt{x}=0, \mathtt{x}=1\}$ the abstract execution ends with $\langle\cdot\rangle_{\mathsf{CE}}$.

The abstract execution with $\Pi = \{\mathtt{err} = 0, \mathtt{x} = 0\}$ starts with the abstract computation described in Example 2, and proceeds as described in Fig. 7.

4 Correspondence between Concrete and Collecting Semantics

In this part we focus on proving the correctness of the K definition of *SIMP* collecting semantics under predicate abstraction with respect to the K definition of the concrete semantics. In other words, we investigate if our K description of model checking with predicate abstraction can be soundly used to prove certain properties for *SIMP* programs.

We revise first some basics of predicate abstraction, namely the Galois connection. In the context of predicate abstraction, the Galois connection is defined

$$\langle\langle\langle 1 : asg(\mathtt{x},0)\_\rangle_{\mathsf{k}\sharp}\ \langle(\mathtt{err}=0)\bot_{(\mathtt{x}=0)}\rangle_{\mathsf{state}\sharp}\ \langle\cdot\rangle_{\mathsf{path}}\ \rangle_{\mathsf{trace}\sharp}\ \langle\cdot\rangle_{\mathsf{store}\sharp}\ \langle\mathtt{err}=0\rangle_{\mathsf{inv}}\ \rangle_{\top\sharp}$$

$$\xrightarrow{R5} \langle\langle\langle 2 : asg(\mathtt{err},\mathtt{x})\_\rangle_{\mathsf{k}\sharp}\ \langle(\mathtt{err}=0)(\mathtt{x}=0)\rangle_{\mathsf{state}\sharp}\ \langle 1\rangle_{\mathsf{path}}\ \rangle_{\mathsf{trace}\sharp}$$
$$\langle (\!| 1, (\mathtt{err}=0)\bot_{(\mathtt{x}=0)} |\!) \rangle_{\mathsf{store}\sharp}\ \langle\mathtt{err}=0\rangle_{\mathsf{inv}}\ \rangle_{\top\sharp}$$

$$\xrightarrow{R5} \langle\langle\langle \mathit{while}(3 : cnd(\mathtt{y}\le 0), K)\_\rangle_{\mathsf{k}\sharp}\ \langle(\mathtt{err}=0)(\mathtt{x}=0)\rangle_{\mathsf{state}\sharp}\ \langle 1,2\rangle_{\mathsf{path}}\ \rangle_{\mathsf{trace}\sharp}$$
$$\langle (\!| 1, (\mathtt{err}=0)\bot_{(\mathtt{x}=0)} |\!), (\!| 2, (\mathtt{err}=0)(\mathtt{x}=0) |\!) \rangle_{\mathsf{store}\sharp}\ \_\rangle_{\top\sharp}$$

where K is $4 : asg(\mathtt{x},\mathtt{x}+1) \curvearrowright 5 : asg(\mathtt{y},\mathtt{y}+\mathtt{x}) \curvearrowright 6 : asg(\mathtt{x},-1+\mathtt{x})$
$\curvearrowright 7 : cnd(\neg(\mathtt{x}=0)) \curvearrowright \mathit{if}(8 : asg(\mathtt{err},1),\cdot)$

$$\overset{R9}{\rightharpoonup} \langle\langle\langle 3 : cnd(\mathtt{y}\le 0) \curvearrowright \mathit{if}(K \curvearrowright \mathit{while}(3 : cnd(\mathtt{y}\le 0), K),\cdot)\_\rangle_{\mathsf{k}\sharp}\ \_\rangle_{\mathsf{trace}\sharp}$$
$$\langle (\!| 1, (\mathtt{err}=0)\bot_{(\mathtt{x}=0)} |\!), (\!| 2, (\mathtt{err}=0)(\mathtt{x}=0) |\!) \rangle_{\mathsf{store}\sharp}\ \_\rangle_{\top\sharp}$$

$$\xrightarrow{R6} \langle\langle\langle \mathit{skip} \curvearrowright\ K\_\rangle_{\mathsf{k}\sharp}\ \langle(\mathtt{err}=0)(\mathtt{x}=0)\rangle_{\mathsf{state}\sharp}\ \langle 1,2,3\rangle_{\mathsf{path}}\ \rangle_{\mathsf{trace}\sharp}$$
$$\langle\langle \mathit{skip}\curvearrowright \cdot \curvearrowright \lfloor 9\rfloor\rangle_{\mathsf{k}\sharp}\ \langle(\mathtt{err}=0)(\mathtt{x}=0)\rangle_{\mathsf{state}\sharp}\ \langle 1,2,3\rangle_{\mathsf{path}}\ \rangle_{\mathsf{trace}\sharp}$$
$$\langle (\!| 1, (\mathtt{err}=0)\bot_{(\mathtt{x}=0)} |\!), (\!| 2, (\mathtt{err}=0)(\mathtt{x}=0) |\!), (\!| 3, (\mathtt{err}=0)(\mathtt{x}=0) |\!) \rangle_{\mathsf{store}\sharp}\ \_\rangle_{\top\sharp}$$

$$\overset{R8}{\rightharpoonup} \langle\langle\langle K\_\rangle_{\mathsf{k}\sharp}\ \langle(\mathtt{err}=0)(\mathtt{x}=0)\rangle_{\mathsf{state}\sharp}\ \langle 1,2,3\rangle_{\mathsf{path}}\rangle_{\mathsf{trace}\sharp}$$
$$\langle\langle \mathit{skip} \curvearrowright \lfloor 9\rfloor\rangle_{\mathsf{k}\sharp}\ \langle(\mathtt{err}=0)(\mathtt{x}=0)\rangle_{\mathsf{state}\sharp}\ \langle 1,2,3\rangle_{\mathsf{path}}\ \rangle_{\mathsf{trace}\sharp}\ \ \_\rangle_{\top\sharp}$$

$$\overset{R8}{\rightharpoonup} \langle\langle\langle 4 : asg(\mathtt{err},\mathtt{x})\_\rangle_{\mathsf{k}\sharp}\ \langle(\mathtt{err}=0)(\mathtt{x}=0)\rangle_{\mathsf{state}\sharp}\ \langle 1,2,3\rangle_{\mathsf{path}}\rangle_{\mathsf{trace}\sharp}$$
$$\langle\langle \lfloor 9\rfloor\rangle_{\mathsf{k}\sharp}\ \langle(\mathtt{err}=0)(\mathtt{x}=0)\rangle_{\mathsf{state}\sharp}\ \langle 1,2,3\rangle_{\mathsf{path}}\ \rangle_{\mathsf{trace}\sharp}\ \ \_\rangle_{\top\sharp}$$

$$\xrightarrow{R1} \langle\langle\langle 4 : asg(\mathtt{x},\mathtt{x}+1)\_\rangle_{\mathsf{k}\sharp}\ \langle(\mathtt{err}=0)(\mathtt{x}=0)\rangle_{\mathsf{state}\sharp}\ \langle 1,2,3\rangle_{\mathsf{path}}\rangle_{\mathsf{trace}\sharp}$$
$$\langle (\!| 1, (\mathtt{err}=0)\bot_{(\mathtt{x}=0)} |\!), (\!| 2, (\mathtt{err}=0)(\mathtt{x}=0) |\!), (\!| 3, (\mathtt{err}=0)(\mathtt{x}=0) |\!) \rangle_{\mathsf{store}\sharp}\ \_\rangle_{\top\sharp}$$

$$\xrightarrow{R5} \langle\langle\langle 5 : asg(\mathtt{y},\mathtt{y}+\mathtt{x})\_\rangle_{\mathsf{k}\sharp}\ \langle(\mathtt{err}=0)\ \neg(\mathtt{x}=0)\rangle_{\mathsf{state}\sharp}\ \langle 1,2,3,4\rangle_{\mathsf{path}}\rangle_{\mathsf{trace}\sharp}$$
$$\langle\_ (\!| 4, (\mathtt{err}=0)(\mathtt{x}=0) |\!) \rangle_{\mathsf{store}\sharp}\ \_\rangle_{\top\sharp}$$

$$\xrightarrow{R5} \langle\langle\langle 6 : asg(\mathtt{x},-1+\mathtt{x})\_\rangle_{\mathsf{k}\sharp}\ \langle(\mathtt{err}=0)\ \neg(\mathtt{x}=0)\rangle_{\mathsf{state}\sharp}\ \langle 1,2,3,4,5\rangle_{\mathsf{path}}\rangle_{\mathsf{trace}\sharp}$$
$$\langle\_ (\!| 5, (\mathtt{err}=0)\ \neg(\mathtt{x}=0) |\!) \rangle_{\mathsf{store}\sharp}\ \_\rangle_{\top\sharp}$$

$$\xrightarrow{R5} \langle\langle\langle 7 : cnd(\neg(\mathtt{x}=0)) \curvearrowright \mathit{if}(8 : asg(\mathtt{err},1),\cdot)\_\rangle_{\mathsf{k}\sharp}\langle(\mathtt{err}=0)\bot_{(\mathtt{x}=0)}\rangle_{\mathsf{state}\sharp}\langle\_6\rangle_{\mathsf{path}}\rangle_{\mathsf{trace}\sharp}$$
$$\langle\_ (\!| 6, (\mathtt{err}=0)\ \neg(\mathtt{x}=0) |\!) \rangle_{\mathsf{store}\sharp}\ \_\rangle_{\top\sharp}$$

$$\xrightarrow{R6} \langle\langle\langle \mathit{skip}\curvearrowright 8 : asg(\mathtt{err},1)\ \_\rangle_{\mathsf{k}\sharp}\langle(\mathtt{err}=0)\ \neg(\mathtt{x}=0)\rangle_{\mathsf{state}\sharp}\langle\_7\rangle_{\mathsf{path}}\rangle_{\mathsf{trace}\sharp}$$
$$\langle\langle \mathit{skip} \curvearrowright \cdot \curvearrowright \mathit{while}(3 : cnd(\mathtt{y}\le 0), K)\_\rangle_{\mathsf{k}\sharp}\langle(\mathtt{err}=0)(\mathtt{x}=0)\rangle_{\mathsf{state}\sharp}\langle\_7\rangle_{\mathsf{path}}\rangle_{\mathsf{trace}\sharp}$$
$$\langle\_ (\!| 7, (\mathtt{err}=0)\ \bot_{(\mathtt{x}=0)} |\!) \rangle_{\mathsf{store}\sharp}\ \_\rangle_{\top\sharp}$$

$$\overset{R8}{\rightharpoonup} \langle\langle\langle \mathit{skip}\curvearrowright 8 : asg(\mathtt{err},1)\ \_\rangle_{\mathsf{k}\sharp}\langle(\mathtt{err}=0)\ \neg(\mathtt{x}=0)\rangle_{\mathsf{state}\sharp}\langle\_7\rangle_{\mathsf{path}}\rangle_{\mathsf{trace}\sharp}$$
$$\langle\langle \mathit{while}(3 : cnd(\mathtt{y}\le 0), K)\_\rangle_{\mathsf{k}\sharp}\langle(\mathtt{err}=0)(\mathtt{x}=0)\rangle_{\mathsf{state}\sharp}\langle\_7\rangle_{\mathsf{path}}\rangle_{\mathsf{trace}\sharp}$$
$$\langle\_ (\!| 3, (\mathtt{err}=0)(\mathtt{x}=0) |\!)\_\rangle_{\mathsf{store}\sharp}\ \ \_\rangle_{\top\sharp}$$

$$\xrightarrow{R3} \langle\langle\langle \mathit{skip}\curvearrowright 8 : asg(\mathtt{err},1)\ \_\rangle_{\mathsf{k}\sharp}\langle(\mathtt{err}=0)\ \neg(\mathtt{x}=0)\rangle_{\mathsf{state}\sharp}\langle\_7\rangle_{\mathsf{path}}\rangle_{\mathsf{trace}\sharp}\ \ \_\rangle_{\top\sharp}$$

$$\overset{R8}{\rightharpoonup} \langle\langle\langle 8 : asg(\mathtt{err},1)\ \_\rangle_{\mathsf{k}\sharp}\ \langle(\mathtt{err}=0)\ \neg(\mathtt{x}=0)\rangle_{\mathsf{state}\sharp}\ \langle\_7\rangle_{\mathsf{path}}\rangle_{\mathsf{trace}\sharp}\ \ \_\rangle_{\top\sharp}$$

$$\xrightarrow{R5} \langle\langle\langle \mathit{while}(3 : cnd(\mathtt{y}\le 0), K)\_\rangle_{\mathsf{k}\sharp}\ \langle \mathit{false}\rangle_{\mathsf{state}\sharp}\ \langle\_8\rangle_{\mathsf{path}}\rangle_{\mathsf{trace}\sharp}$$
$$\langle\_ (\!| 7, (\mathtt{err}=0)\ \neg(\mathtt{x}=0) |\!) \rangle_{\mathsf{store}\sharp}\ \_\rangle_{\top\sharp}$$

$$\overset{R9}{\rightharpoonup} \langle\langle\langle 3 : cnd(\mathtt{y}\le 0) \curvearrowright \mathit{if}(\dots)\_\rangle_{\mathsf{k}\sharp}\ \langle \mathit{false}\rangle_{\mathsf{state}\sharp}\ \langle 1,2,3,4,5,6,7,8\rangle_{\mathsf{path}}\rangle_{\mathsf{trace}\sharp}\ \_\rangle_{\top\sharp}$$

$$\xrightarrow{R4} \langle\langle\langle\cdot\rangle_{\mathsf{k}\sharp}\ \langle\cdot\rangle_{\mathsf{state}\sharp}\ \langle 1,2,3,4,5,6,7,8,3\rangle_{\mathsf{path}}\rangle_{\mathsf{trace}\sharp}\ \_\rangle_{\top\sharp}\ \rightharpoonup\ \langle 1,2,3,4,5,6,7,8,3\rangle_{\mathsf{CE}}$$

Fig. 7. Example of a K abstract execution

as $\mathcal{P}(\mathcal{S}) \ {}^{\alpha}\!\rightleftarrows_{\gamma} \ \mathcal{L}(\Pi)$ where $\mathcal{S} = \{\sigma : V \mapsto \mathbf{Z}\}$ is the set of all states for a *SIMP* program. The abstraction-concretization pair $\langle \alpha, \gamma \rangle$ is defined as follows:

$$\alpha(S) := \sqcap\{\varphi \mid (\forall \sigma \in S)\ \sigma \models \varphi\}, \text{ for any subset of states } S \subseteq \mathcal{S}$$
$$\gamma(\varphi) := \{\sigma \in \mathcal{S} \mid \sigma \models \varphi\}, \text{ for any formula } \varphi \in \mathcal{L}(\Pi)$$

It is easy to verify that $\langle \alpha, \gamma \rangle$ forms a Galois connection. Also, it is standard that $post^{\sharp}$ is a sound approximation of the strongest postcondition (i.e. if $\sigma \models \varphi$ and $\sigma' \in post(\sigma)$ then $\sigma' \models post^{\sharp}(\varphi)$). More on these can be found in [6].

In what follows we prove a similar property about K executions of programs in concrete semantics and collecting semantics under predicate abstraction, respectively. In other words, we check that any concrete execution of a program can be retrieved from the meta-execution, and we state what conditions need to be satisfied such that we can derive from the meta-execution the validity of the property of interest for the given *SIMP* program. We assume as given the *SIMP* program $pgm = \texttt{vars}\ xs; s$, the finite set of predicates Π, and the invariant AGϕ.

Theorem 1. *Any* K *execution in collecting semantics under predicate abstraction is finite.*

This theorem essentially ensures the termination of the program verification method described in the previous section.

Lemma 1. *For any* $\langle \_ \langle \Sigma_1 \rangle_{\mathsf{store}^\sharp}\ \_ \rangle_{\top^\sharp} \overset{*}{\rightarrow} \langle \_ \langle \Sigma_2 \rangle_{\mathsf{store}^\sharp}\ \_ \rangle_{\top^\sharp}$*, a fragment of execution in collecting semantics, we have* $\Sigma_1 \subseteq \Sigma_2$.

This is easy to see from the fact that any rule (R1-9) produces transitions that preserve the ascending inclusion of the $\mathsf{store}^\sharp$ terms.

Lemma 2. *If the* K *execution in collecting semantics encounters a transition that does not change the* $\mathsf{store}^\sharp$ *cell, as* $\langle \_ \langle \Sigma_1 \rangle_{\mathsf{store}^\sharp}\ \_ \rangle_{\top^\sharp} \overset{Ri}{\rightarrow} \langle \_ \langle \Sigma_1 \rangle_{\mathsf{store}^\sharp}\ \_ \rangle_{\top^\sharp}$ *where* $i = 7, 8, 9$*, then the execution evolves either into a terminal configuration, with the rules* $(R1-4)$*, or into a configuration* $\langle \_ \langle \Sigma_2 \rangle_{\mathsf{store}^\sharp}\ \_ \rangle_{\top^\sharp}$ *where* $\Sigma_1 \subset \Sigma_2$*, with the rules* $(R5-6)$.

This lemma ensures that any structural rule enables a computational rule, and, consequently, there is no execution in collecting semantics with a suffix that does not increment the content of the $\mathsf{store}^\sharp$ cell.

Proof of Theorem 1 (sketch). The proof follows from Lemma 1 and Lemma 2, coupled with the fact that there is an upper bound for any $\mathsf{store}^\sharp$ term (because any *SIMP* program has a finite number of labels, and Π has a finite number of predicates, hence $\mathcal{L}(\Pi)$ has a finite number of elements). □

Theorem 2. *If the concrete execution initialized with a* $\langle \sigma_0 \rangle_{\mathsf{state}}$ *evolves into a concrete configuration with* $\langle \sigma \rangle_{\mathsf{state}}$*, namely* $\langle \langle s \rangle_{\mathsf{k}} \langle \sigma_0 \rangle_{\mathsf{state}} \rangle_{\top} \overset{*}{\rightarrow} \langle \_ \langle \sigma \rangle_{\mathsf{state}} \rangle_{\top}$*, and if* $\sigma_0 \models \Gamma_0$ *(i.e.* σ_0 *is contained in the subset of abstract states denoted by* Γ_0*), then the abstract execution starting with the abstract state* $\langle \Gamma_0 \rangle_{\mathsf{state}^\sharp}$ *evolves into an abstract configuration with* $\langle \Gamma \rangle_{\mathsf{state}^\sharp}$*, namely* $\langle \_ \langle k^\sharp(s) \rangle_{\mathsf{k}^\sharp} \langle \Gamma_0 \rangle_{\mathsf{state}^\sharp}\ \_ \rangle_{\top^\sharp} \overset{*}{\rightarrow} \langle \_ \langle \Gamma \rangle_{\mathsf{state}^\sharp}\ \_ \rangle_{\top^\sharp}$*, such that* $\langle \_ \langle \sigma \rangle_{\mathsf{state}}\ \_ \rangle_{\top} \models \langle \_ \langle \Gamma \rangle_{\mathsf{state}^\sharp}\ \_ \rangle_{\top^\sharp}$ *holds true.*

This theorem states that any K execution in the concrete semantics is sinked into a K execution in the collecting semantics under predicate abstraction (in case we take Γ_0 to be *true*, then $\sigma_0 \models \Gamma_0$ for any initial concrete state σ_0).

Remark 1. By $\langle \_ \langle \sigma \rangle_{\mathsf{state}} \_ \rangle_{\mathsf{T}} \models \langle \_ \langle \Gamma \rangle_{\mathsf{state}^\sharp} \_ \rangle_{\mathsf{T}^\sharp}$ we understand that $\sigma \models \Gamma$, and that the abstract computation $\langle k^\sharp \rangle_{\mathsf{k}^\sharp}$ from the trace$^\sharp$ cell containing $\langle \Gamma \rangle_{\mathsf{state}^\sharp}$ is the abstract computation of the program fragment obtained from the cell $\langle k \rangle_{\mathsf{k}}$ in $\langle \_ \langle \sigma \rangle_{\mathsf{state}} \_ \rangle_{\mathsf{T}}$.

Lemma 3. *For any $\langle \_ \langle \sigma \rangle_{\mathsf{state}} \_ \rangle_{\mathsf{T}} \rightarrow \langle \_ \langle \sigma' \rangle_{\mathsf{state}} \_ \rangle_{\mathsf{T}}$ a concrete transition in a concrete execution, if there is an abstract configuration $\langle \_ \langle \Gamma \rangle_{\mathsf{state}^\sharp} \_ \rangle_{\mathsf{T}^\sharp}$ such that $\langle \_ \langle \sigma \rangle_{\mathsf{state}} \_ \rangle_{\mathsf{T}} \models \langle \_ \langle \Gamma \rangle_{\mathsf{state}^\sharp} \_ \rangle_{\mathsf{T}^\sharp}$, then there is $\langle \_ \langle \Gamma' \rangle_{\mathsf{state}^\sharp} \_ \rangle_{\mathsf{T}^\sharp}$ an abstract configuration satisfying the following two properties:*
(1) $\langle \_ \langle \Gamma \rangle_{\mathsf{state}^\sharp} \_ \rangle_{\mathsf{T}^\sharp} \xrightarrow{*} \langle \_ \langle \Gamma' \rangle_{\mathsf{state}^\sharp} \_ \rangle_{\mathsf{T}^\sharp}$ and
(2) $\langle \_ \langle \sigma' \rangle_{\mathsf{state}} \_ \rangle_{\mathsf{T}} \models \langle \_ \langle \Gamma' \rangle_{\mathsf{state}^\sharp} \_ \rangle_{\mathsf{T}^\sharp}$.

Proof (Lemma 3). The proof goes by case analysis over the rules in the concrete semantics. For example, let us consider that the concrete transition from the hypothesis, $\langle \_ \langle \sigma \rangle_{\mathsf{state}} \_ \rangle_{\mathsf{T}} \rightarrow \langle \_ \langle \sigma' \rangle_{\mathsf{state}} \_ \rangle_{\mathsf{T}}$, is the result of the application of the assignment rule $x := a$. Then, in the abstract semantics, the transition is made via application of the rule (R5), and in the next configuration the state$^\sharp$ cell contains $\Gamma' = post^\sharp_\phi(\Gamma, asg(x, a))$. However, from the definition of $post^\sharp_\phi$ we see that $\Gamma[x'/x] \wedge (x = a[x'/x]) \Rightarrow \Gamma'$. But, from hypothesis we have $\sigma \models \Gamma$, so $\sigma' \models \Gamma \wedge (x = a[x'/x])$, hence $\sigma' \models \Gamma'$. □

Proof of Theorem 2 (sketch). The proof uses induction on the length of the concrete derivation and Lemma 3. □

Theorem 3. *For any pgm, Π, and $AG\phi$, if we have the abstract execution*

$$pgm^\phi_\Pi \rightharpoonup \langle \langle \langle k^\sharp(s) \rangle_{\mathsf{k}^\sharp} \ \langle \sqcap\{\varphi \mid \phi \Rightarrow \varphi\} \rangle_{\mathsf{state}^\sharp} \ \langle \cdot \rangle_{\mathsf{path}} \rangle_{\mathsf{trace}^\sharp} \ \langle \cdot \rangle_{\mathsf{store}^\sharp} \ \langle \phi \rangle_{\mathsf{inv}} \rangle_{\mathsf{T}^\sharp} \xrightarrow{*} \langle \langle - \rangle_{\mathsf{store}^\sharp} \ \langle \phi \rangle_{\mathsf{inv}} \rangle_{\mathsf{T}^\sharp} \rightharpoonup \langle \cdot \rangle_{\mathsf{CE}}$$

then, for all $\langle \sigma_0 \rangle_{\mathsf{state}}$ and $\langle \sigma \rangle_{\mathsf{state}}$ concrete states from a concrete execution $pgm \rightharpoonup \langle \langle s \rangle_{\mathsf{k}} \langle \sigma_0 \rangle_{\mathsf{state}} \rangle_{\mathsf{T}} \xrightarrow{*} \langle \_ \langle \sigma \rangle_{\mathsf{state}} \rangle_{\mathsf{T}}$, if $\sigma_0 \models \phi$, then $\sigma \models \phi$ holds true.

This theorem says that if the K execution in collecting semantics under predicate abstraction terminates without finding any counterexample, then the property ϕ is an invariant for any concrete execution of the program *pgm*.

Proof of Theorem 3 (sketch). We observe that since all trace$^\sharp$ terms disappear in the final state of the abstract execution, then it means that rules $(\mathrm{R2,4})_\boxtimes$ were never executed. We can apply Theorem 2 (we know that $\sigma_0 \models \sqcap\{\varphi \mid \phi \Rightarrow \varphi\}$ because $\sigma_0 \models \phi$ and $\phi \Rightarrow \sqcap\{\varphi \mid \phi \Rightarrow \varphi\}$). Hence, there exists a valid abstract state Γ such that $\sigma \models \Gamma$. Moreover, because any intermediate $\langle \Gamma \rangle_{\mathsf{state}^\sharp}$ is a $post^\sharp_\phi$

result, it means that $\Gamma \Rightarrow \phi$. From these two, namely $\sigma \models \Gamma$ and $\Gamma \Rightarrow \phi$, we conclude that $\sigma \models \phi$. □

It is notorious that model checking with abstraction is not a complete procedure. Essentially, this comes from the false negative answers the model checking with abstraction can issue. Nevertheless, the incompleteness of abstract model checking gives rise to a bundle of work known as "abstraction refinement".

5 Conclusions and Future Work

In this paper we study the embedding of predicate abstraction model checking into the K framework. This work makes two contributions: first, it shows that model checking with predicate abstraction can be incorporated as a formal analysis approach following the very same definitional style used for concrete semantics of programming languages in K; second, it shows how to relate the concrete semantics and the predicate abstracted semantics (i.e. collecting semantics), and proves that the latter is correct for the original language.

In near future we plan to give a K definition for symbolic executions of programs, such that we could embed also predicate abstraction CEGAR into the K framework. Another line to pursue is enriching the class of properties we want to verify. However, it is well known that under predicate abstraction we are forced to limit the properties of interest to safety properties. To overcome this, we should investigate the transition predicate abstraction which enables verification of liveness properties. Ultimately, these steps would lead to an automated and founded system in K for defining program reasoning techniques.

Acknowledgments. We would like to give our special thanks to Professor Dorel Lucanu for his insights in organizing and standardizing this work. Our deep gratitude goes to Professor Grigore Roşu for providing us with the latest K documentation and observations to improve our work. Also, we thank to the anonymous reviewers for useful and much appreciated comments and corrections.

References

1. Alba-Castro, M., Alpuente, M., Escobar, S.: Automatic certification of Java source code in rewriting logic. In: Leue, S., Merino, P. (eds.) FMICS 2007. LNCS, vol. 4916, pp. 200–217. Springer, Heidelberg (2008)
2. Clarke, E.M., Grumberg, O., Peled, D.A.: Model Checking. The MIT Press, Cambridge (1999)
3. Cousot, P., Cousot, R.: Abstract interpretation: a unified lattice model for static analysis of programs by construction or approximation of fixpoints. In: Symposium on Principles of Programming Languages, pp. 238–252. ACM Press, New York (1977)
4. Das, S., Dill, D.L.: Counter-example based predicate discovery in predicate abstraction. In: Aagaard, M.D., O'Leary, J.W. (eds.) FMCAD 2002. LNCS, vol. 2517, pp. 19–32. Springer, Heidelberg (2002)

5. Eker, S., Meseguer, J., Sridharanarayanan, A.: The Maude LTL model checker. In: Gaducci, F., Montanari, U. (eds.) Workshop on Rewriting Logic and Its Applications, Electronic Notes in Theoretical Computer Science, vol. 71. Elsevier, Amsterdam (September 2002)
6. Graf, S., Saidi, H.: Construction of abstract state graphs with PVS. In: Grumberg, O. (ed.) CAV 1997. LNCS, vol. 1254, pp. 72–83. Springer, Heidelberg (1997)
7. Henzinger, T.A., Jhala, R., Majumdar, R., Sutre, G.: Lazy abstraction. SIGPLAN Notices 37(1), 58–70 (2002)
8. Hills, M., Roşu, G.: KOOL: An application of rewriting logic to language prototyping and analysis. In: Baader, F. (ed.) RTA 2007. LNCS, vol. 4533, pp. 246–256. Springer, Heidelberg (2007)
9. Martí-Oliet, N., Meseguer, J.: Rewriting logic as a logical and semantic framework. Electronic Notes in Theoretical Computer Science 4 (1996)
10. McMillan, K.L.: Lazy abstraction with interpolants. In: Ball, T., Jones, R.B. (eds.) CAV 2006. LNCS, vol. 4144, pp. 123–136. Springer, Heidelberg (2006)
11. Meredith, P., Hills, M., Roşu, G.: An executable rewriting logic semantics of K-scheme. In: Dube, D. (ed.) Proceedings of the 2007 Workshop on Scheme and Functional Programming (SCHEME 2007), Technical Report DIUL-RT-0701, pp. 91–103. Laval University (2007)
12. Meseguer, J.: Conditional rewriting logic as a unified model of concurrency. Theoretical Computer Science 96(1), 73–155 (1992)
13. Meseguer, J., Palomino, M., Martí-Oliet, N.: Equational abstraction. In: Baader, F. (ed.) CADE 2003. LNCS (LNAI), vol. 2741, pp. 2–16. Springer, Heidelberg (2003)
14. Meseguer, J., Roşu, G.: The rewriting logic semantics project. Theoretical Computer Science 373(3), 213–237 (2007)
15. Nielson, F., Nielson, H.R., Hankin, C.: Principles of Program Analysis. Springer, New York (1999)
16. Palomino, M.: A predicate abstraction tool for Maude. Documentation and tool, `http://maude.sip.ucm.es/~miguelpt/bibliography.html`
17. Roşu, G., Havelund, K.: A rewriting-based framework for computations – preliminary version. Tech. Rep. Department of Computer Science UIUCDCS-R-2007-2926 and College of Engineering UILU-ENG-2007-1827, University of Illinois at Urbana-Champaign (2007)
18. Roşu, G., Schulte, W.: Matching logic — extended report. Tech. Rep. Department of Computer Science UIUCDCS-R-2009-3026, University of Illinois at Urbana-Champaign (January 2009)
19. Roşu, G., Şerbănuţă, T.F.: An overview of the K semantic framework. Journal of Logic and Algebraic Programming (submitted)
20. Şerbănuţă, T.F., Roşu, G.: K-Maude: A rewriting based tool for semantics of programming languages. In: Proceedings of the 8th International Workshop on Rewriting Logic and its Applications (WRLA 2010). LNCS, vol. 6381, pp. 104–122. Springer, Heidelberg (2010)
21. Şerbănuţă, T.F., Roşu, G., Meseguer, J.: A rewriting logic approach to operational semantics. Information and Computation 207, 305–340 (2009)

Concurrent Rewriting Semantics and Analysis of Asynchronous Digital Circuits

Michael Katelman[1], Sean Keller[2], and José Meseguer[1]

[1] Department of Computer Science
University of Illinois at Urbana-Champaign
Urbana, IL 61801, U.S.A.
{katelman,meseguer}@uiuc.edu

[2] Department of Computer Science
California Institute of Technology
Pasadena, CA 91125, U.S.A.
sean@async.caltech.edu

Abstract. Modern asynchronous digital circuits are highly concurrent systems composed largely of customized gates, and can be elegantly modeled using the language of *production rules* (PRs). One of the present limitations of the state of the art in asynchronous circuit design is that no formal executable semantics of asynchronous circuits has yet been given at the PR level. The primary contribution of this paper is to define, using rewriting logic and Maude, an executable formal semantics of asynchronous circuits at the PR level under three common *timing assumptions*. Our semantics provides a circuit designer with a PR-level circuit interpreter and with a *decision procedure* for checking key circuit properties, including *hazard-freedom* and *deadlock-freedom*. We describe several reductions and optimizations that can be used to reduce the state space of circuits in our formal semantics and investigate the impact of these reductions experimentally. The analysis scales up to circuits of over 100 PRs in spite of the high levels of concurrency involved.

1 Introduction

Asynchronous digital circuits have been employed to design low-power, high-performance microprocessors, *e.g.* [1], as well as in emerging applications such as systems-on-chip (SOCs), *e.g.* [2], soft-error tolerant systems, *e.g.* [3], and nano-electronics, *e.g.* [4]. The critical property that makes asynchronous circuits advantageous in these applications is enormous immunity to both intrinsic and extrinsic timing variation. Unfortunately, there are very few commercially supported asynchronous EDA (electronic design automation) tools, making the design and implementation of asynchronous circuits more challenging than that of synchronous ones.

At the highest level of asynchronous circuit design one finds a variety of languages, *e.g.* CHP [5], Tangram [6], Balsa [7], and, more recently, VHDL with handshaking packages [8], all of which are derived from Hoare's CSP (communicating sequential processes) [9] and exhibit a number of syntactic and semantic

P.C. Ölveczky (Ed.): WRLA 2010, LNCS 6381, pp. 140–156, 2010.

commonalities. All of these high-level languages allow the designer to describe separate asynchronous processes that run *concurrently* and, in lieu of a global clock, communicate with each other by way of local synchronization via handshaking. In order to generate actual hardware, however, programs in *all of these languages* must ultimately be *"compiled"* or *"synthesized"* into networks of gates or transistors. The low-level language of *production rules* [10] is a suitable target language for this synthesis step.

In addition to the choice of design language, asynchronous circuit designers use different "timing" simplifications to abstract away the notion of the relative delays between gates/wires. These *timing assumptions* directly affect the difficulty in forming a timing-closure during the place-and-route step, and they range from the most permissive assumption of *delay-insensitivity*, where timing closure is easy, to much more restrictive assumptions of relative delays, where timing closure is substantially more difficult to achieve. When constructing new circuits and templates, it is important to be able to prove that the new design operates correctly under a particular timing assumption of interest.

Sets of production rules are suitable for analyzing asynchronous circuits designed in any of the high-level languages above. That is, from any of the aforementioned languages it is possible to synthesize a production rule set (PRS) directly corresponding to a network of transistors, and to impose the relevant timing assumption; production rule sets are truly universal. Furthermore, it is exactly because of the low-level nature of a PRS that the representation is suitable to address the subtle *timing assumptions* involved in asynchronous circuits, as well as detect *hazards* and other problems, such as *deadlock*, that may exist in the circuit.

To the best of our knowledge, no *formal executable semantics* has previously been given for asynchronous circuits at the PRS level. Because of the intrinsically concurrent semantics of rewriting logic, a rewriting semantics is a perfect candidate for this task. Therefore, in this paper we provide the first ever formal executable semantics of asynchronous circuits at the PRS level using rewriting logic and Maude [11]. The semantics comes in three flavors, corresponding to three common timing assumptions: delay-insensitivity (DI), speed-independence (SI), and quasi-delay-insensitivity (QDI). By virtue of Maude being executable, we are automatically furnished with a *simulator* for asynchronous circuits at the PRS level; and, via Maude's breadth-first search capabilities, it is possible to formally analyze asynchronous circuits and *verify* a range of correctness criteria, such as deadlock-free and hazard-free operation.

Without imposing any constraints on the structure of an asynchronous circuit, it becomes difficult to give strong guarantees about its behavior. For this reason, after giving a formal definition of an asynchronous circuit at the PRS level and explaining *interference and instability hazards* in Section 2.1, we define a "proper" PRS format that enforces a conceptual distinction between *wires* and *gates* in Section 2.2. Our formal, executable semantics for PRS is then given in Section 3. Specifically, we first give an *unconstrained semantics* (delay-insensitive) without hazards in Section 3.2, and then we extend this basic semantics to account for interference and instability hazards in Section 3.3. We then provide semantics for PRS under two other timing assumptions; *speed-independence* in Section 4.1, and *quasi-delay-insensitivity* in Section 4.2. A crucial contribution of our semantics, under each of the timing assumptions, is that it directly provides a *decision procedure* for a very wide class of properties; including hazard-free and deadlock-free operation, but also much more generally. Through Maude's search command [11, Ch. 12], general invariants can be verified, and with Maude's LTL (linear temporal logic) model checker [11, Ch. 13] one can verify LTL formulas against an asynchronous circuit.

The QDI timing assumption is the most realistic timing assumption used for modern design [4], but is quite complex [12]. In our semantics, decidability of hazard-freedom and deadlock-freedom is achieved by model checking, via search, and, for dealing with QDI directly, requires the development of a finitary encoding of the timing assumption that makes model checking possible. However, by a reduction based on a theorem of [12], we are able to reduce the hazard-freedom check under the QDI assumption to the same check under the SI timing assumption, which has a considerably smaller state space. For deadlock-freedom and other properties the encoding is still necessary since the theorem of [12] only addresses hazards. Based on these results and on an additional optimization, we then show that our semantics is able to analyze in practice a number of well-known small asynchronous circuits of up to 130 PRs, both for hazard-freedom and for deadlocks. We consider this a highly nontrivial result, because of the very high levels of concurrency involved and the detailed PRS level of description. We end the paper with a discussion of related work and of some future directions. All source code and circuits used for experimentation are available at [13].

2 Production Rule Sets

The goal of this section is to formally define production rules, and to give some intuition about the circuit abstraction that production rules are designed to represent. In addition, we describe *hazard-free* circuit operation, one of the correctness criteria under consideration, and its connection to *timing assumptions*.

Section 2.1 reviews the basic concepts of production rules and the associated model of computation. Section 2.2 describes a set of structural constraints on production rule sets that are needed to accurately define the two timing assumptions that we are most concerned with.

2.1 Overview

To simplify the exposition we assume a fixed set Y of variables from which circuits will draw node names; we let $T_{\mathbb{B}}(Y)$ denote the set of propositional formulas over Y; and for any $g \in T_{\mathbb{B}}(Y)$ we let

$$\text{vars}(g) = \{x \in Y \mid x \textit{ occurs in } g\}.$$

The following definition of production rule is derived from [10].

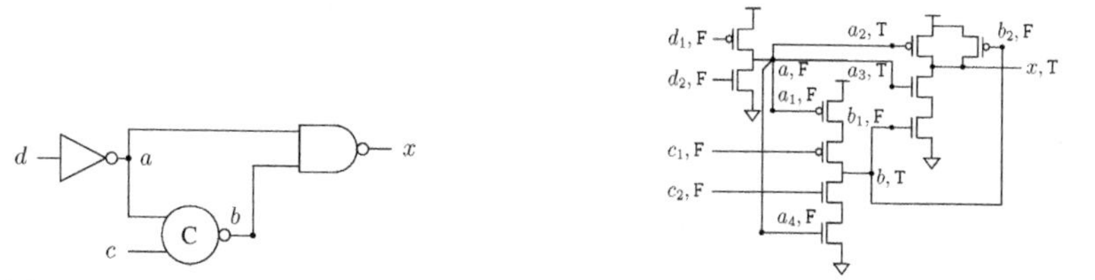

(a) Gate-level representation. (b) CMOS-level representation.

$$\neg d \mapsto a \uparrow \qquad \neg a \wedge \neg c \mapsto b \uparrow \qquad \neg a \vee \neg b \mapsto x \uparrow$$
$$d \mapsto a \downarrow \qquad a \wedge c \mapsto b \downarrow \qquad a \wedge b \mapsto x \downarrow$$

(c) PRS representation (without wires).

Fig. 1. Various representations of the same circuit segment

Definition 1. *A* production rule *is a triple*

$$(g, x, d) \in T_{\mathbb{B}}(Y) \times Y \times \{\uparrow, \downarrow\}.$$

Typically, we denote a production rule (g, x, d) using the more suggestive notation $g \mapsto xd$. A *production rule set* (PRS) is any *finite* set of production rules.

Intuitively, we want a PRS to define a circuit at the level of *switching*, where transistors function as perfect switches but the wires connected to the output of a gate may take arbitrarily long to transition. Any wire connected to the output of an enabled gate may switch, and may do so in any order, non-deterministically. A gate x is *enabled* whenever either (*i*) its has value 0 in the current state and there is a rule $g^+ \mapsto x \uparrow$ with g^+ evaluating to logical-true in the current state, or (*ii*) x is 1 in the current state and there is a rule $g^- \mapsto x \downarrow$ with g^- evaluating to logical-true in the current state.

Fig. 1 shows a segment of digital logic presented in three different formats: (a) as an interconnection of gates, (b) as a CMOS transistor diagram, and (c) as a PRS. The gate marked with a "C" is state-holding and is known as a C-element; when both inputs are 0 it is enabled to switch to 1, when both inputs are 1 it is enabled to switch to 0, and when the inputs are different it holds its current state. Therefore, if the current state of the circuit is

$$a \mapsto 0, b \mapsto 0, c \mapsto 0, d \mapsto 0, x \mapsto 1 \tag{1}$$

then there are 2^4 possible next states: any subset of $\{a, b, c, d\}$ can switch, but the nand gate x is not enabled.

There are two types of *hazards* that must be avoided for an asynchronous circuit to function correctly. Of course, hazard-free operation is only a *necessary* condition for correct operation, the circuit must still satisfy its functional specification.

The first kind of hazard is the *interference hazard.* It occurs when a gate is simultaneously being pulled both up and down. That is, there are two rules

$g^+ \mapsto x \uparrow, g^- \mapsto x \downarrow$ such that both g^+ and g^- evaluate to logical-true in the current state. Note that the PRS given in Fig. 1 does not contain any hazards of this form because all pairs of guards g^+, g^-, as above, are mutually exclusive. However, the custom gates used in practice often *do* have pull-up and pull-down networks which can be simultaneously conducting, *i.e. interfering*.

The second kind of hazard is the *instability hazard.* This occurs when a gate is enabled to switch, but before it actually switches the inputs to the gate change and disable it. For example, in Fig. 1 and state (1) above, if the immediately next state of the system is given by the mappings

$$a \mapsto \mathtt{0}, b \mapsto \mathtt{0}, c \mapsto \mathtt{0}, d \mapsto \mathtt{1}, x \mapsto \mathtt{1} \tag{2}$$

then the inverter a becomes disabled and an instability hazard occurs.

2.2 "Proper" PRS

Asynchronous circuits must be designed so as to avoid hazards. This is done by careful structuring and by imposing some constraints on the relative delays through wire and transistor paths, called *timing assumptions.* The simplest assumption, called *delay insensitivity* (DI), is the degenerate case where arbitrary switching is allowed as described above. Most timing assumptions used in practice typically impose constraints on *forks, i.e.* on connections from the output of a gate to the inputs of multiple subsequent gates.

In order to give a precise definition of common timing assumptions imposed on forks, we will enforce that our PRSs be structured as a set of *gates* which are connected explicitly by *wires.* By *explicitly* we mean that the wire connecting the output x of a gate to the input w of another gate is specified by a pair of production rules $x \mapsto w \uparrow, \neg x \mapsto w \downarrow$; intuitively, w is a *wire branch* of x. Such well-structured PRSs are called *proper* [12].

Definition 2. *Let P be a PRS and $x \in Y$; we denote by $O_{x,P}$ the subset of P*

$$\{g \mapsto x'd \in P \mid x' = x\};$$

when $O_{x,P} \neq \emptyset$ we call it the x operator *with respect to P, and x an* operator variable. *In cases where P is known from context, we may simply write O_x.*

Definition 3. *Let P be a PRS. We say that P has* simple operators *if and only if for all operator variables x in Y, O_x is of the form $\{g^+ \mapsto x \uparrow, g^- \mapsto x \downarrow\}$.*

Definition 4. *Let P be a PRS having simple operators. Then, for each operator variable y in Y;*

1. *we call O_y a* wire *whenever $O_y = \{x \mapsto y \uparrow, \neg x \mapsto y \downarrow\}$ for some $x \in Y$; and*
2. *we call O_y a* gate *otherwise.*

One of the main reasons for structural constraints is to enforce a regular forking structure; a fork being a set of wires with the same variable in the guard

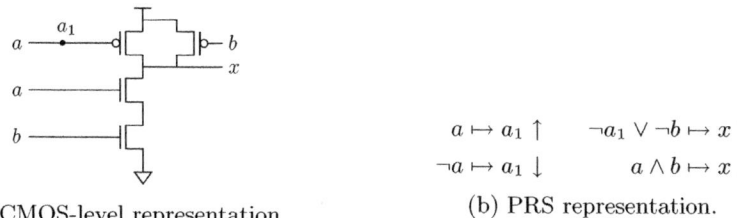

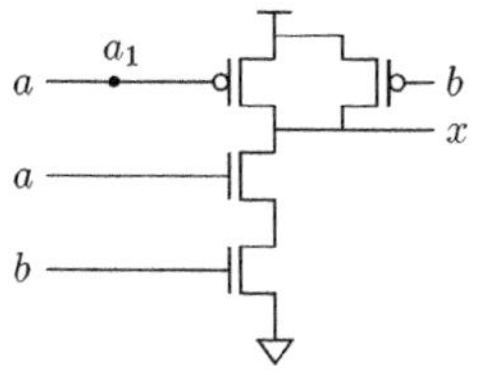

(a) CMOS-level representation.

$$a \mapsto a_1 \uparrow \qquad \neg a_1 \vee \neg b \mapsto x \uparrow$$
$$\neg a \mapsto a_1 \downarrow \qquad a \wedge b \mapsto x \downarrow$$

(b) PRS representation.

Fig. 2. CMOS nand gate with a superfluous wire

of their corresponding production rules. The goal of these requirements is to guarantee a one-to-one correspondence between inter-gate forks in the physical circuit and the wires in its description as a PRS. Formalizing the requirements necessitates a clean way of expressing variable sharing within and between the operators, as such sharing can imply forking.

Definition 5. *Let P be a PRS with simple operators. We associate to P a relation $\longrightarrow_P \subseteq Y \times Y$ defined for all x, y as follows:*

$$x \longrightarrow_P y \Leftrightarrow x \in \text{vars}(g^+) \cup \text{vars}(g^-), \quad \textit{where } \{g^+ \mapsto y\uparrow, g^- \mapsto y\downarrow\} = O_y.$$

Again, when P is clear from context we simply write $\longrightarrow$.

Given a PRS and its associated $\longrightarrow$ relation, we write $\cdot \longrightarrow x, x \longrightarrow \cdot$ to denote the sets $\{y \in Y \mid y \longrightarrow x\}, \{y \in Y \mid x \longrightarrow y\}$ respectively; this notion usefully extends to multiple arrows and multiple dots in the obvious way. *E.g.*, consider the modified nand gate in Fig. 2 having a superfluous wire, *i.e.*, a and a_1 are both wires, with a_1 connected directly to a. Then,

$$\cdot \longrightarrow x = \{a, a_1, b\}$$
$$b \longrightarrow \cdot = \{x\}$$

Superfluous wires like O_{a_1} in Fig. 2 will *not be allowed in a proper PRS*; the output of gates will always connect to wires, and the output of wires will always connect to the input of gates. In addition, a proper PRS will always be a closed system, and all forks must have a regular structure.

Definition 6. *Let P be a PRS. We say that P is* proper *whenever it satisfies all of the following conditions:*

1. *has simple operators and all guards are in disjunctive normal form.*
2. *for all $x \in Y$, $x \longrightarrow \cdot \neq \emptyset$ iff $\cdot \longrightarrow x \neq \emptyset$; i.e. P is closed.*
3. *for all $x, y \in Y$, whenever $x \longrightarrow y$ exactly one of O_x, O_y is a gate; i.e. gates and wires alternate.*
4. *for all gates O_x, O_y, $|x \longrightarrow \cdot \longrightarrow y| \leq 1$; i.e. P has no intra-operator forks.*
5. *for all wires O_w, $|w \longrightarrow \cdot| = 1$; i.e. all inter-operator forks are explicit.*

3 Executable Semantics of PRS

This section presents the executable semantics of PRS. The formalization is in rewriting logic and uses the notation of the rewriting logic language Maude [11]. Section 3.1 defines the data types associated with PRS; Section 3.2 then presents the semantics of PRS *without* hazards or timing assumptions; Section 3.3 deals with hazards and completes the PRS semantics under the DI assumption. The SI and QDI timing assumptions are added in Section 4. The full source code of the semantics, for each of the timing assumptions, as well as the circuits used for experimentation in Section 4 are available at [13] (*note*: [13] and the presentation in this paper have a few small, mostly syntactic, differences).

3.1 Syntax

The syntax of PRS is quite straightforward: it consists of types for the guard expressions, actions, production rules, and sets of production rules. Recall that in a proper PRS, the guards of each production rule are required to be in disjunctive normal form.

```
sorts    Literal Clause Guard .
subsort  Variable < Literal < Clause < Guard .

op   ~_ : Variable              -> Literal .
op  _&_ :   Clause    Clause -> Clause .
op  _|_ :    Guard     Guard ->  Guard .
```

The "action" part of each production rule is everything *but* the guard; that is, the transition variable and a direction for the transition ($x \uparrow, x \downarrow$).

```
sort Action . ops _+ _-  : Variable -> Action .
```

We compose production rules using mix-fix syntax similar to the textual representation used for production rules, *i.e.* "$g \mapsto x \uparrow$".

```
sort ProductionRule . op  [_->_] : Guard Action -> ProductionRule .
```

The resulting data type for production rule sets is then obtained by instantiating the parameterized set module with the appropriate view as follows.

```
fmod SYNTAX is
  pr SET{ProductionRule} * (sort Set{ProductionRule} to PRS,
                              op _,_   to __) .
endfm
```

In Maude, we can now specify the nand gate above as follows, where we take quoted identifiers (*e.g.* `'x`) as a subsort of `Variable`.

```
op nand : -> PRS .
eq nand = [ ~ 'a | ~ 'b -> 'x +]
          [   'a &   'b -> 'x -] .
```

3.2 Unconstrained Semantics without Hazards

We have previously defined a formal semantics for PRS using just notions from naive set theory [12]. The *executable* semantics given here builds on that work, but certain details are omitted which are already spelled out in detail there. For expository purposes we first define those portions of the semantics that are concerned with normal operation, that is, *without hazards*; we then add hazards separately in Section 3.3.

PRS semantics operates as follows: at each step, for each operator O_x make a non-deterministic choice about whether to change the value of x. The possible values that x can be changed to are dictated by the rules comprising O_x and the current state of the circuit. For example, when the state of nand gate above is (note, the nand gate is *enabled*)

$$a \mapsto \texttt{1}, b \mapsto \texttt{1}, x \mapsto \texttt{1},$$

a non-deterministic choice is made to either update the value of x to 0, meaning that the gate has responded to its inputs and switched, or to maintain the value 1, indicating that the voltage on x is not yet low enough to be read as a 0 on any of the wires connected to x.

The *state* of a PRS is represented as a mapping $\chi : Y \longrightarrow \{\texttt{0},\texttt{1}\}$; and a *configuration* is just a PRS and its current state. In rewriting logic

```
sorts Value Configuration .
ops   1 0 :                            ->          Value .
op   <_,_> : PRS Map{Variable,Value} -> Configuration .
```

Map is as defined in Maude [11, §9.13.1]. The top-level rewrite rule is a conditional rule that gathers up all of the non-deterministic "intra-step" level changes for each operator and commits them all at once at the top, yielding a single "step".

```
 op [*_,_*] : PRS Map{Variable,Value} -> [Map{Variable,Value}] .
crl < P, CHI > => < P, CHI' > if [* P, CHI *] => CHI' .          (⋆)
```

Note that the [*_,_*] operator is only given a kind, and not a sort. This makes it so that only fully determined states are reflected at the top. Also, note that in [13] the [*_,_*] operator uses slightly different syntax (<<_,_>>).

The rewrite rules defining the intra-step semantics are given in Fig. 3. Given a configuration $\langle P, \chi\rangle$, the essential idea is to: (*i*) generate a list of all variables occurring in P, and (*ii*) iterate over these variables, making a choice about whether to update the associated value or not. An equation removes the extra intra-step operator symbols when there are no more variables to iterate over, yielding a term of the appropriate sort to be used as the new state value in ($\star$).

A separate intermediate operator is introduced ([*_,_,_,_*]) to iterate over the variables and gather up all of the gate changes. Like [*_,_*], the auxiliary operator [*_,_,_,_*] is defined at the kind level. The A+ and A- predicates simply define when a gate is being pulled-up and pulled-down, respectively;

```
--- empty is the identity for Map
eq [* P, CHI *] = [* mkListOfVars(P), P, CHI, CHI *] .
eq [* empty, P, CHI, CHI' *] = CHI' .

--- choose to switch Y to 1
crl [* (Y YS), P, CHI, CHI' *] => [* YS, P, CHI, CHI'[1 / Y] *]
if A+(Y, P, CHI) .

--- choose to switch Y to 0
crl [* (Y YS), P, CHI, CHI' *] => [* YS, P, CHI, CHI'[0 / Y] *]
if A-(Y, P, CHI) .

--- choose to hold Y
rl [* (Y YS), P, CHI, CHI' *] => [* YS, P, CHI, CHI' *] .
```

Fig. 3. Rewrite rules defining the semantics of PRS, omitting hazards from consideration

```
eq A+  (Y, P, CHI) = eval(pullUpG  (Y, P), CHI) == 1 and
                     eval(pullDownG(Y, P), CHI) == 0 .
eq A-  (Y, P, CHI) = eval(pullUpG  (Y, P), CHI) == 0 and
                     eval(pullDownG(Y, P), CHI) == 1 .
```

that is, `pullUpG` evaluates to the guard of the pull-up rule for the variable given, `pullDownG` the guard of the pull-down rule, and `eval` evaluates a guard relative to a mapping $\chi : Y \longrightarrow \{0, 1\}$ as usual for the Boolean operators.

3.3 Hazards

Automatically checking that an asynchronous circuit is *hazard-free* is one of the primary goals of this paper, so the expression of hazards is a crucial aspect of the PRS semantics. We now extend the rewriting logic definition given in the previous section so as to account for both types of hazards: *interference* and *instability* hazards.

Hazards ultimately get expressed by setting the current value of a gate to a special, intermediate value `X`. Therefore, we must extend the sort `Value` by adding:

```
op X : -> Value .
```

Interference hazards are very simple to characterize; they occur whenever both the guard of the *pull-up* rule ($g^+ \mapsto x \uparrow$) and the guard of the *pull-down* rule ($g^- \mapsto x \downarrow$) for an operator are simultaneously enabled. Then,

```
eq interfering(Y, P, CHI) =
   eval(pullUpG  (Y, P), CHI) == 1 and
   eval(pullDownG(Y, P), CHI) == 1 .
```

The second type of hazard, instability hazards, are more complicated because they are characterized with respect to a pair of successive states

$$\chi, \chi' : Y \longrightarrow \{0, 1, X\}.$$

An instability hazard is triggered whenever a gate is enabled to switch in χ, but does not switch in stepping from χ to χ' while at the same time the inputs to the gate *do* change and the gate is no longer enabled in χ'. This can be specified as follows.

```
eq unstable(Y, P, CHI, CHI') =
    (A+(Y, P, CHI) and eval(Y, CHI') =/= 1 and    --- was enabled
     not A+(Y, P, CHI')) or                       --- no longer enabled
    (A-(Y, P, CHI) and eval(Y, CHI') =/= 0 and    --- was enabled
     not A-(Y, P, CHI')) .                        --- no longer enabled
```

Instability hazards persist from the first point when the `unstable` predicate above becomes true until it is either expressed as an `X` on the output of the gate, or the gate becomes enabled again.

To track instability hazards we add a new component to the state context. Specifically, in addition to χ we also keep a set $I \subseteq Y$ of variables that can switch to `X`

```
sort State .
op (_;_) : Map{Variable,Value} Set{Variable} -> State .
op <_,_> : PRS State -> Configuration .
```

Similarly, we need to change the top-level rewrite rule ($\star$) to include the set of pending hazards and generate the updated I set

```
crl < P, (CHI;I) > => < P, (CHI';I') > if [* P, CHI, I *] => (CHI';I') .
```

and add a new intra-step conditional rule to those in Fig. 3 for the expression of hazards (*note*, the expansion of `[*_,_*]` and `[*_,_,_,_*]` for I):

```
crl [* (Y YS), P, CHI, I, CHI' *] => [* YS, P, CHI, I, CHI'[X / Y] *]
 if Y in I .
```

Of course, the rules in Fig. 3 need to be modified slightly to carry the I component and we must also modify the equations for `[*_,_,_,_*]` to compute the updated I set when there are no more variables left to update during the intra-step computation. The test used to determine whether a given variable x should be included in I' is given in Fig. 4. The full set I' is calculated by performing this test for every variable in the PRS; full details are in our Maude source [13].

4 QDI, SI, and Decidability

It has been shown that the set of hazard-free circuits is extremely limited under the unconstrained, or delay-insensitive, semantics given in Section 3 (see [14]). Therefore, most designs are engineered using stronger timing assumptions than delay-insensitivity. Two common timing assumptions used in practice are the *quasi-delay-insensitivity* (QDI) assumption [10,4,12], and the *speed-independence* (SI) assumption [15,16]. The QDI timing assumption is strictly weaker than the SI assumption, and therefore has benefits in terms of forming a timing closure.

```
--- top-level predicate
eq I'Pred(Y, P, CHI, I, CHI') =
   I+Pred(Y, P, CHI, CHI') or                        --- Y in I+
   (Y in I and not I-Pred(Y, CHI, CHI')) .  --- Y in I minus I-

--- auxiliary predicates
eq I+Pred(Y, P, CHI, CHI') =
   interfering(Y, P, CHI')       --- hazard type 1
   or unstable(Y, P, CHI, CHI') --- hazard type 2
   or A*(Y, P, CHI') .           --- propagation of hazards.
eq I-Pred(Y,    CHI, CHI') = CHI[Y] =/= CHI'[Y] .
```

Fig. 4. Update calculation for the I component of the PRS state

It is the timing assumption that we are most interested in from the standpoint of pragmatic asynchronous circuit design.

The question becomes: *can we decide interesting properties of PRS under QDI operation?* The answer is *yes*, both directly using a finitary encoding of the assumption into our executable semantics and, in the case of verifying hazard-free operation, indirectly using a result from [12]. Although in this paper we experiment only with hazard-free and deadlock-free operation, it is possible to use our semantics to verify a wide range of properties, under any of the timings assumptions we have considered. Indeed, Maude's search command permits general safety properties to be verified, and with Maude's LTL model checker one can verify propositional LTL formulas against the circuit. Without the finitary encoding, the formalization of [12] does not permit this general decidability in QDI.

In Section 4.1 we define PRS semantics under the SI timing assumption, and show that it can easily be imposed using the infrastructure from Section 3. Then, in Section 4.2 we describe the QDI timing assumption at a high level and explain how it can be encoded in a finitary way within our rewriting semantics. A detailed discussion of QDI is made in [12], and is omitted here for space reasons.

4.1 Speed-Independence

The speed-independence timing assumption [15,16] forces all wires connected to a gate to *switch in unison*. Equivalently, it must never be the case that two wires connected to the same gate have a different value

```
ceq SI([Y -> W1 +] [~ Y -> W1 -] [Y -> W2 +] [~ Y -> W2 -] P, CHI)
     = false if CHI[W1] =/= CHI[W2] .
 eq SI(P, CHI) = true [owise] .
```

To enforce the SI timing assumption, we simply add the SI predicate to the conditions of the top-level rewrite rule ($\star$).

```
crl  < P, (CHI;I) > => < P, (CHI';I') >
      if [* P, CHI, I *] => (CHI';I') /\ SI(CHI') .
```

4.2 Quasi-Delay-Insensitivity

The QDI timing assumption is substantially more complicated than the SI timing assumption. The basic idea in QDI is that different branches of a fork may have different values, unlike SI, but once an *acknowledgment path* from one branch of the fork reaches a gate on an unacknowledged branch, the unacknowledged branch must have switched. When a gate switches, it *acknowledges* any inputs in the guard of the corresponding production rule which are part of an enabled disjunctive clause. This is why guards are assumed to be in disjunctive normal form in a proper PRS: to facilitate the definition of acknowledgment. The full formal definition of acknowledgment, and its transitive extension, is given in [12].

Given a state $\chi : Y \longrightarrow \{0, 1, X\}$, the function `ackableOf` gathers up all of the variables that would be acknowledged if the gate corresponding to the given guard were to switch.

```
op ackableOf : Guard Map{Variable,Value} -> Set{Variable} .
eq ackableOf(C | G, CHI) = if eval(C, CHI) == 1
                             then varsOf(C)
                             else empty fi
                           , ackableOf(G, CHI) .
eq ackableOf(C     , CHI) = if eval(C, CHI) == 1
                             then varsOf(C)
                             else empty fi .
```

Encoding the QDI timing assumption as defined in [12] and checking hazard and deadlock-free operation by search, *i.e.* by explicit enumeration of all states, is not possible because the resulting state space is infinite. The basic reason for this is that sequences of acknowledgments are tagged with indices from ω. QDI is imposed on an *execution sequence* given as a set of configurations $\{c_i\}_{i\in\omega}$; where $c_i \longrightarrow^1 c_{i+1}$ for all $i \in \omega$ using the rewrite rule $(\star)$ from Section 3.2. The assumption is satisfied in $\{c_i\}_{i\in\omega}$ if there are no indices $j, k \in \omega$ such that there is a sequence of acknowledgments starting at some gate, O_x, in c_j and ending at another gate, say O_y, at c_k, and where an unacknowledged branch from O_x also reaches O_y. A mathematically precise definition is given in [12].

Careful study of the QDI assumption yields up a solution, however, because between j and k, *acknowledged wires connected to x cannot switch.* If one does, it kills the unacknowledged branches of the fork at index j, and the QDI assumption cannot be violated. Therefore, to encode the QDI assumption in a *finitary* way we record, for every fork, the set of nodes which transitively acknowledge the *current value* on that fork. That is, we endow the configuration state with a new component called an *acknowledgment map*

```
op (_;_;_) : Map{Variable,Value} Set{Variable} Map{Variable,VariableSet}
               -> State .
```

that records for each fork, f, the set of nodes acknowledging f. The details of the update to the acknowledgment map are omitted for space reasons, but the key idea is applying `ackableOf` to every operator that switches during the current step (defined by rule $(\star)$ above) and adding the new transitive acknowledgments into the map. A separate function handles the case when a gate switches and propagates a *new* value to its wires; we then simply clear the mapping for the fork, except for those wires that are part of the fork and equal to the current value of the gate. The QDI timing assumption is then defined as

```
eq QDI(P,            empty, CHI) = true .
eq QDI(P, (F |-> YS,ACK), CHI) =
    if empty =/= intersection(
           unackToGate(F, YS, P) --- unacknowledged branches of F
         ,   ackToGate(   YS, P) ---   acknowledged branches of F
        ) then false else QDI(P, ACK, CHI) fi .
```

That is, for every fork we take the intersection of (*i*) those gates connected to unacknowledged branches of f and (*ii*) the set of gates connected to nodes which transitively acknowledge f and we make sure that these two sets contain no gates in common. If so, QDI is violated. Additional details are in the source code [13].

4.3 Decidability and Experimental Results

One of the compelling advantages of a rewriting logic semantics for PRS is the possibility of using rewriting logic tools such as Maude to *simulate and model check* a PRS against important correctness criteria. Of course, in most cases decidability is contingent upon having a *finite* state space, which is why the encoding for QDI given above is crucial. This section reports on experiments in Maude to decide hazard-free and deadlock-free operation for several small asynchronous circuits. We successfully check circuits up-to 130 production rules, after which the state space so large that the check did not finish after several days on a machine with 2.33GHz (Intel E5410)/8GB RAM/64-bit Linux.

Using the encoding above, we decide hazard-free operation for QDI by enumerating all possible configurations from an initial, *reset* configuration and checking for hazards

```
op hazard! : Configuration -> Bool [frozen] .
eq hazard!(< P, (Y |-> X, CHI ; I ; ACK) >) = true .
eq hazard!(< P, (CHI; I ; ACK) >) = false [owise] .
```

In Maude, we simply use the search command [11, §12]

```
search [1] in QDI-SEMANTICS : initCnfg =>* C:Configuration
                             such that hazard!(C:Configuration) .
```

delay-insensitivity and speed-independence assumptions are handled similarly.

Fig. 5 presents the results of checking hazard-free operation for several small asynchronous circuits according to our semantics and the implementation of the timing assumptions described above. Clearly, the SI timing assumption yields a significantly smaller state space and is therefore easier to check for hazard-free operation. The main result of [12] allows us to reduce the QDI check to checking hazard-free operation in the same circuit under SI. As shown in [17], though verifying hazard-free operation under SI is NP-complete.

Circuit Name	Size	DI	SI	QDI
3InverterRing	12rl	yes	yes 12st – 20ms – 55, 254rw	yes 17st – 32ms – 100, 313rw
ClosedBuffer	26rl	no	yes 20st – 192ms – 513, 522rw	yes 59st – 1, 040ms – 2, 561, 312rw
Toggle	28rl	no	yes 28st – 128ms – 392, 887rw	yes 139st – 944ms – 2, 648, 172rw
PCHBAndFixed	66rl	no	yes 681st – 21, 081ms – 55, 564, 688rw	yes 2, 679st – 213, 869ms – 409, 224, 700rw
1BitFullAdderFixed	118rl	–	–	–
PCHBAndToggle	130rl	–	–	–

Fig. 5. Analysis of hazard-freedom for several small circuits. Benchmark Machine: Xeon @2.33GHz (E5410), 8GB RAM, 64-Bit Linux. Key: rl=rules, st=states, ms=milliseconds, rw=rewrites, DI=delay insensitive, SI=speed independent, QDI=quasi-delay insensitive.

Theorem 1 (main result of [12]). *Let P be a proper PRS. P is hazard-free under QDI if and only if it is hazard-free under SI.*

This reduction clearly results in faster checking of hazard-free operation; *e.g.* the PCHBAndFixed circuit runs an order of magnitude faster under the SI assumption. But, it is disappointing that the reduction does not provide enough leverage to successfully check either 1BitFullAdderFixed or PCHBAndToggle.

To gain additional leverage, we tried a number of strategies. First, we modified our semantics to get rid of all of the conditional rewriting rules, resulting in just three unconditional rewrite rules. The cost of this transformation is that we had a large increase in state space considered by Maude, because the intra-step semantics needed to occur at the top. This resulted in much slower run times since Maude now had to check for duplicate states in a much larger top-level state space. We also tried an intermediate approach, with just a single conditional rewrite rule at the top; the performance of this solution was about the same as the original, highly conditional system.

Finally, we applied a transformation to remove all wires from the PRS and wire the output of each gate directly to subsequent gates.

Conjecture 2. *Let P be a proper PRS and let $P\restriction_{gate}$ be the PRS obtained by removing wires from P and connecting the output of each gate directly to the inputs of subsequent gates by modifying the guards. P is hazard-free under SI if and only if $P\restriction_{gate}$ is hazard-free under the unconstrained semantics.*

We conjecture that this transformation is sound, although we have not yet developed a detailed formal proof. This provided enough leverage to successfully check hazard-free operation for the PCHB-And connected to a circuit that toggles the input, and also to do the same check for the 1-bit full-adder with fixed inputs. The additional model checking results are given in Fig. 6.

Circuit Name	Size	NO-WIRE
3InverterRing	12rl	yes 6st – < 1ms – 9,514rw
ClosedBuffer	26rl	yes 10st – < 1ms – 37,596rw
Toggle	28rl	yes 12st – < 1ms – 42,546rw
PCHBAndFixed	66rl	yes 114st – 1ms – 4Mrw
1BitFullAdderFixed	118rl	yes 1,800st – 239ms – 453Mrw
PCHBAndToggle	130rl	yes 2,844st – 159ms – 299Mrw

(a) Hazard-freedom check when we omit all wires.

Circuit Name	Size	SI
Toggle	28rl	yes 28st – < 1ms – 393Krw
ToggleMod	69rl	no 459st – 13ms – 31Mrw

(b) Deadlock-freedom check on the toggle circuit.

Fig. 6. Hazard-freedom check omitting wires, and deadlock-freedom check for a toggle circuit

In addition, we checked the toggle circuit for *deadlock-freedom*, and a modified toggle circuit with a deadlock. The results are given in Fig. 6. For this, we used the following Maude command.

```
search [1] in SI-SEMANTICS : initCnfg =>! C:Configuration .
```

The full Maude code for the semantics under all of three timing assumptions, the system with unconditional rules, and the transformation to remove wires, can be found at [13], as can all of the PRSs used in the experiments above.

5 Related Work and Conclusions

A sophisticated method for formally verifying hazard-free operation in asynchronous circuits is given in [18]. The methods developed there rely on the existence of *two* designs of the circuit; one *high-level* and one *low-level*. Both designs are given as specialized automata, and while a full enumeration of the reachable state space in the high-level design is necessary, a careful analysis shows how to avoid doing the same for the low-level design. This yields a more efficient analysis of hazard-free operation since the high-level design has a smaller state space than the more detailed, low-level design.

It is not entirely clear how to apply the methods of [18] to PRS. As used in this paper, PRS is a very low-level representation of a circuit and no method has been developed to generate the high-level design required by [18] from PRS. Even with the high-level design, it is likely that if the enumeration of reachable states is done naively, it will result in a combinatorial explosion that precludes analysis of complex, modern circuits. We have some ideas about how this could be done efficiently with a combination of CHP, PRS, and common circuit templates; but this is left for future work.

A use of symbolic model checking for checking hazard-free operation of SI circuits is given in [19]. However, comparison with the results presented here is difficult, because the circuits analyzed were only described informally. By contrast, our analysis is based on a formal semantics for PRS and can be applied directly to any circuit given in PRS, and also used in more general kinds of analysis. The other tool that we know of to verify hazard-free operation is called `prlint` (see [17]). However, the tool is not readily available and we were unable to acquire a version capable of running on a modern Linux workstation. In practice, statistical methods based on Monte Carlo methods are normally used to analyze large circuits.

To the best of our knowledge, this work provides the first formal executable semantics of asynchronous circuits at the PRS level. The realization of our semantics in Maude also provides a PRS-level simulator and a model checker to verify other properties such as deadlock freedom. Another important contribution of this work is the *finitary encoding* of the QDI timing assumption. Although Theorem 1 makes it unnecessary for deciding hazard-free operation, the *finitary encoding* is instrumental in making it feasible to model check other interesting properties.

The primary current limitation of our semantics and model checking methods is of course the high level of concurrency intrinsic in the operation of an asynchronous circuit, and the resulting state space explosion. Although the reduction and optimization methods we have presented here have allowed us to scale the model checking up to the level of over one hundred PRs, new reduction methods are needed to scale up to bigger asynchronous circuits. The investigation of new reduction methods, intuitively collapsing many uninteresting "intermediate states" of the asynchronous computation and possibly taking the form of a stuttering bisimulation reduction, is left for future research.

References

1. Martin, A.J., Nyström, M., Wong, C.G.: Three Generations of Asynchronous Microprocessors. IEEE Design & Test of Computers 20(6), 9–17 (2003)
2. Martin, A., Nyström, M.: Asynchronous Techniques for System-on-Chip Design. Proceedings of the IEEE 94(6) (2006)
3. Jang, W., Martin, A.J.: A Soft-error-tolerant Asynchronous Microcontroller. In: 13th NASA Symposium on VLSI Design (2007)
4. Martin, A.J., Prakash, P.: Asynchronous Nano-Electronics: Preliminary Investigation. In: Proceedings of the 2008 14th IEEE International Symposium on Asynchronous Circuits and Systems, pp. 58–68. IEEE Computer Society, Los Alamitos (2008)
5. Martin, A.J.: Programming in VLSI: From communicating processes to delay-insensitive circuits. In: Hoare, C.A.R. (ed.) Developments in Concurrency and Communication, pp. 1–64. Addison-Wesley, Reading (1990)
6. van Berkel, K., Kessels, J., Roncken, M., Saeijs, R., Schalij, F.: The VLSI-programming language Tangram and its translation into handshake circuits. In: Proceedings of the Conference on European Design Automation, Amsterdam, The Netherlands, pp. 384–389. IEEE Computer Society Press, Los Alamitos (1991)
7. Bardsley, A., Edwards, D.: Compiling the language Balsa to delay insensitive hardware. In: Hardware Description Languages and their Applications, pp. 89–92. Chapman & Hall, Ltd., Boca Raton (1997)

8. Renaudin, M., Vivet, P., Robin, F.: A Design Framework for Asynchronous/Synchronous Circuits Based on CHP to VHDL Translation. In: Proceedings of the 5th International Symposium on Advanced Research in Asynchronous Circuits and Systems, vol. 135. IEEE Computer Society, Los Alamitos (1999)
9. Hoare, C.A.R.: Communicating sequential processes. ACM Commun. 21(8), 666–677 (1978)
10. Martin, A.J.: Compiling communicating processes into delay-insensitive VLSI circuits. Distributed Computing 1(4) (1986)
11. Clavel, M., Durán, F., Eker, S., Lincoln, P., Martí-Oliet, N., Meseguer, J., Talcott, C.: All About Maude - A High-Performance Logical Framework. How to Specify, Program and Verify Systems in Rewriting Logic. LNCS, vol. 4350. Springer, Heidelberg (2007)
12. Keller, S., Katelman, M., Martin, A.J.: A Necessary and Sufficient Timing Assumption for Speed-Independent Circuits. In: 15th IEEE International Symposium on Asynchronous Circuits and Systems, ASYNC 2009 (2009)
13. Katelman, M., Keller, S., Meseguer, J.: Source Code for an Executable Formal Semantics of Production Rule Sets in Maude; with Examples (2010), `http://hdl.handle.net/2142/14863`
14. Martin, A.J.: The limitations to delay-insensitivity in asynchronous circuits. In: AUSCRYPT 1990: Proceedings of the sixth MIT conference on Advanced research in VLSI, pp. 263–278. MIT Press, Cambridge (1990)
15. Muller, D.E., Bartky, W.S.: A theory of asynchronous circuits. In: Proceedings of an International Symposium on the Theory of Switching, pp. 204–243. Harvard University Press, Cambridge (1959)
16. Miller, R.E.: Switching Theory. Sequentical Circuits and Machines, vol. II. John Wiley & Sons, Inc., Chichester (1965)
17. Cook, J.N.: Produciton Rule Verification for Quasi-Delay-Insensitive Circuits. Master's thesis, California Institute of Technology (1993)
18. Beerel, P.A., Burch, J.R., Meng, T.H.Y.: Sufficient Conditions for Correct Gate-Level Speed-Independent Circuits. In: Proceedings of the International Symposium on Advanced Research in Asynchronous Circuits and Systems, pp. 33–43 (1994)
19. Yenigün, H., Levin, V., Peled, D., Beerel, P.A.: Hazard-Freedom Checking in Speed-Independent Systems. In: Pierre, L., Kropf, T. (eds.) CHARME 1999. LNCS, vol. 1703, pp. 317–320. Springer, Heidelberg (1999)
20. Rosenblum, L.Y., Yakovlev, A.: Signal Graphs: From Self-Timed to Timed Ones. In: International Workshop on Timed Petri Nets, pp. 199–206. IEEE Computer Society, Los Alamitos (1985)
21. Petri, C.A.: Kommunikation mit Automaten. PhD thesis, Technische Universität Darmstadt (1962)
22. Agerwala, T.: Putting Petri nets to work. IEEE Computer 12(12), 85–94 (1979)
23. Papadantonakis, K.: Design Rules for Non-Atomic Implementation of PRS. Technical Report CaltechCSTR:2005.001, California Institute of Technology (2005)

A Formal Pattern Architecture for Safe Medical Systems

Mu Sun, José Meseguer, and Lui Sha

University of Illinois at Urbana-Champaign

Abstract. Design patterns have demonstrated major practical uses for cost savings and modular design in software engineering. For safety-critical systems, however, such patterns should also provide formal guarantees that critical safety properties are met. We leverage the power of rewriting logic and parameterization available in Real-Time Maude to add a formal basis for analysis of a novel safety pattern for medical devices. We demonstrate practicality and applicability of our pattern by instantiating it to a pacemaker specification, and we validate our pattern by verifying the safety invariant in the pacemaker instantiation.

1 Introduction

In life, we naturally use patterns as a powerful form of abstraction that not only serves to concisely represent the large amounts of information around us, but also provides a reasoning mechanism for situations that we have not yet encountered. In the same spirit, engineers have also realized, that after successfully designing many similar systems, a common part of these designs can be extracted as a pattern. Future engineers can then use these patterns as a starting point and benefit from the tried and true experience of successful designs in the past.

Patterns have been enormously useful in software engineering after being introduced through the gang of four book [4]. However, current design patterns are used mostly to ensure modularity, portability, scalability, and maintainability of code. There have been many works formalizing software design patterns including: [6] using high level temporal behaviors, [3] using UML-based semantics, [10] using the concept of responsibilities and rewards, etc. However, most of these focus on formalizing structural constraints and interaction properties between different objects. For safety-critical systems, we need patterns with provable safety properties for the system as a whole. For example, even if we verify the correct use of the *observer pattern* [4] in a system design, this does not provide any guarantees on the safety of the system. We need to have a clear way to attach to patterns formal conditions for their applicability and formal guarantees for their behavior when such conditions are met. Clearly, to harness the notion of patterns in safety-critical systems which require verifiable properties, a more precise notion of pattern is needed.

We show that the notion of a safety-pattern can be captured by parameterization in rewriting logic as supported by Maude [1]. In particular, we present

P.C. Ölveczky (Ed.): WRLA 2010, LNCS 6381, pp. 157–173, 2010.

in detail a pattern for medical devices, which is actually applicable to a wide range of safe medical device operations. To do this, we first formalize a notion of medical device safety, *SR-safety*, characterized by stress and relax events, and second, we specify a *Command-Shaper Pattern* in the form of an object wrapper which is able to manipulate input commands to a medical device in order to satisfy a given SR-safety definition.

We present a condensed description of all the elements necessary for our pattern specification. We first highlight some important safety requirements for various medical devices that we have considered (Section 2) along with high level intuitions behind our *Command-Shaper Pattern* (Section 3). We provide a brief background on Real-Time Maude and parameterization (Section 4) before delving into the specification details. We then formally define our notion of medical device safety (Section 5), which sets up the ground work to finally present our pattern (Section 6). We also provide an example instantiation to a pacemaker system which we have validated via timed search (Section 7).

We focus on the specification and instantiation of our medical device pattern. This requires us to leave out some details including the correctness proof for our pattern and the distributed emulation of the pattern. These complementary topics and the full specification can be found in our technical report [11].

2 Safety of Life-Critical Medical Devices

In this paper we describe in detail a safety pattern called the *Command-Shaper Pattern* that we have found applicable to a wide range of medical devices. Before we can talk about a generic safety pattern, we must first discuss a generic notion of safety for medical devices.

When studying medical device operation, we have found a recurring pattern of *command restrictions*. Consider the following three examples:

- Infusion pumps for pain medication are normally incorporated into Patient Controlled Analgesia (PCA) systems, where the patient can demand additional bolus doses of drugs with the push of a button. If the patient pushes the button too often without safety checks, this will clearly lead to depression of the nervous system, and even death. The PCA needs some safety mechanism to make sure that not too many bolus doses are administered.
- Pacemakers normally need to adapt the heart rate to the degree of patient activity. However, pacemaker activity sensors often pick up false positives during bumpy car rides. Pacing a heart at high rates for a prolonged period of time could lead to patient discomfort or even cardiac arrest. Thus, pacemakers must have safety mechanisms to prevent them from pacing too fast for too long.
- A ventilator machine may need to be turned off temporarily for another piece of equipment to work on a patient. Sometimes humans forget to turn ventilator machines back on, potentially causing brain damage to the patient due to oxygen deprivation. The ventilator should have time triggers to make sure that it does not turn off too often or for too long.

Intuitively, all these examples illustrate a common theme with medical devices: human bodies are normally self stabilizing, and our bodies can normally be placed under some stress temporarily, provided they are given sufficient time to recover.

Thus, for the medical device examples presented, all of the device states can be partitioned into two classes: *stressed states* and *relaxed states*. Stressed states are states where the patient cannot stay for too long (heart pacing too fast, holding breath for too long, etc.), since otherwise permanent physical harm may result for the patient. Relaxed states are states that allow the patient to recover over time from a previous period of stressed states.

Although the partitioning of device states into stressed and relaxed states is common to many devices, it should be noted that not all devices can be placed into this category. For example, a glucose-insulin pump does not have any static relaxed states. There always exist patient contexts where any potential device state: *infuse insulin*, *infuse glucose*, or *do nothing* could be considered an unsafe action. Such devices, which depend on external context and sensor information for their safety, are not addressed by the pattern that we present in this paper.

3 Command-Shaper Pattern for Safety Monitoring

The key idea of the command-shaper pattern is that *commands from external devices should only be taken as suggestions.* Figure 1 shows this pattern applied in the form of a wrapper around a pacing module in a cardiac pacemaker. If a command is detected to be deviating or unsafe, the command-shaper can either ignore the command, or more generally, modify the command into a safe variant. To do this, we must first come up with a reasonably general definition of which commands are safe and, also, of how to respond to the actions of commands that are unsafe.

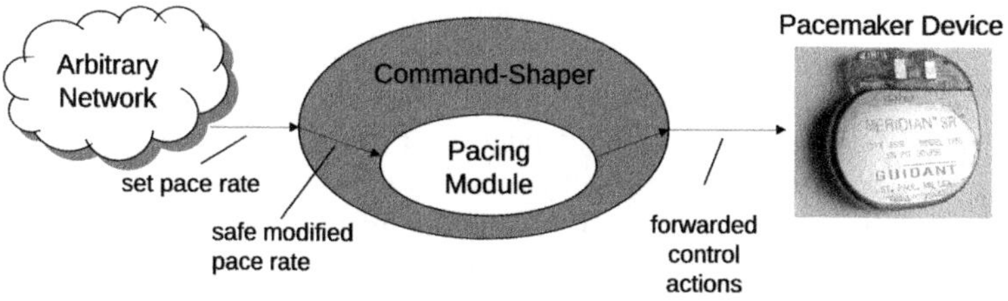

Fig. 1. Command-Shaper Wrapper Pattern for the Pacemaker

The notion of safety in medical systems is of course very broad: ranging from biological conditions to electronic interference. We consider an important aspect of medical device safety which applies to medical device controllers. This notion of safety is definable via stressed and relaxed states, which we call *SR-safety*. Consider the heart rate at which a pacemaker is pacing the patient's heart shown in Figure 2. We assume that the doctors customized the pacemaker's parameters to adapt to the patient's normal expected heart rhythm. For this patient a safe minimal heart rate is 60 bpm when inactive. Furthermore, a critical heart rate is defined to be at 100 bpm, so that any heart rate above 100 bpm will be considered *stressed*, and any heart rate at or below 100 bpm will be considered *relaxed*. The key idea is that the pacemaker can pace at relaxed rates indefinitely without compromising patient safety, but the pacemaker should not pace at stressed rates for prolonged periods of time. These constraints can expressed as upper bounds on stress durations over time. Of course, the maximum allowable stress duration at any instant in time may depend on relax durations also. For example, if the pacemaker's pace drops below 100 bpm after a long stress duration, it will need to stay in the relaxed region for some time before allowing the pace to become stressed again. In this pacemaker example, we have abstracted away the device states into *stress* and *relax* regions, and the safety property can be defined as a predicate on the history of stress and relax intervals. We call this type of safety property *SR-safety*.

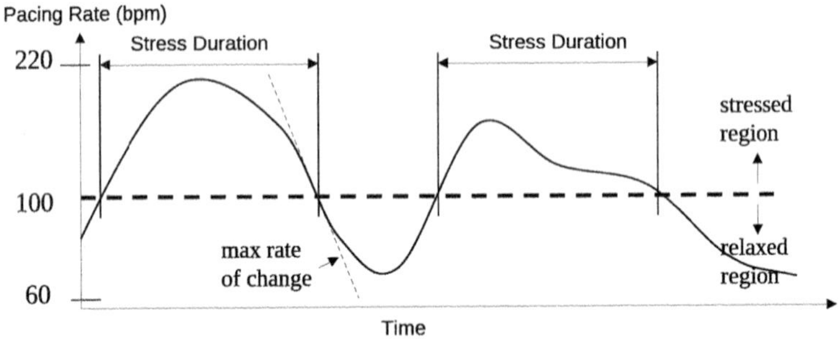

Fig. 2. Characterizing Medical Device Safety

There is also another important point for safe device operation: device states reflecting continuous physiological parameters should change *gradually* in order for the patient to slowly adapt to the effects. For example, in a pacemaker, if the pacing rate of 100 bpm immediately drops to 60 bpm over 1 second, the patient may start to feel a bit light-headed. Patient safety in this case requires constraining how fast the pacing rate can change over time shown by the slope in Figure 2. The maximum rate of change could in general be a function of the current state and whether the change is increasing or decreasing.

After the notion of safety is defined, a system can be designed to detect if an issued command to a device is safe or not. Any command that takes the device into a relaxed state is safe by default. The only commands we need to worry about are the commands that take the device into a stressed state or keep the device in a stressed state. From the definition of SR-safety we know that, once the device enters a stressed state, the time that it remains in stressful states must be shorter than some maximum duration. Figure 3 intuitively shows how this type of safety can be enforced in the system design. Given a maximum stress duration, we can construct an envelope such that any state in the envelope can transition to a relaxed state (satisfying the rate of change constraints) before the maximum stress duration ends. Now to check whether a new command is safe or not, we just need to perform a one step look-ahead to detect whether the command will keep the device state in the envelope. If the look-ahead device state is outside the stress-envelope, then the command should be ignored and a default safe command should be issued to gradually transition back to a relaxed state. This describes the essence of the operations performed by the command-shaper pattern to enforce SR-safety.

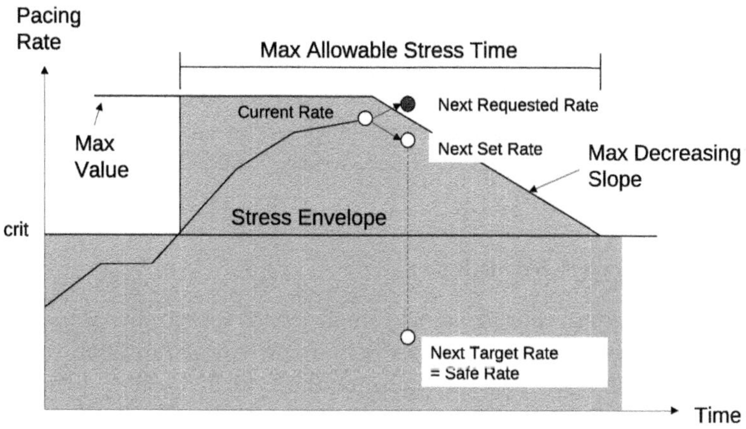

Fig. 3. Stress Envelope for Enforcing Safety

4 Real-Time Maude and Parameterized Modules

We have informally covered the concepts of medical device SR-safety and the command-shaper pattern. To formally define these concepts, we have used the Maude rewriting logic framework [1]. In this section we briefly cover the important constructs we used from Real-Time Maude and parameterized modules. We assume the reader is familiar with basic Maude constructs including modules (`mod`), sorts (`sort`), operators (`op`), unconditional and conditional equations (`eq` and `ceq`) and unconditional and conditional rules (`rl` and `crl`).

4.1 Full Maude and Real-Time Maude

Full Maude [2] is a Maude interpreter written in Maude, which in addition to the Core Maude constructs provides syntactic constructs such as object oriented modules. Object oriented modules implicitly add in sorts `Object` and `Msg`. Furthermore, OO-modules add a sort called `Configuration` which consists of a multiset of terms of sort `Object` or `Msg`.

Objects are represented as records:

```
< objectID : classID | AttributeName : Attribute, ... >
```

Rewriting logic rules are then used to describe state transitions of objects based on consumption of messages. For example, the following rule expresses the fact that a pacemaker object consumes a message to set the pacing period to T:

```
rl setPeriod(pm, T)
   < pm : Pacing-Module | pacing-period : PERIOD >
   => < pm : Pacing-Module | pacing-period : T > .
```

Real-time Maude [8] is a real time extension of Maude in Full Maude. It adds syntactic constructs for defining timed modules. Timed modules automatically import the TIME module, which defines the sort Time (which can be chosen to be discrete or continuous) along with various arithmetic and comparison operations on Time. Timed modules also provide a sort System which encapsulates a Configuration and implicitly associates with it a time stamp of sort Time. After defining a time-advancing strategy, Real-time Maude provides timed execution (trew), timed search (tsearch), which performs search on a term of sort System based on the time advancement strategy, and timed and untimed LTL model checking commands.

4.2 Parameterized Modules

Modules in Maude use initial model semantics for execution. Maude also supports theories which are given a loose semantics. Normally theories are instantiated via views to other theories or to modules. In particular, a theory can be instantiated to any module whose initial model satisfies all equational, membership, and rewrite sentences of the theory. For example, if we defined a theory SAFE-STATE and a module SAFE-PACEMAKER-DURATION satisfying all the sentences of the theory (after renaming), then we can define a view from SAFE-STATE to SAFE-PACEMAKER-DURATION. Below we show one equational sentence in the SAFE-STATE theory that is satisfied by the module SAFE-PACEMAKER-DURATION through view Safe-PD.

```
(fth SAFE-STATE ...
   eq min-val <=risk safe-val = true .
... endfth)
(fth SAFE-PACEMAKER-DURATION ...
   eq D <=risk D' = D >= D' .
   eq min-risk-dur = 100 .
   eq safe-dur = 75 .
... endfth)
(view Safe-PD from SAFE-STATE to SAFE-PACEMAKER-DURATION is ...
   op min-val to min-risk-dur .
   op safe-val to safe-dur .
... endv)
```

Parameterized modules are modules which take theories as input parameters and define operations (parametrically) in terms of the input theories.

Parametrized modules are instantiated by providing views to concrete modules for the corresponding input theories. Once instantiated, the parametrized module is given the free extension semantics for the initial models of the targets of the input views. For example, a parameterized module `PATTERN{X :: SAFE-STATE}` can be instantiated with the view `Safe-PD` using the syntax `PATTERN{Safe-PD}`.

Patterns as Parameterized Specifications. Parameterized modules are very powerful constructs as defining a parameterized module really defines a wide range of modules, one for each possible correct instantiation of its input theories. This means that any theorems we prove about a parameterized module should hold no matter what instantiation of the input theories is given. This has a nice correspondence with design patterns (design structures that can be reused within different contexts). If a design pattern can be formalized as a parameterized module and we prove a safety property for it, then any time we apply the pattern to a system (assuming the context satisfies all the preconditions specified by the input theories), we can be sure that the safety property holds in the instantiated system also. We present the specification of our command-shaper pattern using parameterization in this paper. A detailed the proof of its safety properties can be found in our technical report [11].

5 Preliminary Definitions and the Safety Theory

To aid discussion, this and following sections contain various snippets from our formal specifications in Maude. More detailed specifications can be found in [11], and the full specification can also be downloaded from:
https://netfiles.uiuc.edu/musun/www/medical_pattern/specification.zip.

5.1 Formal Models Stress-Relax Event Streams

The medical device safety model starts by defining the notion of time advancement. We use the conventions of defining `tick` and `mte` described in the Real-Time Maude documentation [9] to ensure deterministic timed rewriting. The detailed definitions can be found in module `TICK-MTE-SEM` (see [11] Section 4.3).

We have already discussed earlier that SR-Safety for a medical device can be specified as a predicate on the history of stress and relaxed intervals. This history can be captured by recording the time instances when the device state changes from a relaxed state to stressed state (a `!stress` event) and when the device state changes from a stressed state to a relaxed (a `!relax` event). This idea is shown in Figure 4, where logging two types of events over time gives us all the information of when the device is in stressed states and in relaxed states.

We formally define (see [11] Section 4.3) the log of events as a parameterized module `EVENT-LOG` (parameterized on the set of events). To illustrate how event logs are represented, Figure 5 provides a graphical representation of the term `events`. The clock `c` for each event keeps track of the time elapsed until the next

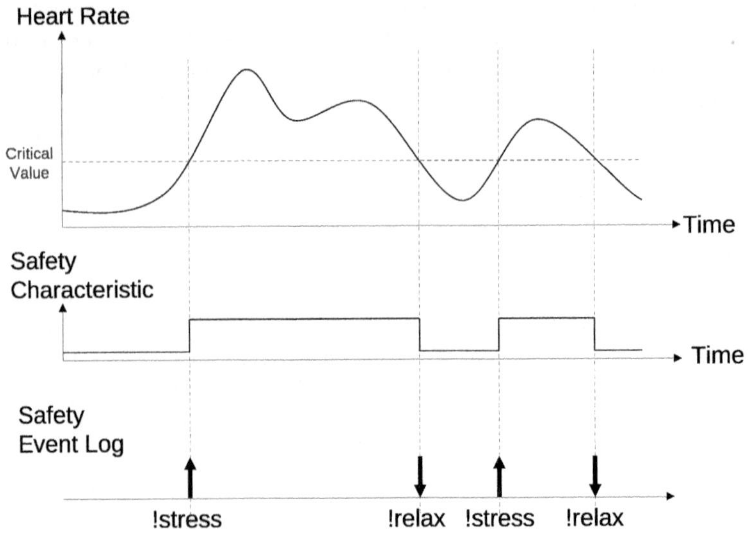

Fig. 4. Stress Relax Event Log

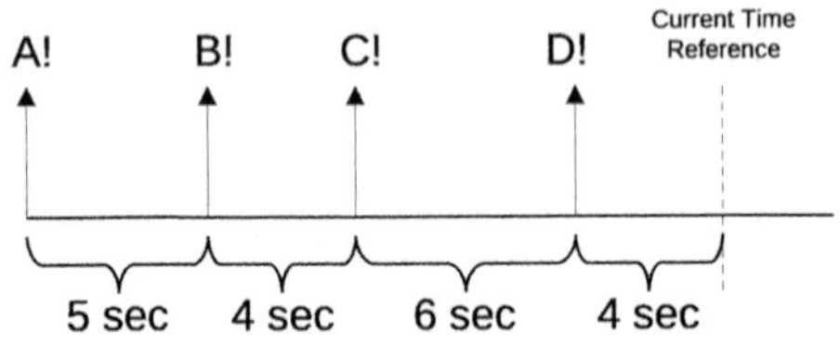

Fig. 5. Event Example

event (or the current time if there is not next event). Notice that only the clock for the latest event needs to be running, since the time interval between two consecutive events will remain fixed.

```
events = E(D!, c(4, run)) E(C!, c(6, stop))
         E(B!, c(4, stop)) E(A!, c(5, stop)) nil .
```

Now for the *stress-relax log* shown in Figure 4, we assume that the system initially starts at a value below the threshold. This means that a stress-relax log imposes additional structure on top of event logs:

1. The first event logged must be a !*stress* event.
2. Events must alternate between !*stress* and !*relax* events over time.

These constraints, and the notion of a stress-relax log, are both captured by the parameterized instantiation STRESS-RELAX-LOG (see [11] Section 4.3).

5.2 Formal Definition of Safety

We now formally define the notion of *SR-safety* motivated in Section 3. For devices that have an *SR-safety* property (recall Figure 2), we start by characterizing device states by a set *Val* of values and an order relation $\leq_{risk}$) as shown in the theory SAFE-STATE.

```
(fth SAFE-STATE is
   pr TOTAL-ORDER * (sort Elt to Val, op _<=_ to _<=risk_) ...
   ops min-val max-val crit-val safe-val : -> Val .
   op period : -> Time . --- period of wrapper dispatch
   op del : Val Val -> Val .
   op tdel-min : Val Val -> TimeInf .
   op safe? : Stress-Relax-Log -> Bool .
   op norm : Stress-Relax-Log -> Stress-Relax-Log .
... endfth)
```

Intuitively, the total order $\leq_{risk}$ relates the relative safety of two states. Furthermore, in terms of risk there are three key constants $v_{min} \leq_{risk} v_{crit} \leq_{risk} v_{max}$ with the given ordering corresponding to the minimum, critical, and maximum values in Figure 2 respectively. That is, value v is considered *relaxed* when $v_{min} \leq_{risk} v \leq_{risk} v_{crit}$ and is considered *stressed* when $v_{max} \geq_{risk} v >_{risk} v_{crit}$. In addition, we define v_{safe} ($v_{min} \leq_{risk} v_{safe} \leq_{risk} v_{crit}$) as a default safe state to eventually transition to when input commands to the device are unsafe.

The theory SAFE-STATE defines period indicating a constant period for execution dispatch. For two states V, V', the operator del(V, V') defines the value maximally changed from V towards the direction of V' in one period. tdel-min(V, V') defines the minimum amount of time it will takes to change from V to V'. Operators del and tdel-min provide a discrete way of defining bounds on the rate of change for the device state.

We want to have a generic way of characterizing the amount of time that devices may safely remain in relaxed and stressed states. This is captured by a predicate safe? on Stress-Relax-Log. For the Stress-Relax-Log, it is assumed that a !stress event is recorded when the device state rises above v_{crit}, and a !relax event is recorded when the state falls to or below v_{crit}. The safe? predicate is left generic, so that arbitrarily complex conditions on time bounds can be defined on stress and relax intervals. However, safe? should satisfy some monotonicity assumptions. These assumptions include the fact that staying in a stress situation for a longer amount of time cannot make a device safe when it was unsafe before. Also, staying in a relax situation for a longer amount of time should keep the device safe if it was safe before. We show a subset of the monitonicity properties below.

```
ceq [stress-safe] :
safe?(E(!stress, c(T', CA)) L) implies safe?(E(!stress, c(T, CA)) L)
       = true if T' ge T .
eq [relax-safe] : safe?(E(!relax, c(T, CA)) L) = safe?(L) .
```

Finally, we define a norm operator which is really a modeling construct used for efficiency of representation. Instead of maintaining an unbounded list of events that grows over time, many times it is possible to contract or normalize the history of events to a bounded-sized list.

6 Command Shaper Pattern as a Parameterized Specification

The *Command Shaper Pattern* is expressed as a wrapper object around an existing device object. To the external environment its interface looks exactly like interface for the internal wrapped device, but it also has the capability to ignore unsafe commands.

The precise definition of what types of objects can be wrapped is provided by the theory WRAPPED-OBJECT (see [11] Section 5.1). A wrapped object is of class Wrapped and has a single attribute set-val for its state. We classify messages into input and output messages. This is so that the wrapper knows which messages to forward to the external configuration. The message set-val, to set the device state, is assumed to be the only message of sort InMsg.

```
(oth WRAPPED-OBJECT is inc SAFE-STATE .
   class Wrapped | set-val : Val .
   sort InMsg OutMsg . ...
   msg set-val : Oid Val -> InMsg .
... endoth)
```

6.1 Parameterized Wrapper Object

The wrapper class structure is parameterized by the theory WRAPPED-OBJECT. The definition of the wrapper is split up into three modules for ease of readability. The first part EPR-WRAPPER (see [11] Section 5.2) defines the wrapper class with the appropriate accessor and modifier operations for each attribute.

```
class EPR-Wrapper{X} | inside : NEConfiguration,
   next-val : X$Val, val : X$Val, disp : Timer,
   stress-intervals : Stress-Relax-Log .
```

The EPR-Wrapper class has four attributes. The inside attribute defines the internal wrapped configuration. This is assumed to contain an instance of an object of the instantiated class Wrapped and possibly various messages of type InMsg and OutMsg. The next-val and val attributes describe the next requested (target) state and the current state, respectively. The last attribute stress-intervals is the log of stress and relaxed events used to evaluate safety.

Safety Envelope Calculations for the Wrapper. The parameterized module WRAPPER-AUX describes auxiliary operations based on the operations defined in the theory SAFE-STATE:

```
(fmod WRAPPER-AUX{X :: WRAPPED-OBJECT} is inc EPR-WRAPPER{X} .
   op cap : X$Val -> X$Val .
   op stress? : X$Val -> Bool .
   ops toStress? toRelax? : X$Val X$Val -> Bool .
   op inEv? : X$Val Stress-Relax-Log -> Bool .
... endfm)
```

The operator cap changes the state to be within the min and max risk range, if it was originally outside of this range. The predicates toStress? and toRelax? describe when a value has crossed the crit-val threshold in the more risky direction or less risky direction, respectively. The last predicate inEv?, which is an abbreviation for *inside the envelope*, is the most important predicate. It detects whether a configuration can persistently satisfy the safe? predicate in the future by performing a look ahead to see the shortest time it will take to reach a relaxed state.

```
ceq [inev-unreachable] : inEv?(V, L) = false
   if tdel-min(V, crit-val) == INF /\ stress?(V) .
ceq [inev-stress] : inEv?(V, L) = safe?(L) and
   safe?(log(tick(L, tdel-min(V, crit-val) plus period), !relax))
   if stress?(V) /\ tdel-min(V, crit-val) :: Time .
ceq [inev-relax] : inEv?(V, L) = safe?(L) if not stress?(V) .
```

It is easy to see that inEv? is a stronger predicate than safe?. All the equations have the form $inEv?(V, L) = false$ or $inEv?(V, L) = safe?(L) \wedge term$. Essentially, inEv? strengthens safe? enough for it to become stable over time. That is, if $inEv?(V, L) = true$, then there always exists a controllable path of operation for the system to remain safe (recall the intuition provided in Figure 3). Equations *inev-unreachable* and *inev-stress* specify that if the device is in a stress state, then it is in the envelope iff it can transition to a relaxed state before the safe? predicate is violated. Equation *inev-relax* says that relaxed states are always in the envelope.

Wrapper Execution. Finally, with all the auxiliary functions and the wrapper object fully defined, EPR-WRAPPER-EXEC (see [11] Section 5.4) describes how a wrapper object executes. Intuitively, the wrapper should filter and correct improper settings for the state, so that the system always remains safe. Aside from defining standard timed behavior via mte and tick and delivering and forwarding messages to and from the internal wrapped object. The most important part of the EPR-WRAPPER-EXEC specification states that if the next state value V' is outside the extended safety envelope, then the wrapper object will ignore it and use a safe value safe-val as the next target value.

```
ceq next-val(< O : EPR-Wrapper{X} |
   val : V, next-val : V', stress-intervals : L >)
   = del(V, safe-val)
   if not inEv?(del(V, V'),
      log(L, log-entry(V, del(V, V')))) .
```

Notice that the statement in the condition performs a look-ahead by precomputing `del(V, V')` and logging any new events. If the precomputed next state and event log are not in the envelope, then the next requested state `V'` is ignored, and a default `safe-val` is used as a target for transitioning to the next state (again recall Figure 3). Of course, under normal operation, when commands are actually safe to perform, the wrapper will not change the behavior of the device.

6.2 Pattern Instantiation

At this point we have fully defined the formal *Command Shaper Pattern*. Furthermore, any instantiation of the pattern has provably correct properties for safety [11]. In this subsection we demonstrate an instantiation of the pattern to a cardiac pacemaker. Once instantiated, the specifications also become executable, so we are able to use model checking to validate that our specifications are safe.

Pacemaker Instantiation. The pacemaker system represents a very generic application of the *Command Shaper Pattern*. It preserves the structure of the pattern without introducing any collapsing of terms or degeneracies, so it is a good test case to completely cover most constructs of the pattern.

We assume that the instantiation is customized for a specific patient, so the specific patient safety properties are as follows (since pacing periods are modeled more naturally than a pacing rate, we set the constraints on pacing periods):

1. Only pacing periods in the range between 500ms (120 bpm) and 1000ms (60 bpm) are considered valid.
2. Any pacing period below 660ms (above 90bpm) is considered stressful.
3. The pacemaker should not pace continuously at stressful rates for more than 1 minute.
4. Once the pacemaker's pacing rate drops down from stressful rates, the pacing rate should remain relaxed for a duration proportional to twice the previous stress interval.
5. The pacing period can be updated at most once every second
6. An updated pacing period can increase the period by at most 30ms from the previous pacing period, or it can decrease the period by at most 20ms from the previous pacing period.

These requirements are captured in the module `SAFE-PACEMAKER-DURATION` (see [11] Section 5.5) with each time unit representing 10ms. Requirements 1 and 2 constraining the pacing periods are easily specified.

```
eq min-risk-dur = 100 . --- x 10ms = 60 bpm
eq safe-dur = 75 . --- x 10ms = 80 bpm
eq crit-dur = 66 . --- x 10ms = 90 bpm
eq max-risk-dur = 50 . --- x 10ms = 120 bpm
```

Requirements 5 and 6 constraining the rate of change can also be easily specified. We have to consider two cases based on whether the pacing period is increasing or decreasing.

```
eq period = 100 . --- x 10ms = 1s
eq risk-dec-max(D) = 2 . --- x 10ms = 20ms
eq risk-inc-max(D) = 3 . --- x 10ms = 30ms ...
ceq del(D, D') = del-inc-risk(D, D') if (D <=risk D') ...
ceq del-inc-risk(D, D') = D' if (D - D' <= risk-inc-max(D)) ...
ceq del-dec-risk(D, D') = D' if (D' - D <= risk-dec-max(D)) ...
```

Finally, requirements 3 and 4 are a bit more verbose to specify due to the generality and flexibility of the `safe?` predicate. Essentially, we have to go through the log of events to check whether some time duration in the past has violated safety.

```
eq max-stress-interval = 6000 . --- x 10ms = 1 min
eq min-relax-interval(T) = 2 * T . ...
eq safe?(E(!stress, C) L) = safe?(L) and
   value(C) <= max-stress-interval .
eq safe?(E(!stress, C'') E(!relax, C) E(!stress, C') L) =
   safe?(E(!stress, C') L) and
   value(C'') <= max-stress-interval and
   value(C) >= min-relax-interval(value(C')) . ...
ceq norm(E(!stress, C) L) = E(!stress, C) nil if
   safe?(E(!stress, C) L) .
```

In addition to requirements 3 and 4, we also include a definition of a normalization function `norm` which will throw away all history aside from the current stress duration if everything is already safe. This makes sense since for the defined pacemaker safety properties, there is no need to keep track of previous stress durations that the patient has already had sufficient time to recover from.

After safety has been defined for the pacemaker, we can specify how the internal (wrapped) pacing module behaves in `WRAPPED-PACING-MODULE` (see [11] Section 5.5). The core functionality is just to send a `pace` event whenever the pacing timer expires.

```
rl [reset-next-pace] :
< O : Pacing-Module | nextPace : t(0), period : T >
    => < O : Pacing-Module | nextPace : t(T) > pace  .
```

We have defined `WRAPPED-PACING-MODULE` to satisfy all the sentences in the theory `WRAPPED-OBJECT`, so we can now define the view `Safe-Pacer` (see [11] Section 5.5) from `WRAPPED-OBJECT` to `WRAPPED-PACING-MODULE` which specifies some renaming of sorts and operators. With a view defined for the pattern's input theory, we can finally instantiate the wrapper to an executable system specification:

```
(tomod PARAM-PACEMAKER is
   pr EPR-WRAPPER-EXEC{Safe-Pacer} . ...
   eq wrapper-init =
      < pacing-module : EPR-Wrapper{Safe-Pacer} |
         inside : < pacing-module : Pacing-Module |
                       nextPace : t(0), period : safe-dur >,
         val : safe-dur, next-val : safe-dur, disp : t(period),
         stress-intervals : (nil).Event-Log{Stress-Relax} > .
... endtom)
```

An important thing to notice is that the object ID of the wrapper module is exactly the same as the internal module being wrapped. This is needed for modularity and allows the wrapper object to be used anywhere the original (internal) object can be used, receiving the exact same set of input messages. The model also includes a set of initial delayed messages `msgs-init` to simulate external input and an external environment model `extern-init` which includes the pacemaker lead being shocked.

7 Verification of an Instantiation

Our *Command Shaper Pattern* is provably safe [11], so naturally, all instantiations should satisfy the necessary safety properties. However, since we already have executable instantiations available for certain medical devices, we can also use timed search or model checking as an extra level of validation for the correctness of our pattern given certain initial states. However, we cannot say anything about the completeness of these verification results without some additional requirements on the system specification.

7.1 Verification Completeness for Compositional Nested Systems

In general, Real-Time Maude provides sound but incomplete model checking for system specifications [7]. That is, any counterexample found will be a real counterexample, but some real counterexamples may be missed. However, if time advancement strategies and propositions satisfy the properties of time robustness and tick invariance (i.e. no important system states are missed due to the time advancement strategy), then the timed model checking results are sound and complete [7]. In [7], Theorem 14 provides a simple criterion for verifying that flat object-oriented specifications are time-robust. However, for many practical application we have nested or wrapped objects, to which Theorem 14 does not apply. We provide a proof sketch of a refined time-robust criterion for nested configurations as a Theorem in [11] which gives sufficient conditions for ensuring completeness of model checking in configurations with nested objects. In this section we are concerned with checking the non-reachability of unsafe states. This can be verified just using timed search, and with time robustness and tick invariance of the safety property, we are guaranteed not to miss intermediate states where the safety property may be violated.

Verifying the Pacemaker. Given the module PARAM-PACEMAKER described in Section 6.2, we can immediately model check that the safety log of events in the wrapper satisfies the defined safety properties of the pacemaker using a timed search.

```
Maude> (tsearch [1] in PARAM-PACEMAKER : {init} =>*
   {C:Configuration
   < pacing-module : EPR-Wrapper{Safe-Pacer} |
     A:AttributeSet, stress-intervals : L::Stress-Relax-Log >}
   such that not safe?(L::Stress-Relax-Log) in time <= 10000 .)
...
No solution
```

This tells us that the pattern instantiation does indeed perform what it is meant to do with the given initial state and up to a given time. However, safety is defined for the patient and not the device, so is patient safety the same as device safety? For pacemaker operation, patient safety is close to device safety but there may be some time delays. For example, if the pacemaker's rate changes between heart beats, then the patient would not notice it until the next heart beat. Thus, to capture the actual events affecting the patient, we have created a model of a pacemaker lead (the bioelectrical element that actually stimulates heart contractions by creating an activation potential on muscle tissue). In our case, the model of the lead is quite simple: whenever it receives a `!shock` event (from the pacing module), it will log the event. Thus, the model of the lead effectively keeps track of all the heart beats stimulated by the pacemaker PM-LEAD (see [11] Section 6.2). With a lead model, it is now necessary to define patient safety in terms of heart beat intervals. These corresponding safety properties are somewhat more tedious to specify, since we are at a much lower level of abstraction, but it is still reasonably straightforward.

```
(omod PM-LEAD-SAFETY-PROP is pr PM-LEAD ...
   eq max-period = 100 . --- x 10ms = 60 bpm
   eq min-period = 50 . --- x 10ms = 120 bpm
   eq crit-period = 66 . --- x 10ms = 90 bpm ...
   eq stressed?(T) = T < crit-period ...
   eq dec-max = 3 . --- x 10ms = 20ms
   eq inc-max = 2 . --- x 10ms = 30ms ...
   eq max-stress-dur = 6000 . --- x 10ms = 1 min
   eq min-relax-dur(T) = 2 * T . ...
   --- periods must be within range ...
   eq range-safe?(T) = T >= min-period and T <= max-period .
   --- periods cannot change too fast ...
   eq log-change-safe?(E E' L) =
      (not stopped?(E) or change-safe?(elapsed(E), elapsed(E')))
      and log-change-safe?(E' L) ...
   --- periods cannot remain stressed too often ...
   ceq log-stress-safe?(E L, T, T') =
      log-stress-safe?(L, T, T' + elapsed(E))
      and stress-safe?(T, T' + elapsed(E))
      if stressed?(elapsed(E)) and stopped?(E) ...
endom)
```

Now, we can perform model checking using the patient's safety requirements, and check that the system indeed still satisfies the true safety requirements.

```
Maude> (tsearch [1] in PARAM-PACEMAKER : {init} =>*
    {C:Configuration < lead : Lead | A:AttributeSet >}
    such that not pm-safe?(< lead : Lead | A:AttributeSet >) in time
    <= 10000 .
...
No solution
```

The reason that safety still holds despite delays is because the wrapper pattern implicitly assumed that a delay of at most one period may be required for actuation (in module `WRAPPER-AUX` equation `inev-stress` there is a `plus period` in the evaluation of `inEv?`). In the pacemaker the period was 1 second, and each heart beat was at most 1 second apart (60 bpm), so the delay in pacing could not exceed 1 second.

8 Conclusions

In the world of medical device plug-and-play [5], it is essential that any medical devices inside a system adapt to a diverse and varied environment without compromising safety. Throughout this paper we have described in detail a command-shaper pattern to ensure that certain safety properties hold for a subclass of medical devices (including ventilators, infusion pumps, and pacemakers [11]). For devices that can be partitioned into a set of stressed and relaxed states, we can provably guarantee the safety requirements for the device state: always stay in a valid state, always change the devices state based on a bounded rate of change, and do not remain in stressed states for too long. A main concern of medical device plug-and-play is that connecting more devices may lead to more points of failure. We have shown through our pattern that some essential safety properties can be isolated inside individual devices independent of the network communication. Thus, by introducing a provably safe medical design pattern, we are one step closer towards the goal of safe and reliable medical systems.

Acknowledgements. We would like to thank all the people of the UIUC MD PnP group and the UIUC Maude Seminar group who have provided many valuable insights and suggestions towards improving this work. This work was supported in part by ONR grant N000140810896 and by NSF grants 0720482 and 0834709.

References

1. Clavel, M., Durán, F., Eker, S., Lincoln, P., Martí-Oliet, N., Meseguer, J., Talcott, C.: All About Maude - A High-Performance Logical Framework. LNCS, vol. 4350. Springer, Heidelberg (2007)
2. Durán, F., Meseguer, J.: The Maude specification of Full Maude. Technical report, SRI International (1999)

3. France, R.B., Kim, D.-K., Ghosh, S., Song, E.: A UML-Based Pattern Specification Technique. IEEE Trans. Softw. Eng. 30(3), 193–206 (2004)
4. Gamma, E., Helm, R., Johnson, R., Vlissides, J.: Design patterns: elements of reusable object-oriented software. Addison-Wesley Longman Publishing Co., Inc., Boston (1995)
5. Medical Devices and Medical Systems - Essential Safety Requirements for Equipment Comprising the Patient-Centric Integrated Clinical Environment (ICE), `http://mdpnp.org/uploads/ICE_Part_I_draft_21Dec2008_N30_web.pdf`
6. Mikkonen, T.: Formalizing Design Patterns. In: ICSE 1998: Proceedings of the 20th International Conference on Software Engineering, Washington, DC, USA, pp. 115–124. IEEE Computer Society, Los Alamitos (1998)
7. Ölveczky, P.C., Meseguer, J.: Abstraction and completeness for Real-Time Maude. Electronic Notes in Theoretical Computer Science 176(4), 5–27 (2007)
8. Ölveczky, P.C., Meseguer, J.: Semantics and pragmatics of Real-Time Maude. Higher-Order and Symbolic Computation 20(1-2), 161–196 (2007)
9. Ölveczky, P.: Real-Time Maude 2.3 Manual (August 2007)
10. Soundarajan, N., Hallstrom, J.O.: Responsibilities and Rewards: Specifying Design Patterns. In: ICSE 2004: Proceedings of the 26th International Conference on Software Engineering, Washington, DC, USA, pp. 666–675. IEEE Computer Society, Los Alamitos (2004)
11. Sun, M., Meseguer, J., Sha, L.: A Formal Pattern Architecture for Safe Medical Systems, `https://netfiles.uiuc.edu/musun/www/medical_pattern/techrep.pdf`

On the Behavioral Semantics of Real-Time Domain Specific Visual Languages

José E. Rivera, Francisco Durán, and Antonio Vallecillo

GISUM/Atenea Research Group. Universidad de Málaga, Spain
{rivera,duran,av}@lcc.uma.es

Abstract. Domain specific visual languages (DSVLs) are becoming commonplace for specifying systems at a high-level of abstraction, using a notation very close to the problem domain and quite intuitive for domain experts. Usually, DSVLs are defined only in terms of their abstract and concrete syntaxes, with no precise semantics—something that may hamper the use of tools to simulate or analyze the produced models. In this paper we show how rewriting logic, and in particular Real-Time Maude, can be effectively used to provide semantics to real-time DSVLs, and how these Maude specifications can be automatically generated from the visual specifications. The use of Real-Time Maude provides additional interesting benefits, such as being able to simulate the DSVL specifications or to conduct formal analysis on them.

1 Introduction

Domain specific visual languages (DSVLs) are becoming essential elements of Model-Driven Engineering (MDE). DSVLs are normally defined in terms of their abstract and concrete syntaxes. The abstract syntax of a DSVL is defined as a metamodel, which describes the concepts of the language, the relationships between them, and the structuring rules that constrain the combination of model elements according to the domain rules. The concrete syntax specifies how the domain concepts included in its metamodel are represented, and is usually defined as a mapping between the metamodel and a textual or graphical notation. Explicit and formal specification of model semantics is receiving more attention recently, since the lack of explicit behavioral semantics strongly hampers the development of simulation and formal analysis tools. This is particularly important in certain domains that require rigorous and precise specifications (e.g., safety-critical real-time and embedded systems).

There are different ways of providing semantics to DSVLs, from operational (interpreting the language as sequences of computational steps) to translational approaches (providing a mapping to another language with precisely defined semantics) [6]. One way of specifying the dynamic behavior of a DSVL is by describing the evolution of the modeled artifacts along some time model. This can be done, for instance, using model transformations supporting in-place update [8].

In this paper we investigate the use of rewriting logic [11], and specifically Real-Time Maude [13,12], for giving semantics to real-time DSVLs. More precisely, we give semantics to the DSVLs that can be defined with *e-Motions* [1], a tool for defining

P.C. Ölveczky (Ed.): WRLA 2010, LNCS 6381, pp. 174–190, 2010.

the behavior of visual languages. In particular, we present a mapping from these time-dependent behavioral specifications into Real-Time Maude specifications. The use of Maude as a target semantic domain brings very interesting benefits, because it enables, e.g., simulation of the real-time specifications and the conduction of reachability analysis and model checking.

The expressiveness of Real-Time Maude enables many possible ways of specifying real-time systems, and selecting the one that best fits our DSVL is not always obvious; we present here an encoding of the constructs of a real-time DSVL into Real-Time Maude that aims at allowing the simulation and analysis of specifications. We have defined a mapping from the *e-Motions* specifications to the corresponding Real-Time Maude specifications, that provides the semantics in Maude of any DSVL defined with *e-Motions*. Such an automatic mapping has been implemented as a set of ATL [22] model transformations from the e-Motions metamodel to the Real-Time Maude metamodel.

After this introduction, Section 2 presents our proposal for defining real-time DSVLs. Section 3 presents how the language constructs can be represented in Real-Time Maude. Section 4 describes the current tool support. Finally, Sections 5 and 6, respectively, compares our work with related proposals, and draws some conclusions and outlines some future research activities.

2 Real-Time DSVLs with e-Motions

Let us introduce a modeling language for mobile phone networks (MPNs), which will serve as the motivating example to show how we provide semantics to real-time DSVLs. A more detailed version of this example was originally presented in [15] to illustrate the use of *e-Motions* to define real-time DSVLs.

The Structure of the System. The MPN metamodel is shown in Fig. 1 (a). An MPN has cell phones and antennas. Antennas provide coverage to cell phones, depending on their relative distance. A cell phone is identified by its number, and can perform calls to other phones of its contact list. Dialed and received calls are registered. Phone's attribute bps represents the battery consumption per time unit. Fig. 1 (b) shows an MPN example using a visual concrete syntax. This model consists of three cell phones and one antenna. The position of each element is dictated by its position on a grid. All phones are initially off, and their contacts are represented by arrows between them.

Dynamic Behavior. The dynamic behavior of the system is described as an in-place model transformation, which is composed of a set of rules. Each one of these rules represents a possible *action* of the system. These rules are of the form $l : [\mathrm{NAC}]^* \times \mathrm{LHS} \rightarrow \mathrm{RHS}$, where l is the rule's label (its name); LHS (left-hand side), RHS (right-hand side), and NAC (negative application conditions) are model patterns that represent certain (sub-)states of the system. The LHS and NAC patterns express the precondition for the rule to be applied, whereas the RHS one represents its postcondition, i.e., the effect of the corresponding action. Thus, a rule can be applied, i.e., *triggered*, if an occurrence (or match) of the LHS is found in the model and none of its NAC patterns occurs. Generally, if several matches are found, one of them is non-deterministically selected

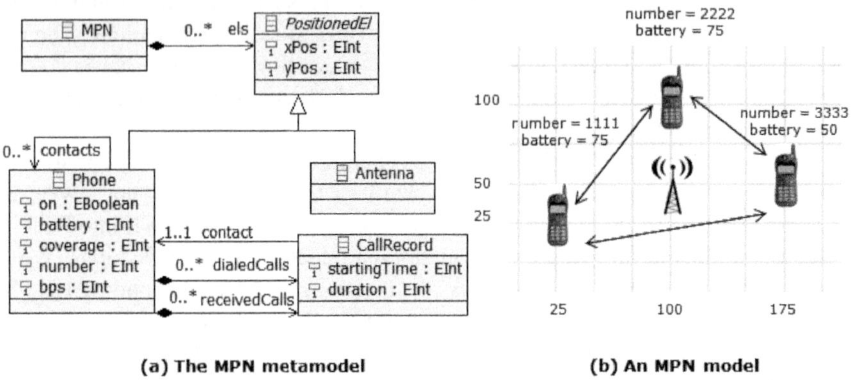

Fig. 1. A Mobile Phone Network Model and Metamodel

and applied, producing a new model where the match is substituted by the appropriate instantiation of its RHS pattern (the rule's *realization*). The model transformation proceeds by applying the rules in a non-deterministic order, until none is applicable—although this behavior can be usually modified by some execution control mechanism (e.g., Maude's strategies [9] were used in [18] to control model transformations).

Time-Dependent Behavior. One natural way to model time-dependent behavior quantitatively consists in decorating the rules with the time they consume, i.e., by assigning to each action the time it takes. Thus, we define **atomic** rules as in-place transformation rules of the form $l : [\mathrm{NAC}]^* \times \mathrm{LHS} \xrightarrow{t} \mathrm{RHS}$, where t expresses the duration of the action modeled by the rule.

As normal in-place transformation rules, an atomic rule can be triggered whenever an occurrence (or match) of its LHS, and none of its NAC patterns, is found in the model. Then, the action specified by such rule is scheduled to be realized after t time units. At that time, the rule is applied by substituting the match by its RHS and performing the attribute computations. Since actions have now a duration, elements can be then engaged in several actions at the same time. The triggering of an atomic rule is only forbidden if another occurrence of the same rule is already being executed with the same participants. Note that the states of the elements that participate in a timed action may vary during the action's time elapse, since the states of the elements may be modified by other actions. The only condition for the final application of an atomic rule is that the elements involved are still there; otherwise the action will be aborted. If we want to make sure that something happens (or does not happen) during the execution of an action, we can make use of *action execution* objects to model the corresponding exceptional behavior (see below).

As first examples, Fig. 2 shows the *SwitchOn* and *BatteryOff* atomic rules. When a phone is off, it can be switched on if it has enough battery (see the condition specified in the WITH clause). This action takes ten time units. Whenever a phone is on and has no battery, it is switched off—as modeled by the *BatteryOff* rule, whose duration is one time unit.

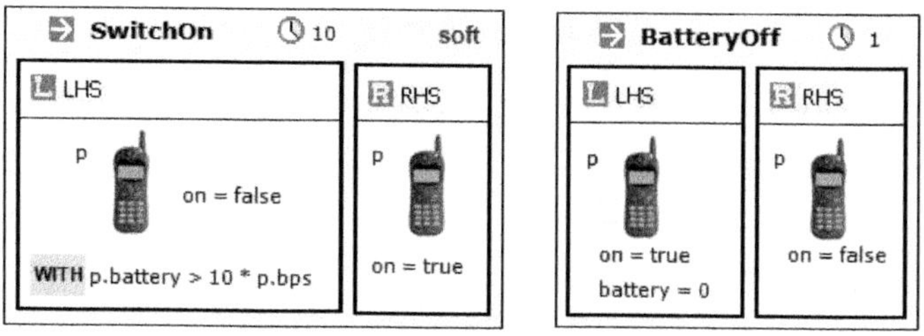

Fig. 2. The *SwitchOn* and *BatteryOff* atomic rules

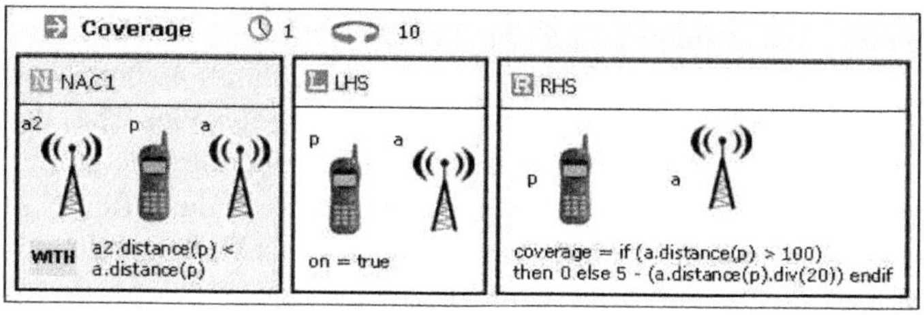

Fig. 3. The *Coverage* atomic rule

We distinguish normal rules, which are triggered as soon as possible, from **soft** rules, which are not necessarily triggered immediately, but in a non-deterministic moment in time in the near future. The *SwitchOn* action in Fig. 2 is modeled by a soft rule (notice the soft label in the rule's header): we allow phones to be switched on at a non-deterministic moment in time. However, mobile phones must go off as soon as they run out of battery.

Another essential aspect for modeling time-dependent behavior is **periodicity**. Atomic rules admit a parameter that specifies an amount of time after which the triggering of the action is periodically attempted. Normal periodic rules are tried to be triggered at the beginning of the period, while soft periodic rules can be triggered at any time within the period (only once per period). Fig. 3 shows the *Coverage* rule, which specifies the way in which antenna coverage changes. Coverage is updated every ten time units (notice the loop icon in the header of the *Coverage* rule). Each cell phone is covered by the closest antenna: as specified in its NAC pattern, the rule cannot be applied if there exists another antenna closer to the phone. To compute the distance between the two objects, we have the following **helper** (OCL operation):

```
context Antenna::distance(p : Phone): Integer
body: (self.xPos   p.xPos).abs() + (self.yPos  p.yPos).abs()
```

Helper invocations in LHS and NAC patterns are computed at the triggering of actions, while helper invocations in RHS patterns are computed in their finalization. Thus, note that the distance between the antenna and the phone may vary on these two different moments of time.

In addition to atomic rules, we also count on rules to model actions that are continuously progressing, perhaps without a specific duration. Think for instance of an action

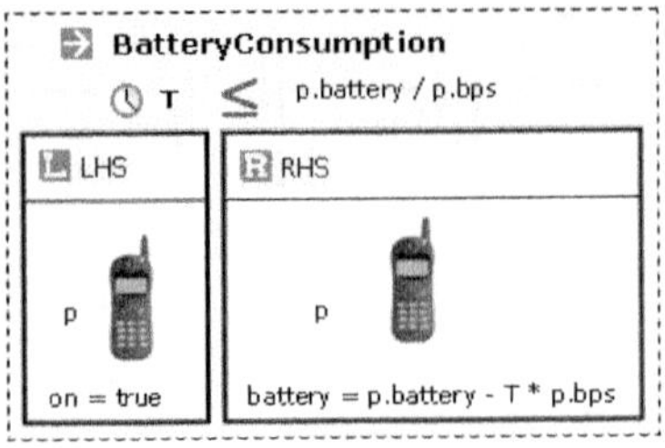

Fig. 4. The *BatteryConsumption* ongoing rule

that models the consumption of a phone battery, whose level decreases continuously with time. Actions of this kind do not have a specific duration, but are required to be continuously updated. **Ongoing** rules model this kind of behavior. They do not have any *a priori* duration time: they progress with time while the rule preconditions (LHS and NACs) hold, or until the time limit (*maximal duration*) of the action is reached. Note that rule preconditions act as a kind of *invariants* for this kind of actions.

For example, Fig. 4 shows the *BatteryConsumption* ongoing rule, which models phone battery consumption (note the use of dotted box to distinguish ongoing actions from atomic ones). According to this rule, the battery power is decreased bps battery units per time unit. To explicitly identify the state in which a phone runs out of battery, and not to decrease the battery power below zero, we limit the duration of the rule (see the expression on the righthand side of symbol $\leq$). Since ongoing actions progress with time, phones' battery will always be updated whenever an atomic rule is tried to be triggered.

Action Executions. In standard in-place transformation approaches, LHS, RHS, and NAC patterns are defined in terms of system states. This is a strong limitation in those situations in which we need to refer to actions currently under execution, or to those that have been executed in the past. For example, we can be interested in knowing whether an object is currently performing a given action to take some decision, e.g., to interrupt the action (e.g., in case an exception occurs), or not to allow the object to perform another action. In general, the inability of being able to model and deal with action occurrences hinders the specification of some useful action properties, unless some unnatural changes are introduced in the system model—such as extending the system state with information about the actions currently happening.

In order to be able to model both state-based and action-based properties, we propose extending model patterns with **action executions** to specify action occurrences. These action executions represent actions that are currently happening or that were previously performed (by using the past attribute). Action executions describe the type of the action (given by the name of the atomic rule that represents the action), its identifier, its starting and ending time, and the set of objects involved in such an action.

For example, Fig. 5 shows the *Call* atomic rule, which models the behavior of a call from a cell phone to one of its contacts. To make a call, both phones must be on and have coverage. The two NAC patterns forbid the execution of the rule whenever

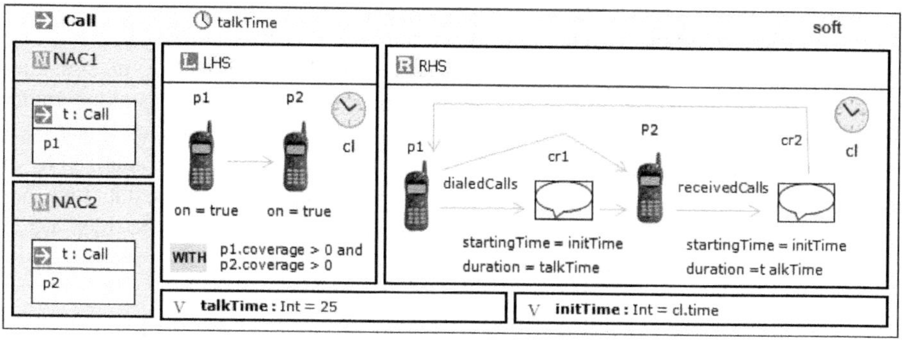

Fig. 5. The *Call* atomic rule

one of the phones is participating in another call (we do not allow call waiting). This is explicitly described with two action execution elements that specify that phone p1 (or p2, respectively) is participating in a *Call* action. At the end of the talk, the call is registered in both phones (as a dialed call in phone p1 and as a received call in phone p2) including the duration of the call (talkTime) and its starting time (initTime). The values of talkTime and initTime are computed, as every **variable** value, when the rule is triggered. The context of a user-defined variable is the rule in which it is defined. We assume that a conversation will take 25 time units. The starting time is computed by using a special kind of object, named **clock**, that represents the current global time elapse (using its attribute time). A unique and read-only clock instance is provided by the system to model time elapse through the underlying platform.

3 The Encoding in Real-Time Maude

In the previous sections, we have introduced the e-Motions language to define time-dependent behavior of DSVLs in an intuitive and informal manner, by means of descriptions of its main features in natural language. However, this lack of rigorous definitions may lead to imprecisions and misunderstandings that might hinder the proper usage and analysis of the language. In this section, we provide a precise semantics to models, metamodel, and this behavioral language by defining how they are represented in terms of Real-Time Maude constructs.

3.1 Real-Time Maude

Real-Time Maude [13,12] is a rewriting-logic-based specification language and formal analysis tool that extends the Maude system [7] to support the formal specification and analysis of *real-time systems*. Real-Time Maude provides support for symbolic simulation through timed rewriting, and time-bounded temporal logic model checking and search for reachability analysis.

Rewriting logic [11] is a logic of change that can naturally deal with states and non-deterministic concurrent computations. A rewrite logic theory is a tuple $(\Sigma, E \cup A, R)$, where $(\Sigma, E \cup A)$ is a *membership equational logic* [4] theory with Σ its signature, E a set of conditional equations, A a set of equational axioms such as associativity, commutativity and identity, so that rewriting is performed *modulo* A, and R is a set of labeled conditional rules. In rewriting logic, a distributed system is axiomatized by an equational theory $(\Sigma, E \cup A)$, describing its set of states as an algebraic data type, and a collection of conditional rewrite rules, specifying its dynamics. Rewrite rules, which are written:

$$\text{crl } [l] : t \Rightarrow t' \text{ if } C \ .$$

with l the rule label, t and t' terms, and C a condition, describe the local, concurrent transitions that are possible in the system, i.e., when a part of the system state fits the pattern t, then it can be replaced by the corresponding instantiation of t'. The guard C acts as a blocking precondition, in the sense that a conditional rule can only be fired if its condition is satisfied. The form of conditions is EqC_1 /\ ... /\ EqC_n where each of the EqC_i is either an ordinary equation $t = t'$, a *matching equation* $t := t'$, a sort constraint $t : s$, or a term t of sort `Bool`, abbreviating t = `true`. In the execution of a matching equation $t := t'$, the variables of the term t, which may not appear in the lefthand side of the corresponding conditional equation, become instantiated by *matching* the term t against the canonical form of the bounded term t'. See [7] for further details on Maude.

Real-Time Maude provides a sort Time to model the time domain, which can be either discrete or dense time (users can also define their own time domains). It also provides a sort TimeInf to extend the time domain with an infinity value INF. Moreover, there is a predefined constructor {_} of sort GlobalSystem, and an extended form of rewrite rules, known as *tick rules*, with syntax:

$$\text{crl } [l] : \{t\} \Rightarrow \{t'\} \text{ in time } \tau \text{ if } C \ .$$

where τ is a term of sort Time that denotes the *duration* of the rewrite, and that affects the *global time elapse*.

3.2 Encoding Models and Metamodels in Real-Time Maude

The representation of models used here is inspired by the Maude representation of object-oriented systems [7]. It was first introduced in [20] and further developed in [18,1]. Boronat and Meseguer use a similar representation in [2].

We represent models in Real-Time Maude as structures of sort @Model of the form $mm\{obj_1 \ obj_2 \ ... \ obj_N\}$, where mm is the name of its metamodel and obj_i are the objects that constitute the model. An object is a record-like structure of the form $< o : c \mid a_1 : v_1, ..., a_n : v_n >$ (of sort @Object), where o is the object identifier (of sort Oid), c is the class the object belongs to (of sort @Class), and $a_i : v_i$ are attribute-value pairs (of sort @StructuralFeatureInstance). Given the appropriate definitions for all classes, attributes and references in its corresponding metamodel (see below), the following Real-Time Maude term describes the MPN model shown in Fig. 1 (b):

```
op initModel : -> @Model .
eq initModel = MPN_MM {
  < 'mpn : MPN | els : Set{'a, 'p1, 'p2, 'p3} >
  < 'a : Antenna | xPos : 100, yPos : 50 >
  < 'p1 : Phone | dialedCalls : OrderedSet[], receivedCalls : OrderedSet[],
    contacts : OrderedSet['p2 ; 'p3], number : 1111, on : false, battery : 75,
    coverage : 0, bps : 1, xPos : 25, yPos : 25 >
  < 'p2 : Phone | dialedCalls : OrderedSet[], receivedCalls : OrderedSet[],
    contacts : OrderedSet['p1 ; 'p3], number : 2222, on : false, battery : 75,
    coverage : 0, bps : 1, xPos : 100, yPos : 100 >
  < 'p3 : Phone | dialedCalls : OrderedSet[], receivedCalls : OrderedSet[],
    contacts : OrderedSet['p1 ; 'p2], number : 3333, on : false, battery : 50,
    coverage : 0, bps : 1, xPos : 175, yPos : 50 > } .
```

Note that quoted identifiers are used as object identifiers; references are represented as object attributes by means of object identifiers; and OCL collections (Set, OrderedSet, Sequence, and Bag) are supported by means of mOdCL [19].

Metamodels are encoded using a sort for every metamodel element: sort @Class for classes, sort @Attribute for attributes, sort @Reference for references, etc. Thus, a metamodel is represented by declaring a constant of the corresponding sort for each metamodel element. More precisely, each class is represented by a constant of a sort named after the class. This sort, which will be declared as subsort of @Class, is defined to support class inheritance through Maude's order-sorted type structure. The following Maude specification describes a fragment of the MPN metamodel depicted in Fig. 1 (a):

```
(mod MPN_MM is extending ECORE_MM .
  op MPN_MM : -> @Metamodel .
  op MPN_Pack : -> @Package .

  sort MPN .
  subsort MPN < @Class .
  op MPN : -> MPN .
   els : -> @Reference .

  sort PositionedEl .
  subsort PositionedEl < @Class .
  op PositionedEl : -> PositionedEl .
   op xPos : -> @Attribute .
   op yPos : -> @Attribute .

  sort Antenna .
  subsort Antenna < PositionedEl .
  op Antenna : -> Antenna .

  sort Phone .
  subsort Phone < PositionedEl .
  op Phone : -> Phone .
   op on : -> @Attribute .
   op battery : -> @Attribute .
   op coverage : -> @Attribute .
   op number : -> @Attribute .
   op bps : -> @Attribute .
  ...
 endm)
```

Other properties of metamodel elements, such as whether a class is abstract or not, the opposite of a reference (to represent bidirectional associations), or attributes and reference types, are expressed by means of equations defined over the constant that represents the corresponding metamodel element. Classes, attributes and references are qualified with their containers' names, so that classes with the same name belonging to different packages, as well as attributes and references of different classes, are distinguished. These qualifications are omitted here to improve readability. See [18] for further details.

3.3 Encoding e-Motions Timed Rules in Real-Time Maude

Since tick rules affect the global time, in Real-Time Maude time elapse is usually modeled by one single tick rule, and the system dynamic behavior by instantaneous transitions [13]. This single tick rule models time elapse by using two functions: the delta function, that defines the effect of time elapse over every model element, and the mte (maximal time elapse) function, that defines the maximum amount of time that can elapse before any action is performed. Then, time advances non-deterministically by any amount T, which must be equal or less than the maximum time elapse of the system.

```
var MODEL : @Model .                    var T : Time .

crl [tick] : {MODEL} => {delta(MODEL, T)} in time T if T <= mte(MODEL) [nonexec] .
```

The delta and mte functions are applied over the whole model. However, we want DSVL objects to be completely unaware of time. In the same way that DSVL designers describe the structural aspects of the language (in terms of its metamodel) separately from its dynamic behavior (defined by means of in-place transformations) we think that time and action concerns, which are behavioral aspects, should also be defined separately from the DSVL metamodel, and then added to the specification.

With this goal in mind, we introduce classes whose instances represent time and action properties: a Clock instance will represent the current time elapse (time attribute); and ActionExec objects will gather all the information related to a rule execution and include a timer to the finalization of the corresponding action. This representation allows us to: (a) reason about and refer to actions, and (b) define time elapse only over these special objects, making DSVL objects completely unaware of it (see below). We show in what follows a more specific encoding of the main features of our approach.

Atomic Rules. Atomic rules can be naturally represented as two Real-Time Maude instantaneous rules, one modeling its *triggering* and one modeling its actual *realization*.

The triggering rule. When a rule's precondition is satisfied, an AtomicActionExec object is created. AtomicActionExec objects represent atomic rules' executions, each one acting as a countdown (timer attribute) to the finalization of the action. They gather all the information needed for its instantiation, such as the rule's name (action attribute), the identifiers of the elements involved in the action (participants attribute), the starting time (startingTime attribute), the ending time (endingTime attribute), and the variable definitions (variables attribute). Initially, the timer is set to the given duration of the rule, and its ending time is left undefined. For instance, the following Maude rule corresponds to the encoding of the *SwitchOn* action's triggering rule (see Fig. 2):

```
vars p CLK@ CNT@ ACTEXC@ OR1@ : Oid .
vars p@SFS ACTEXC@@SFS SFS : Set{@StructuralFeatureInstance} .
var  OBJSET@ : Set{@Object} .   var  PHONE : Phone .       var  @CNT@ : Nat .
vars @TIME@ DURATION@ : Time .  var  ON@p@ATT : OCL-Type .  var  MODEL@ : @Model .

crl [SwitchOn@Triggering] :
  MM@ { < p : PHONE | on : ON@p@ATT, p@SFS >
        < CLK@ : Clock | time : @TIME@ >
        < CNT@ : Counter | value : @CNT@ >
        OBJSET@ }
  =>
  MM@ { < p : PHONE | on : ON@p@ATT, p@SFS >
        < CLK@ : Clock | time : @TIME@ >
        < CNT@ : Counter | value : (@CNT@ + 2) >
        < ACTEXC@ : AtomicActionExec | action : "SwitchOn", timer : DURATION@,
          startingTime : @TIME@, endingTime : undefined,
```

```
        participants : Set{OR1@}, variables : Set{} ) >
      < OR1@ : ObjectRole | actualObject : p, role : "p" >
      OBJSET@ }
  if MODEL@ :=   MM@ { < p : PHONE | on : ON@p@ATT, p@SFS >
                      < CLK@ : Clock | time : @TIME@ >
                      < CNT@ : Counter | value : @CNT@ >
                      OBJSET@ }
  /\ not currentExec@SwitchOn(Set{p}, MODEL@)
  /\ eval(p . battery > 10 * p . bps, empty, MODEL@)
  /\ ACTEXC@ := newId(@CNT@)
  /\ OR1@ := newId(@CNT@ + 1)
  /\ ON@p@ATT = eval(false, ctx(self, p), MODEL@)
  /\ DURATION@ := toRat(eval(10, empty, MODEL@)) .
```

Note the use of the MODEL@ variable matched in the condition of the rule to avoid repeating the configuration in the righthand side of the rule. In the rules below, we will write dots to abbreviate their presentations.

Objects of transformation rules' LHS patterns are encoded as Real-Time Maude objects placed in the left-hand side of the Real-Time Maude rule; they are also included in the right-hand side so that they remain as such. LHS conditions and attribute-value pairs are encoded as rule conditions, which are computed by the mOdCL's eval operation [19]. The arguments of the eval operation are an OCL expression, a context, and the model in which such OCL expression is to be evaluated. The Clock object is included to set the starting time of the ActionExec element ACTEXC@ that represents the rule execution. The Counter object is included to compute the identifiers of the new created objects with different natural numbers (see the newId operation). ObjectRole elements represent the participants of the rule execution (with their corresponding roles). Additionally, the currentExec@SwitchOn operation is included to forbid the triggering of the rule whenever another occurrence of the same rule is already being executed with the same set of participants. This operation checks the existence of an AtomicActionExec object that refers to the rule (SwitchOn) with the same participants (Set{p}) and with an undefined endingTime in the model (@MODEL). Although there is no NAC patterns in this case, they are encoded as invocations to predicates in the corresponding rule condition that check whether occurrences of the specified patterns are found in the model.

The realization rule. Once an action's timer is consumed (i.e., there is an AtomicActionExec object whose timer attribute's value is 0) the corresponding action can be performed if none of the action's participants has been deleted. Then, the matching of the LHS is substituted by the corresponding instantiation of the RHS and the attribute values are computed. To keep track of the performed actions, the AtomicActionExec objects are not deleted and their ending times are set. The realization rule of the *SwitchOn* action is as follows:

```
rl [SwitchOn@Realization] :
  MM@ { < p  PHONE | on : ON@p@ATT, p@SFS >
        < ACTEXC@ : AtomicActionExec | action : "SwitchOn",endingTime : undefined,
          timer : 0, participants : Set{OR1@}, variables : Set{}, ACTEXC@@SFS >
        < OR1@ : ObjectRole | actualObject : p, role : "p" >
        < CLK@ : Clock | time : @TIME@ >
        < CNT@ : Counter | value : @CNT@ >
        OBJSET@ }
  =>
  readjust(Set{}, mt-ord,
    MM@{ < p : PHONE | on : eval(true, ctx(self, p), MODEL@), p@SFS >
         < ACTEXC@ : AtomicActionExec | action : "SwitchOn", endingTime : @TIME@,
```

```
      timer : 0, participants : Set{OR1@}, variables : Set{}, ACTEXC@@SFS >
    < OR1@ : ObjectRole | actualObject : p, role : "p" >
    < CLK@ : Clock | time : @TIME@ >
    < CNT@ : Counter | value : @CNT@ >
    OBJSET@ }) .
```

Attribute-value pairs in RHS patterns are encoded as computations in the right-hand side of the rule. The readjust operation deletes objects (first parameter) and links (second parameter) that are specified in the in-place rule to be deleted (an empty set and an empty list in this case, respectively). It also deletes contained objects, dangling references, and current action executions of this set of deleted objects.[1]

Ongoing rules. Ongoing rules are typically used to model actions that progress with time, and their realization will then depend on the time elapsed. Since the delta function defines the effect of time on the model objects, and because of the form of the tick rule, that makes the time elapse to be computed in the mte function before the delta function is applied, we encode the realization of the ongoing rules into the delta function itself. The values required for the calculations in the mte function are provided by corresponding OngoingActionExec objects created in instantaneous rules fired at the beginning of the action.

The initial instantaneous rule. These are encoded as the triggering rules of atomic actions. When the rule precondition is satisfied, an OngoingActionExec object is created, indicating that the corresponding ongoing rule can be executed at that moment of time. It also gathers information about its instantiation, including a new timer that represents a countdown to the rule's upper bound (upperBoundTimer attribute), and the OCL expression that represents the maximal duration (maxDuration attribute). As an example, the following Real-Time Maude specification corresponds to the encoding of the *BatteryConsumption* action's instantaneous rule (see Fig. 4):

```
crl [BatteryConsumption@Triggering] :
  MM@ { < p : PHONE | on : ON@p@ATT, p@SFS >
        < CLK@ : Clock | time : @CLK@ >
        < CNT@ : Counter | value : @CNT@ >
        OBJSET@ }
  =>
  MM@ { < p : PHONE | on : ON@p@ATT, p@SFS >
        < CLK@ : Clock | time : @TIME@ >
        < CNT@ : Counter | value : (@CNT@ + 2) >
        < ACTEXC@ : OngoingActionExec | action : "BatteryConsumption",
          maxDuration : freeze(p . battery / p . bps), variables : Set{},
          startingTime : @TIME@, endingTime : undefined,
          participants : Set{OR1@}, upperBoundTimer : undefined >
        < OR1@ : ObjectRole | actualObject : p, role : "p"  >
        OBJSET@ }
   if MODEL@ := MM@ { < p : PHONE | on : ON, p@SFS > ... OBJSET@ }
   /\ not currentExec@BatteryConsumption(Set{p}, MODEL@)
   /\ ACTEXC@ := newId(@CNT@)
   /\ OR1@ := newId(@CNT@ + 1)
   /\ ON@p@ATT := eval(true, ctx(self, p), MODEL@) .
```

[1] Note that dangling references are only deleted with the *spo* formalization, since *dpo* forbids the application of rules (by means of rule conditions) that may result in dangling references [16]. Both *spo* and *dpo* formalizations are available in e-Motions as alternative options.

Ongoing rules' maximal duration expressions are not evaluated in their triggering rules: they are *frozen* (i.e., maintained with the freeze operation) to be later computed in the mte function to get the real values at that moment of time. Furthermore, in this case the upperBoundTimer attribute is set to undefined, since the *BatteryConsumption* rule's upper bound is not specified.

Finally, note that Real-Time Maude rules are applied in a non-deterministic order. Therefore, the realization rule of an atomic action can be applied, e.g., after the execution of the triggering rule of an ongoing action. This application can make an OngoingActionExec object to represent an invalid action execution: since the state of the system may change from the moment of the generation of an OngoingActionExec object to the moment in which the ongoing action is in fact realized (the moment in which the delta operation is applied), the precondition of the action that represent the OngoingActionExec object may be violated. These *invalid* OngoingActionExec objects will be removed from the specifications, as we shall see below.

The applyOngoingRules operator. In the following time elapse, the delta equation calls the applyOngoingRules function. One applyOngoingRules equation is added *per* ongoing rule. This equation substitutes the LHS matching by its RHS, if applicable, and sets the OngoingActionExec object's ending time and maximal duration, which has already been computed by the mte function.

```
var BATTERY@p@ATT MAXDURATION@ : OCL-Exp .

op applyOngoingRules : @Model TimeInf -> @Model .
ceq applyOngoingRules(
   MM@ {< p : PHONE | on : ON@p@ATT, battery : BATTERY@p@ATT, p@SFS >
        < ACTEXC@ : OngoingActionExec | action : "BatteryConsumption",
          endingTime : undefined, participants : Set{OR1@}, variables : Set{},
          maxDuration : freeze(MAXDURATION@), ACTEXC@@SFS >
        < OR1@ : ObjectRole | actualObject : p, role : "p" >
        < CLK@ : Clock | time : @TIME@ >
        < CNT@ : Counter | value : @CNT@ >
        OBJSET@ },
   T)
 = applyOngoingRules(
     readjust(Set{}, mt-ord,
       MM@ { < p : PHONE | on : ON@p@ATT, battery :
                 eval(p . battery - T * p . bps, ctx(self, p), MODEL@), p@SFS >
             < ACTEXC@ : OngoingActionExec | action : "BatteryConsumption",
                 maxDuration : eval(MAXDURATION@, empty, MODEL@),
                 variables : Set{}, participants : Set{OR1@},
                 endingTime : (@TIME@ plus T), ACTEXC@@SFS >
             < OR1@ : ObjectRole | actualObject : p, role : "p" >
             < CLK@ : Clock | time : @TIME@ >
             < CNT@ : Counter | value : @CNT@ >
             OBJSET@ }),
        T)
  if MODEL@ := MM@ { ...  OBJSET@ }
  /\ ON@p@ATT := eval(true, ctx(self, p), MODEL@) .
--- Remaining applyOngoingRules equations ...
eq applyOngoingRules(MODEL@, T) = deleteOngoingActionExecs(MODEL) [owise] .
```

The applyOngoingRules operation is recursively called until every possible execution of an ongoing action is realized, and therefore the endingTime of the OngoingActionExec objects that represent them set. Remaining OngoingActionExec objects with an undefined endingTime represent current invalid actions, i.e., actions that cannot be performed in that moment of time. These *invalid* applyOngoingRules objects are deleted in the owise equation. This operation needs the current time elapse T, which is provided by the delta operation: (a) to set the rule's ending time, and (b) to perform the attribute computations. Attributes in this kind of actions typically depend on such time elapse. See for instance the battery attribute computation that progresses with time T by means of the expression (p . battery - T * p . bps).

Time elapse. As previously mentioned, time elapse is modeled by using the delta and mte functions. Both functions need to be defined only over time-dependent elements, namely the Clock instance, and AtomicActionExec and OngoingActionExec objects.

The delta function applies ongoing actions (with the applyOngoingRules auxiliary operation), and then decreases AtomicActionExec timers and increases the clock value. The DSVL objects remain unchanged.

```
vars T T' : Time .                vars OBJECT OBJECT' : @Object .

op delta : @Model Time -> @Model [frozen] .
op deltaAux : @Model Time -> @Model [frozen] .
op delta : Set{@Object} Time -> Set{@Object} [frozen] .
eq delta(MODEL@, T) = deltaAux(applyOngoingRules(MODEL@, T), T) .
eq deltaAux(MM@ { OBJSET@ }, T) = MM { delta(OBJSET@, T) } .
eq delta(< O : AtomicActionExec | timer : T', SFS > OBJSET@, T)
  = < O : AtomicActionExec | timer : (T' monus T), SFS > delta(OBJSET@, T) .
eq delta(< O : Clock | time : T', SFS > OBJSET@, T)
  = < O : Clock | time : (T' plus T), SFS > delta(OBJSET@, T) .
eq delta(OBJSET@, T) = OBJSET@ [owise] .
```

Note the use of the owise attribute: we act on time-dependent elements and we leave the rest unaffected. Note as well that the frozen attribute guarantees that no rule is applied on any of the arguments of a delta function.

The mte function is defined as the minimum of (a) timer values of current AtomicActionExec objects, (b) maxDuration and upperBoundTimer values of current OngoingActionExec objects, and (c) the difference between the following beginning of rule period or lower bound and the current time elapse. We make sure that time does not pass if something can happen by adding an extra mte equation for every (non-soft) atomic rule. This equation forbids time to elapse (mte = 0) whenever the rule can be applied and it has not been so. Note the use of the owise attribute.

```
var OCLEXP : OCL-Exp .

op mte : @Model -> TimeInf [frozen] .
op mteAux : @Model @Model -> TimeInf [frozen] .
---- (mte = 0) equation of the SwitchOn rule
ceq mte( MM@ { < p : PHONE | on : ON@p@ATT, p@SFS >
               < CLK@ : Clock | time : @TIME@ >
               < CNT@ : Counter | value : CNT@ >
               OBJSET@ })
  = 0
  if MODEL@ :=  MM@ { < p : PHONE | on : ON@p@ATT, p@SFS > ... OBJSET@ }
  /\ ON@p@ATT := eval(false, (ctx(self, p)), MODEL@)
  /\ eval(p . battery > 10 * p . bps, empty, MODEL@)
  /\ not currentExec@SwitchOn(Set{p}, MODEL@) .
---- Remaining (mte = 0) equations ...
eq mte(MODEL@) = mteAux(MODEL@, MODEL@) [owise] .
eq mteAux(MM@ { < O : AtomicActionExec | timer : T,
                  endingTime : undefined, SFS) > OBJSET@ }, MODEL@)
  = minimum(T, mteAux(MM@ { OBJSET@ }, MODEL@)) .
eq mteAux(MM@ { < O : OngoingActionExec | maxDuration : freeze(OCLEXP),
                  endingTime : undefined, upperBoundTimer : T', SFS > OBJSET@ },
        MODEL@)
```

```
  = minimum(toRat(eval(OCLEXP, empty, MODEL@)),
            minimum(T', mteAux(MM@ { OBJSET@ }, MODEL@))) .
eq mteAux(MM@ { < O : Clock | time : T > OBJSET@ }, MODEL@)
  = minimum(minimum(nextLowerBound(T, rulesInformation),
                    nextPeriod(T, rulesInformation)) monus T,
           mteAux(MM@ { OBJSET@ }, MODEL@)) .
eq mteAux(MODEL@, MODEL@') = INF [owise] .
```

The rulesInformation constant is defined as a model that gathers all the rule properties, such as their periodicity and lower and upper lower bounds. Operations nextPeriod and nextLowerBound make use of it to compute the following beginning of rule's period and lower bound, forcing time to stop in these moments in time. In this way, we allow periodic and time-bounded rules to be applied in their corresponding interval and period, respectively. Additionally, the triggering rules of actions of these kinds will also include conditions to forbid several applications of the same rule with the same participants in the same period (in case of periodic rules) or to be triggered out of its interval of time (in case of time-bounded rules).

Analysis and Simulation. Once the specification of our system is encoded in Real-Time Maude, what we get is a rewriting logic specification of such a system. Since the rewriting logic specification produced is executable, this specification can be used as a prototype of the system, which allows us to simulate and analyze it.

Our model encoding enables, e.g., the use of Real-Time Maude's model simulation, reachability analysis and model checking tools. These tools are the timed versions of Maude's rewriting, search, and model-checking commands (see, e.g., [16,12,18] for examples of the kinds of analyses that can be accomplished on Real-Time Maude and on models and metamodels like the ones considered here). In particular, they extend them to consider the non-deterministic time advance, and to allow to, e.g., include time bounds in the analysis and simulation.

For instance, the Real-Time Maude tsearch command allows us to explore (following a breadth-first strategy up to a specified bound) the reachable state space in a certain time interval from an initial model. This command is useful, e.g., to check safety properties. For example, given variables O of sort Oid, BAT of sort Int, SFS of sort Set{@StructuralFeatureInstance} and OBJSET of sort Set{@Object}, we can check whether, starting from initModel (the model depicted in Fig. 1 (b)), the battery power of any cell phone is decreased below zero:

```
(tsearch {initModel} =>* {MPN_MM { < O : Phone | battery : BAT, SFS > OBJSET }}
  such that BAT < 0 in time < 200 .)

No solution.
```

Although all phones will probably run out of battery before 200 time units, since there is a periodic action the system can run for ever if we do not bound the search. Since no solutions are found, we can state that (starting from initModel, and in 100 time units) the battery power of the phones are never decreased below zero.

We refer the interested reader to [13,12] for details on Real-Time Maude analysis tools, and to [15,18] for examples of use on models as the ones presented here.

4 Tool Support

The representation of the *e-Motions* behavioral specifications in Real-Time Maude provide their precise semantics. However, it is unrealistic to think that average system modelers will write such Real-Time Maude specifications. What we have defined is a mapping between the *e-Motions* and the Real-Time Maude metamodels (i.e., a *semantic mapping* between these two semantic domains) that realizes the automatic generation of the Real-Time Maude specifications corresponding to a DSVL defined with *e-Motions*.

Such mapping to Real-Time Maude has been defined and implemented by means of a set of ATL [22] transformations. In particular, we have specified three model transformations to encode (EMF) models, (EMF) metamodels, and *Behavior* models (conforming to the *e-Motions Behavior* metamodel [15]). For this purpose, we adapted the metamodel of Maude in [14] to cover Real-Time Maude specifications. The tool and the ATL transformations can be downloaded from the *e-Motions* website [1].

At this moment, the simulation and analysis of the DSVL models needs to be performed in the Real-Time Maude environment. We are currently working on the integration of Real-Time Maude analysis tools into the *e-Motions* environment, so that a system modeler can perform the simulation and formal analysis of the visual models inside such environment.

5 Related Work

Maude has already been proposed as a formal notation and environment for specifying and effectively analyzing models and metamodels [20,17,2]. Simulation, reachability and model-checking analysis are then possible on the models using the tools and techniques provided by Maude. In [16] we showed how Maude is also suitable as a semantic domain for standard in-place rules, formalizing graph transformations using rewriting logic. In this paper we have shown how Real-Time Maude can provide a target semantic domain for providing semantics to real-time domain specific visual languages whose behavior is expressed in terms of in-place rules extended with time properties.

There are several approaches that propose in-place model transformations to deal with the behavior of DSVLs, from textual to graphical. Furthermore, several formalizations of graph transformation to perform different kinds of system analysis have been proposed (see [16] for a discussion on these topics). However, none of them includes a quantitative model of time. When time is needed, it is usually modeled in an intrusive way, by adding clocks or timers to the DSVL metamodel. This is, for example, the approach followed in [10], where graph transformation systems are provided with a model of time by representing logical clocks as a special kind of node attributes.

Syriani and Vangheluwe propose in [21] to complement graph grammar rules with the Discrete EVent system Specification (DEVS) formalism to model time-dependent behavior. Although it allows modular designs, this approach requires specialized knowledge and expertise on the DEVS formalism. Furthermore, they do not provide analysis capabilities: system evaluation is accomplished through simulation.

Real-Time Fujaba [5] aims at supporting the model-driven development of correct software for safety-critical, networked, real-time systems. A restricted UML model serves as the basis for model checking. The tool supports the modeling of the system structure by means of UML component diagrams, and the modeling of real-time behavior by means of real-time extended UML state machines.

Boronat and Ölveczky have recently proposed in [3] a collection of built-in timed constructs for defining the timed behavior of model-based systems that are specified with in-place model transformations. These timed constructs can be added to the DSVL metamodel itself, or separately defined in another metamodel (in a non-intrusive way). They also formalize in-place model transformations into Real-Time Maude. In fact, the model of time they use can be considered as a straightforward translation from the Real-Time Maude model of time: opposite to our approach, in which in-place rules are extended with time-related constructs and then transparently encoded in Real-Time Maude with timer objects, they propose handling these timers directly in the model transformation.

6 Conclusions and Future Work

In a previous work [15] we showed how some timed behavioral specifications can be supported, extending in-place rules with a quantitative model of time and with mechanisms that allow designers to state action properties, easing the design of real-time complex systems. In this paper we have shown how it is possible to provide a formal semantics to our visual notation, using a mapping from this timed-dependent behavioral specifications to Real-Time Maude specifications. We are then able to perform the same kind of analysis we were able to perform for time-unaware systems [18,20]. Such an encoding in Maude can be useful to other DSVLs, which can make use of it by simply providing model transformations from their models to it. This mapping will help providing these languages with precise semantics, and also gaining access to Maude's formal environment.

We are currently working on further extensions of our graphical tool to automate the interaction with Real-Time Maude and its analysis tools using the native visual notation of the DSVL. This will make the use of Real-Time Maude completely transparent to users.

Acknowledgements. The authors would like to thank the anonymous referees for their insightful comments and very constructive suggestions. This work has been supported by Spanish Research Projects TIN2008-03107 and P07-TIC-03184.

References

1. Atenea group. The e-Motions tool (2009), `http://atenea.lcc.uma.es/index.php/Main_Page/Resources/E-motions`
2. Boronat, A., Meseguer, J.: An algebraic semantics for MOF. In: Fiadeiro, J.L., Inverardi, P. (eds.) FASE 2008. LNCS, vol. 4961, pp. 377–391. Springer, Heidelberg (2008)
3. Boronat, A., Ölveczky, P.C.: Formal real-time model transformations in MOMENT2. In: Rosenblum, D.S., Taentzer, G. (eds.) FASE 2010. LNCS, vol. 6013, pp. 29–43. Springer, Heidelberg (2010)

4. Bouhoula, A., Jouannaud, J.-P., Meseguer, J.: Specification and proof in membership equational logic. Theoretical Computer Science 236(1), 35–132 (2000)
5. Burmester, S., Giese, H., Hirsch, M., Schilling, D., Tichy, M.: The Fujaba real-time tool suite: model-driven development of safety-critical, real-time systems. In: Proc. of ICSE 2005, pp. 670–671. ACM, New York (2005)
6. Clark, T., Sammut, P., Willans, J.: Applied Metamodelling, Ceteva, 2nd edn. (2004)
7. Clavel, M., Durán, F., Eker, S., Lincoln, P., Martí-Oliet, N., Meseguer, J., Talcott, C.: All About Maude - A High-Performance Logical Framework. LNCS, vol. 4350. Springer, Heidelberg (2007)
8. Czarnecki, K., Helsen, S.: Classification of model transformation approaches. In: OOPSLA 2003 Workshop on Generative Techniques in the Context of MDA (2003)
9. Eker, S., Martí-Oliet, N., Meseguer, J., Verdejo, A.: Deduction, strategies, and rewriting. Electron. Notes Theor. Comput. Sci. 174(11), 3–25 (2007)
10. Gyapay, S., Heckel, R., Varró, D.: Graph transformation with time: Causality and logical clocks. In: Corradini, A., Ehrig, H., Kreowski, H.-J., Rozenberg, G. (eds.) ICGT 2002. LNCS, vol. 2505, pp. 120–134. Springer, Heidelberg (2002)
11. Meseguer, J.: Conditional rewriting logic as a unified model of concurrency. Theoretical Computer Science 96(1), 73–155 (1992)
12. Ölveczky, P.C.: Real-Time Maude 2.3 Manual (2007), `http://www.ifi.uio.no/RealTimeMaude/`
13. Ölveczky, P.C., Meseguer, J.: Semantics and pragmatics of Real-Time Maude. Higher-Order and Symbolic Computation 20(1-2), 161–196 (2007)
14. Rivera, J.E., Durán, F., Vallecillo, A.: A metamodel for maude. Technical report, University of Málaga (2008), `http://atenea.lcc.uma.es/images/e/e0/MaudeMM.pdf`
15. Rivera, J.E., Durán, F., Vallecillo, A.: A graphical approach for modeling time-dependent behavior of DSLs. In: Proc. of VL/HCC 2009. IEEE Computer Society, Los Alamitos (2009)
16. Rivera, J.E., Guerra, E., de Lara, J., Vallecillo, A.: Analyzing rule-based behavioral semantics of visual modeling languages with Maude. In: Gašević, D., Lämmel, R., Van Wyk, E. (eds.) SLE 2008. LNCS, vol. 5452, pp. 54–73. Springer, Heidelberg (2009)
17. Rivera, J.E., Vallecillo, A.: Adding behavioral semantics to models. In: Proc. of EDOC 2007, pp. 169–180. IEEE Computer Society, Los Alamitos (October 2007)
18. Rivera, J.E., Vallecillo, A., Durán, F.: Formal specification and analysis of domain specific languages using Maude. Simulation: Transactions of the Society for Modeling and Simulation International 85(11/12), 778–792 (2009)
19. Roldán, M., Durán, F.: Representing UML models in mOdCL (2008) (manuscript), `http://maude.lcc.uma.es/mOdCL`
20. Romero, J.R., Rivera, J.E., Durán, F., Vallecillo, A.: Formal and tool support for model driven engineering with Maude. Journal of Object Technology 6(9), 187–207 (2007)
21. Syriani, E., Vangheluwe, H.: Programmed graph rewriting with time for simulation-based design. In: Vallecillo, A., Gray, J., Pierantonio, A. (eds.) ICMT 2008. LNCS, vol. 5063, pp. 91–106. Springer, Heidelberg (2008)
22. The AtlanMod Team. ATL, `http://www.eclipse.org/m2m/atl/doc/`

Multiset Rewriting: A Semantic Framework for Concurrency with Name Binding*

Fernando Rosa-Velardo

Dpto. de Sistemas Informáticos y Computación
Universidad Complutense de Madrid
fernandorosa@sip.ucm.es

Abstract. We revise multiset rewriting with name binding, by combining the two main existing approaches to the study of concurrency by means of multiset rewriting: multiset rewriting with existential quantification and constrained multiset rewriting. We obtain ν-MSRs, where we rewrite multisets of atomic formulae, in which some names may be restricted. We prove that ν-MSRs are equivalent to a class of Petri nets in which tokens are tuples of pure names, called $p\nu$-APNs. Then we encode π-calculus processes into ν-MSRs in a very direct way, that preserves the topology of bound names, by using the concept of derivatives of a π-calculus process. Finally, we discuss how the recent results on decidable subclasses of the π-calculus are independent of the particular reaction rule of the π-calculus, so that they can be obtained in the more general framework of ν-MSRs. Thus, those results carry over not only to the π-calculus, but to any other formalism that can be encoded within it, as $p\nu$-APNs.

1 Introduction

Dynamic name generation has been thoroughly studied in the last decade, mainly in the field of security [9,1] and mobility [16]. The paper [9] presents a meta-notation for the specification and analysis of security protocols. This meta-notation involves facts and transitions, where facts are first-order atomic formulae and transitions are given by means of rewriting rules, with a precondition and a postcondition. For instance, the rule

$$A_0(k), Ann(k') \rightarrow \exists x.(A_1(k,x), N(enc(k', \langle x, k\rangle)), Ann(k'))$$

specifies the first rule of the Needham-Schroeder protocol, in which a principal A with key k ($A_0(k)$) decides to talk to another principal, with a key k' that has been announced ($Ann(k')$), for which it creates a nonce x and sends to the network the pair $\langle x, k\rangle$ ciphered under k'. This notation gave rise to the specification language for security protocols MSR [8].

* Work partially supported by the Spanish projects DESAFIOS10 TIN2009-14599-C03-01 and PROMETIDOS S2009/TIC-1465.

P.C. Ölveczky (Ed.): WRLA 2010, LNCS 6381, pp. 191–207, 2010.

In [12] *Constraint Multiset Rewriting Systems* (CMRS) are defined. As in [9], facts are first-order atomic formulae, but the terms that can appear as part of such formulae must belong to a *constraint system*. For instance, the rule $count(x), visit \rightarrow count(x+1), enter(x+1)$ could be used to count the number of visits to a web site. For a comprehensive survey of CMRS see [13]. In CMRS, there is no mechanism for name binding or name creation, so that it has to be simulated using the order in the constraint system (for instance, simulating the creation of a fresh name by taking a value greater than any of the values that have appeared so far). Thus, in an unordered version of CMRS, in which only the equality predicate between atoms is used, there is no way of ensuring that a name is fresh.

It is our goal in this paper to find a minimal set of primitives that allow us to specify concurrent formalisms with name binding. This specification may be achieved by means of some encoding, provided this encoding preserves concurrency and name topology. Let us remark that our goal is therefore different to the one expressed in [18]:

The goal (...) is to express as faithfully as possible a very wide range of concurrency models, each on its own terms, avoiding any encodings or translations.

We combine the features of the meta-notation in [9] and CMRS, obtaining ν-MSRs. On the one hand, we maintain the existential quantifications in [9] to keep a compositional approach, closer to that followed in process algebra with name binding. On the other hand, we restrict terms in atomic formulae to be pure names, that can only be compared with equality or inequality, unlike the arbitrary terms over some syntax, as in [9], or terms in a constraint system, as in CMRS.

The formalism obtained can be seen as a particular instance of the Chemical Abstract Machine [4], in which a configuration is given by a multiset of molecules, atomic formulae in our case. In the terminology of [4], the heating reactions in ν-MSRs are given by a structural congruence that, essentially, deals with name binding, as is usual in process algebras with name binding.

Two of the most well established models for concurrency are Petri nets and process algebra. The π-calculus is the paradigmatic example of process algebra with name binding. Names in the π-calculus can be used to build a dynamic communication topology. To our knowledge, the only approach to dynamic name generation in the field of Petri nets are the ν-APNs [23] and Data Nets [17]. In ν-APNs, tokens are pure names that can move along the places of the net, be used to restrict the firing of transitions to happen only when some names match, and be created fresh. ν-APNs are Well Structured Transition Systems (WSTS) [25,14], but $p\nu$-APNs, its polyadic version, in which tokens are *tuples* of pure names, are not. Actually, $p\nu$-APNs are Turing-complete [24], even if restricted to the binary case, in which tokens are just pairs of names. In Data Nets, tokens are taken from a linearly ordered and dense domain, and whole-place operations (like transfers or resets) are allowed. However, in Data Nets (which are also WSTS), fresh name creation has to be simulated using the linear order, as happens in CMRS. Actually, the paper [2] proves that CMRS and Data Nets are equivalent, even if the former cannot perform whole-place operations (using a language-based comparison, where the criterion for accepting words is a coverability one, instead of the more standard reachability criterion).

We will first prove that ν-MSRs are equivalent to $p\nu$-APNs. We will see that this equivalence is a rather strong one (isomorphism between the transitions systems). As an immediate consequence, we obtain Turing-completeness of ν-MSRs. Moreover, the subclass of monadic ν-MSRs, that are equivalent to ν-APNs, are WSTS, so that coverability, boundedness and termination are decidable for them.

Next, we will see that processes of the π-calculus can be simulated, in a very natural way, by ν-MSRs. This translation is inspired by the results by Meyer about *structural stationary* π-calculus processes, that can be mapped to P/T nets [21].

The search for a subclass of the π-calculus in which some interesting properties (like reachability or termination) that are undecidable in the general model become decidable, is an active field of research [5,21,20,22,3]. Usually, decidability of such properties is achieved by mapping the considered subclass to Petri Nets, or some extension of Petri nets. Most of the restrictions considered in the π-calculus, are restrictions on the dynamic topology of names in all reachable markings. We claim that these properties are independent of the particular reaction rule of the π-calculus, so that they can be specified in ν-MSRs, obtaining the analogous results in a more general framework. Therefore, those results carry over to other formalisms that can be encoded within ν-MSR, as $p\nu$-APNs. As an example, depth-boundedness in π-calculus processes (boundedness of the interdependence of names) can also be defined for ν-MSR terms. Moreover, the proof of Well Structuredness of depth-bounded processes carries over to depth-bounded ν-MSR terms so that, as a corollary, we know that depth-bounded $p\nu$-APNs are also WSTS.

The rest of the paper is organized as follows. Section 2 defines ν-MSRs. In Section 3 the equivalence between ν-MSRs and $p\nu$-APNs is proved. Section 4 presents the encoding of π-calculus terms within ν-MSR. In Section 5 we briefly show how ν-MSRs can be specified using Maude in a straightforward manner. Finally, Section 6 presents our conclusions and some directions for future work.

2 ν-MSRs

We fix a finite set of predicate symbols $\mathcal{P}$, a denumerable set Id of names and a denumerable set Var of variables. We use $a, b, c, \ldots$ to range over Id, $x, y, \ldots$ to range over Var, and $\eta, \eta' \ldots$ to range over $Id \cup Var$.

An atomic formula over $\mathcal{P}$ and Var has the form $p(\eta_1, \ldots, \eta_n)$, where $p \in \mathcal{P}$ and $\eta_i \in Var \cup Id$ for all i. A ground atomic formula has the form $p(a_1, \ldots, a_n)$, where $p \in \mathcal{P}$ and $a_i \in Id$ for all i. We use $X, Y, \ldots$ to range over atomic formulae and $A, B, \ldots$ to range over atomic ground formulae. We denote by $Var(X)$ and $Id(X)$ the set of variables and names appearing in X, respectively. We will write

$\tilde{x}$ and $\tilde{a}$ to denote finite sequences of variables and names, respectively, so that we will sometimes write $p(\tilde{x})$ or $p(\tilde{a})$. Moreover, we will sometimes use set notation with these sequences and write, for instance, $x \in \tilde{x}$ or $\tilde{x}_1 \cup \tilde{x}_2$.

Definition 1. *A ν-MSR term is given by the following grammar:*

$$M ::= \mathbf{0} \mid A \mid M_1 + M_2 \mid \nu a.M$$

We denote by $\mathcal{M}$ the set of ν-MSR terms, and use M, M', $M_1, \ldots$ to range over $\mathcal{M}$. We define $fn : \mathcal{M} \to \mathcal{P}(Var)$ as $fn(\mathbf{0}) = \emptyset$, $fn(A) = Id(A)$, $fn(M_1+M_2) = fn(M_1) \cup fn(M_2)$, and $fn(\nu a.M) = fn(M) \setminus \{a\}$.

Definition 2. *A rule t is an expression of the form*

$$t : X_1 + \ldots + X_n \to \nu\tilde{a}.(Y_1 + \ldots + Y_m)$$

such that if $x \in Var(Y_j)$ for some j then $x \in Var(X_i)$ for some i. A ν-MSR is a tuple $\langle \mathcal{R}, M_0 \rangle$, where M_0 is the initial ν-MSR term and $\mathcal{R}$ is a finite set of rules.

In examples, we will use commas instead of the symbol $+$. For instance, we will write $p(x,y), q(y,y) \to \nu a.q(x,a)$ instead of $p(x,y) + q(y,y) \to \nu a.q(x,a)$. For a rule $t : X_1, \ldots, X_n \to \nu\tilde{a}.(Y_1, \ldots, Y_m)$ we write $pre(t) = \bigcup_{i=1}^{n} Var(X_i)$, $post(t) = \bigcup_{j=1}^{m} Var(Y_m)$, and $Var(t) = pre(t) \cup post(t)$. With these notations, every rule t satisfies $post(t) \subseteq pre(t)$.

We will identify ν-MSR terms up to $\equiv$, defined as the least congruence on $\mathcal{M}$ where α-conversion of bound names is allowed, such that $(\mathcal{M}, +, \mathbf{0})$ is a commutative monoid and:

$$\nu a.\nu b.M \equiv \nu b.\nu a.M \qquad \nu a.\mathbf{0} \equiv \mathbf{0}$$

$$\nu a.(M_1 + M_2) \equiv \nu a.M_1 + M_2 \quad \textit{if} \quad a \notin fn(M_2)$$

The first rule justifies our notation $\nu\tilde{a}.M$. The last rule is usually called name extrusion when applied from right to left. A *mode* for $t : X_1 + \ldots + X_n \to \nu\tilde{a}.(Y_1 + \ldots + Y_m)$ is any substitution $\sigma : Var(t) \to Id$. We use the term mode for analogy with the modes in high level Petri nets, since a rewrite rule can be applied in different modes. We write $pre_t(\sigma) = \sigma(X_1) + \ldots + \sigma(X_n)$, where $\sigma(p(\eta_1, \ldots, \eta_n)) = p(a_1, \ldots, a_n)$, with $a_i = \sigma(\eta_i)$ if $\eta_i \in Var$, or $a_i = \eta_i$ if $\eta_i \in Id$.

In order to define the analogous $post_t(\sigma)$, and to avoid capturing free names, we consider a sequence of pairwise different names $\tilde{b}$ (of the same length as $\tilde{a}$) such that $\sigma(Var(t)) \cap \tilde{b} = \emptyset$. Then, we take $\sigma' = \sigma \circ \{\tilde{a}/\tilde{b}\}$ and $post_t(\sigma) = \nu\tilde{b}.(\sigma'(Y_1) + \ldots + \sigma'(Y_m))$, where $\{\tilde{a}/\tilde{b}\}$ denotes the simultaneous substitution of each $a_i \in \tilde{a}$ by the corresponding $b_i \in \tilde{b}$. Let us define the transition system $(\mathcal{M}, \to, M_0)$, where $\to$ is the least relation such that:

$$(t)\ \frac{\sigma \textit{ mode for } t}{pre_t(\sigma) \to post_t(\sigma)} \qquad \frac{M_1 \equiv M_1' \to M_2' \equiv M_2}{M_1 \to M_2}\ (\equiv)$$

$$(+)\ \frac{M_1 \to M_2}{M_1 + M \to M_2 + M} \qquad \frac{M_1 \to M_2}{\nu a.M_1 \to \nu a.M_2}\ (\nu)$$

Rules $(+)$ and (ν) state that transitions can happen inside a sum or inside a restriction, respectively. Rule $(\equiv)$ is also standard, and formalizes that we are rewriting terms modulo $\equiv$. Then we have a rule schema (t) for each $t \in \mathcal{R}$. For instance, let $t : p(x), q(x) \to \nu b.p(b)$ be a rule in $\mathcal{R}$. Then the rewriting $p(a), q(a) \to \nu b.p(b)$ can take place by taking $\sigma(x) = a$, which satisfies the conditions for modes and $pre_t(\sigma) = p(a), q(a)$ and $post_t(\sigma) = \nu b.p(b)$. Consider now the term $p(b), q(b)$. In order to apply the previous rule, one must necessarily consider the mode given by $\sigma(x) = b$, that does not satisfy $\sigma(Var(t)) \cap \{b\} = \emptyset$. Therefore, we need to first rename b in the right handside of the rule, obtaining (e.g. if we replace b by a) $\nu a.p(a)$.

As in the π-calculus, we can consider several normal forms, that force a certain rearrangement of bound names.

Definition 3. *A term M is in* standard normal form *if there is a set of names $\tilde{a}$ and atomic formulae $A_1, \ldots, A_n$ such that $M = \nu\tilde{a}.(A_1 + \ldots + A_n)$.*

Clearly, every term is equivalent to some term in standard form. To obtain it, it is enough to apply the extrusion rule (from right to left) as much as necessary. The standard form is unique up to commutativity and associativity of $+$, and α-conversion and commutativity of the names in $\tilde{a}$. Moreover, we can prove the following result, that relates the transition relation with the standard normal form.

Proposition 1. *$M_1 \to M_2$ iff $M_i \equiv \nu\tilde{a}_i.(A^i_1 + \ldots + A^i_{n_i} + M)$ for $i = 1, 2$, and there is $t : X^1_1 + \ldots + X^1_{n_1} \to \nu\tilde{a}.(X^2_1 + \ldots + X^2_{n_2})$ in $\mathcal{R}$, σ mode for t and $\tilde{b}$ with $\sigma(Var(t)) \cap \tilde{b} = \emptyset$ such that $\sigma(X^1_j) = A^1_j$, $\sigma(X^2_j)\{\tilde{a}/\tilde{b}\} = A^2_j$ and $\tilde{a}_1 \sqcup \tilde{b} = \tilde{a}_2$.*

Proof. We prove the *if* implication by induction on the rules proving $M_1 \to M_2$.

- If $M_1 = pre_t(\sigma)$ and $M_2 = post_t(\sigma)$ for some rule t and some mode σ for t, then trivially both M_1 and M_2 are in standard form, and $\tilde{a}_1 = \emptyset$ and $\tilde{a}_2 = \tilde{b}$, so that clearly $\tilde{a}_1 \sqcup \tilde{b} = \tilde{a}_2$.
- Let $M_i = M' + M'_i$ with $M'_1 \to M'_2$, so that by the induction hypothesis, $M'_i \equiv \nu\tilde{a}_i.(A^i_1 + \ldots + A^i_{n_i} + M)$ and $\tilde{a}_1 \sqcup \tilde{b} = \tilde{a}_2$. We assume $fn(M') \cap \tilde{a}_i = \emptyset$, or we rename the names in $\tilde{a}_i$ that are free in M, obtaining a term that is equivalent modulo $\equiv$. Let $M \equiv \tilde{c}.M''$ in standard form. As before, we can assume that $\tilde{a}_i \cap \tilde{c} = \emptyset$. Then, $M_i = M'_i + M' \equiv \nu\tilde{a}_i.(A^i_1 + \ldots + A^i_{n_i} + M) + \nu\tilde{c}.M''$, which by the extrusion rule is equivalent to $\nu\tilde{a}_i, \tilde{c}.(A^i_1 + \ldots + A^i_{n_i} + M + M'')$. Moreover, $\tilde{a}_2 \sqcup \tilde{c} = \tilde{a}_1 \sqcup \tilde{b} \sqcup \tilde{c}$.
- The cases for (ν) and $(\equiv)$ are straightforward.

Conversely, $A^1_1 + \ldots + A^i_{n_1} \to \nu\tilde{b}.(A^2_1 + \ldots + A^2_{n_2})$ holds by rule (t). Rules $(+)$ and $(\equiv)$ for the extrusion, tells us that $A^1_1 + \ldots + A^i_{n_1} + M \to \nu\tilde{b}.(A^2_1 + \ldots + A^2_{n_2} + M)$, and by successively applying rule (ν) for all the names in $\tilde{a}_1$, we obtain that $\nu\tilde{a}_1.(A^1_1 + \ldots + A^1_{n_1} + M^1) \to \nu\tilde{a}_2.(A^2_1 + \ldots + A^2_{n_2} + M^2)$. Finally, again by rule $(\equiv)$ we can conclude that $M_1 \to M_2$.

Let us now define in our setting the restricted normal form of a term, which can be seen as the opposite concept to standard form. Intuitively, a term is in restricted form if the scope of its restrictions is minimal, that is, if every expression $\nu a.(A_1 + \ldots + A_m)$ satisfies $a \in fn(A_i)$ for all i, so that no extrusion rule can be applied from left to right.

Definition 4. *Let us define $\hat{\equiv}$ as the least congruence on $\mathcal{M}$ such that $+$ is commutative and associative with* **0** *as identity, and $\rightsquigarrow$ as the least binary relation on $\mathcal{M}$ such that:*

$$\frac{a \notin fn(M_2)}{\nu a.(M_1 + M_2) \rightsquigarrow \nu a.M_1 + M_2} \qquad \frac{M_1 \hat{\equiv} M_1' \rightsquigarrow M_2' \hat{\equiv} M_2}{M_1 \rightsquigarrow M_2}$$

$$\frac{M_1 \rightsquigarrow M_2}{M_1 + M \rightsquigarrow M_2 + M} \qquad \frac{M_1 \rightsquigarrow M_2}{\nu a.M_1 \rightsquigarrow \nu a.M_2}$$

We say M is in restricted form if there is no M' with $M \rightsquigarrow M'$.

The relation $\rightsquigarrow$ is confluent, up to $\hat{\equiv}$. Moreover, if $M \rightsquigarrow M'$ then $M \equiv M'$. We do not have a result analogous to Prop. 1, that is, the restricted normal form is not compatible with the transition relation. However, restricted forms give more insight about the topology of pure names. In particular, they are the basis of the proof that depth-bounded ν-MSR terms yield WSTS.

As in [20], we can use the restricted form to define the so called *fragments* of a term. We say a marking M in restricted form is a *fragment* if it cannot be decomposed as $M = M_1 + M_2$. Obviously, any M in restricted form satisfies $M = F_1 + \ldots + F_n$ with F_i fragments. Intuitively, within a fragment some bound names are shared. Let us consider the following hypergraph interpretation of a ν-MSR term. Given a term M, we consider the hypergraph whose nodes are the atomic formulae in M, and an edge between two such formulae labelled by a name when they share that name. Then, the fragments of M correspond to the connected components of its hypergraph. When the process is depth-bounded, the paths in the hypergraph are also bounded, which can be used to endow depth-bounded ν-MSR terms with a well-structure.

3 ν-MSRs and $p\nu$-APNs

A $p\nu$-APN is a Petri net in which tokens are tuples of pure names. Arcs are labelled by tuples of variables (or multiset of such tuples, if we allow weights) that specify how tokens flow from preconditions to postconditions. Variables are taken from a set Var. Some of the variable in postarcs can be in the set of special variables $\Upsilon \subset Var$ that can only be instantiated to names that do not occur in the current marking, thus creating fresh names. We use ν, ν', $\nu_1, \ldots$ to range over Υ. We take $\mathcal{L} = \bigcup_{i>0} Var^i$, that is, the set of tuples of variables of arbitrary length. We will sometimes use set notation for tuples, so that we will write, for instance, $x \in (x, y)$. Moreover, we will use an arbitrary set Id of names.

Given an arbitrary set A, we will denote by $\mathcal{MS}(A)$ the set of finite multisets of A, that is, the set of mappings $m : A \to \mathbb{N}$ such that the set $S(m) = \{a \in$

$A \mid m(a) > 0\}$ (called support of m) is finite. We denote by $m_1 + m_2$, $m_1 \subseteq m_2$ and $m_1 - m_2$ the multiset addition, inclusion, and substraction, respectively. Given $f : A \rightarrow B$ and $m \in \mathcal{MS}(A)$ then we can define $f(m) \in \mathcal{MS}(B)$ by $f(m)(b) = \sum_{f(a)=b} m(a)$.

Definition 5. *A pν-APN is a tuple $N = (P, T, F)$, where P and T are finite disjoint sets of elements called places and transitions, respectively,*

$$F : (P \times T) \cup (T \times P) \rightarrow \mathcal{MS}(\mathcal{L})$$

is such that for every $t \in T$, $pre(t) \cap \Upsilon = \emptyset$, and $post(t) \setminus \Upsilon \subseteq pre(t)$, where $pre(t) = \bigcup_{p \in P} S(F(p,t))$, $post(t) = \bigcup_{p \in P} S(F(t,p))$ and $Var(t) = pre(t) \cup post(t)$.

Let us denote by $\mathcal{T}$ the set of tuples of names of arbitrary length, that is, $\mathcal{T} = \bigcup_{i>0} Id^i$. The tokens of a pν-APN are taken from $\mathcal{T}$. We will use φ, φ', $\varphi_1, \ldots$ to range over tokens.

Definition 6. *A marking of a pν-APN $N = (P, T, F)$ is any $M : P \rightarrow \mathcal{MS}(\mathcal{T})$.*

We define $Id(M) = \{a \in Id \mid \text{there are } p \in P \text{ and } \varphi \text{ st } a \in \varphi \in M(p)\} \subset Id$, the set of all the names appearing in some token in some place, according to the marking M.

Transitions are fired with respect to a mode, that chooses which tokens are taken from preconditions and which are put in postconditions. Given a transition t of a net N, a mode of t is a mapping $\sigma : Var(t) \rightarrow Id$, that instantiates each variable involved in the firing of t to an identifier. We will use $\sigma, \sigma', \sigma_1 \ldots$ to range over modes. We extend modes to tuples of variables by taking $\sigma((x_1, \ldots, x_n)) = (\sigma(x_1), \ldots, \sigma(x_n))$.

Definition 7. *Let N be a pν-APN, M a marking of N, t a transition of N and σ a mode of t. We say t is enabled with mode σ if $\sigma(\nu) \notin Id(M)$ for all $\nu \in Var(t) \cap \Upsilon$, and $\sigma(F(p,t)) \subseteq M(p)$ for all $p \in P$. The reached state of N after the firing of t with mode σ is the marking M', given by*

$$M'(p) = (M(p) - \sigma(F(p,t))) + \sigma(F(t,p)) \quad \forall p \in P$$

We will write $M \xrightarrow{t(\sigma)} M'$ if M' is reached from M when t is fired with mode σ. We also define the relations $\longrightarrow$ and $\longrightarrow^*$, as usual. Fig. 1 depicts a simple example of a pν-APN and the firing of its only transition. Notice that the transition can be fired because the second component of the pair (a, b) in p_1 matches the name in p_2, as demanded by the labels in the arcs.

In order to capture the intuition that the names in Id are pure, we work modulo $\equiv_\alpha$, which allows consistent renaming of names in markings. Accordingly, the order $\sqsubseteq_\alpha$ that induces coverability for pν-APNs is defined as follows: $M \sqsubseteq_\alpha M'$ if there is an injection $\iota : Id(M) \rightarrow Id(M')$ such that for every place $p \in P$, $\iota(M(p)) \subseteq M'(p)$, where $\iota((a_1, \ldots, a_n)) = (\iota(a_1), \ldots, \iota(a_n))$.

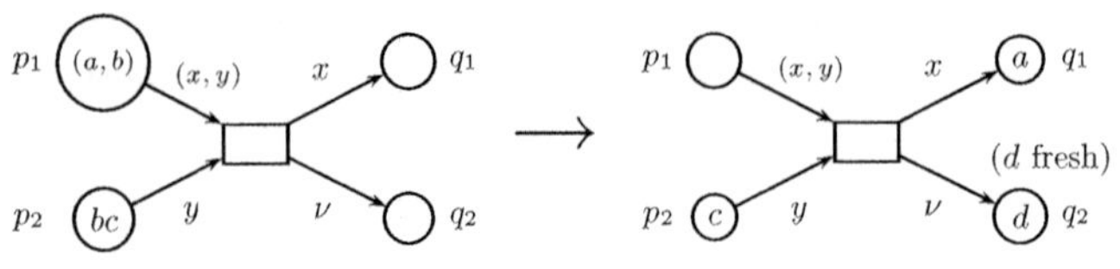

Fig. 1. A simple $p\nu$-APN

Proposition 2. *For any $p\nu$-APN N there is a ν-MSR $F(N)$ (with F computable) such that N and $F(N)$ are isomorphic (as transition systems).*

Proof. Let $N = (P, T, F, M_0)$ be a $p\nu$-APN. For every $t \in T$, if $\tilde{\nu}$ is a sequence formed by the special variables in postarcs of t, let us take any sequence (of the same length) of arbitrary names $\tilde{a}$, and let us define the rule

$$F(t) : \sum_{p \in P} \sum_{\tilde{x} \in F(p,t)} p(\tilde{x}) \rightarrow \nu\tilde{a}. \sum_{p \in P} \sum_{\tilde{x} \in F(t,p)} p(\tilde{x}\{\tilde{a}/\tilde{\nu}\})$$

For every marking M with $\tilde{b} = Id(M)$, we define M^* as the ν-MSR term $\nu\tilde{b}.(\sum_{p \in P} \sum_{\tilde{a}_i \in M(p)} p(\tilde{a}_i))$. Then, we define $\mathcal{R} = \{F(t) \mid t \in T\}$ and $F(N) = \langle \mathcal{R}, M_0^* \rangle$. For two markings M_1 and M_2 with $M_1 \rightarrow M_2$, it holds that $M_1^* \rightarrow M_2^*$. On the other hand, for two ν-MSR terms M_1 and M_2 such that $M_1 \rightarrow M_2$, Prop. 1 tells us that $M_i \equiv \nu\tilde{a}_i(A_1^i + \ldots + A_{n_i}^i + M)$, so that M_i is equivalent to a $M_i'^*$ for some markings M_i' and M_2'. Moreover, $M_1' \rightarrow M_2'$ and the thesis follows.

For instance, consider the $p\nu$-APN in Fig. 1. The previous construction yields the ν-MSR given by the rule $t : p_1(x, y), p_2(y) \rightarrow \nu a.(q_1(x), q_2(a))$. The initial marking is represented by the term $\nu a, b.(p_1(a, b), p_2(b))$, which evolves to $\nu a, c.(q_1(a), q_2(c))$.

Therefore, $p\nu$-APNs can be just thought of as a graphical representation of ν-MSRs. However, since $p\nu$-APNs lack a name binding operator, intuitively they always work with terms in their standard normal form. Indeed, for a marking M, the term M^* is in standard form. Let us now prove the converse result.

Proposition 3. *For any ν-MSR S there is a $p\nu$-APN $G(S)$ (with G computable) such that S and $G(S)$ are isomorphic (as transition systems).*

Proof. Let $S = \langle \mathcal{R}, M_0 \rangle$ be a ν-MSR. We define $G(S) = (\mathcal{P}, \mathcal{R}, F, M_0^*)$ as follows. Let $t : \sum_{i=1}^{n} p_i(\tilde{x}_i) \rightarrow \nu\tilde{a}.(\sum_{i=1}^{m} q_i(\eta_i))$ be a rule in $\mathcal{R}$. We assume for the sake of readability that no names appear in the tuples of the left handside of the rule, and that the only names appearing in the right handside are those in $\tilde{a}$. Let $\tilde{\nu}$ be a sequence (of the same length of $\tilde{a}$) of pairwise different special variables. We define $F(p, t) = \sum_{p = p_i} \tilde{x}_i$ and $F(t, p) = \sum_{p = q_i} \eta_i\{\tilde{\nu}/\tilde{a}\}$. For a ν-MSR term $M \equiv \nu\tilde{a}.(\sum_{i=1}^{n} p_i(\tilde{a}_i))$ we define M^* as the marking given by $M^*(p) = \{\tilde{a}_i \mid p = p_i\}$. As in the previous result, for two terms M_1 and M_2, thanks to Prop. 1, it holds that $M_1^* \rightarrow M_2^*$. Moreover, for two markings M_1 and M_2 such that $M_1 \rightarrow M_2$, $M_i = M_i'^*$ for some terms M_1' and M_2', with $M_1' \rightarrow M_2'$.

In [24] we proved that $p\nu$-APNs are Turing complete. Therefore, Prop. 2 tells us that so are ν-MSRs. It is easy to devise some decidable subclasses of ν-MSRs. For instance, if a ν-MSR S is monadic, that is, if atomic formulae have the form $p(\eta)$, then the $p\nu$-APN $G(S)$ obtained in Prop. 3 is a ν-APN [23], that is, a Petri net in which tokens are pure names. In [25] we proved that coverability is decidable for them, so that they are also decidable for monadic ν-MSRs. Moreover, if we consider a ν-MSR with only binary predicates and so that for every formula $p(a,b)$ there are only finitely many b_i such that $p(a,b_i)$ appears in any reachable term, then $G(S)$ is a restricted binary $p\nu$-APN [24], for which coverability is also decidable. We claim that these results could have been obtained directly for the restricted classes of ν-MSRs, so that the corresponding results for Petri nets could have been obtained as a corollary instead. Finally, let us remark that in the case of ordinary P/T nets (that are a subclass of ν-APNs, in which only one element of *Id* is used) our translation yields a ν-MSR that coincides with the rewriting logic specification obtained in [29].

4 ν-MSRs and the π-calculus

Let us see that ν-MSRs can simulate any π-calculus process. We use the monadic version of the π-calculus used in [21,20,27], with parameterized recursion. The prefixes of the π-calculus are defined by

$$\pi ::= \overline{x}\langle y\rangle \;\mid\; x(y) \;\mid\; \tau$$

The set of the π-calculus processes is defined by

$$P ::= \sum_{i=1}^{n} \pi_i.P_i \;\mid\; P_1 \mid P_2 \;\mid\; \nu a.P \;\mid\; K\lfloor\tilde{a}\rfloor$$

The empty sum (with $n=0$) is denoted as $\mathbf{0}$. As usual, we identify processes up to $\equiv$, which is the least congruence that allows α-conversion of bound names, such that $+$ and $\mid$ are commutative and associative with $\mathbf{0}$ as neutral element, and the following equations hold: $\nu a.0 \equiv 0$, $\nu a.\nu b.P \equiv \nu b.\nu a.P$ and $\nu a.(P \mid Q) \equiv \nu a.P \mid Q$, if $a \notin fn(Q)$, where $fn(P)$ is the set of names that occur *free* in P. If a name in P is not free then it is *bound*. As usual, we omit pending $\mathbf{0}$ in the examples. The reaction relation is defined by the following rules:

$$\tau.P + M \rightarrow P \qquad x(y).P + M \mid \overline{x}\langle z\rangle.Q + N \rightarrow P\{z/y\} \mid Q$$

$$K\lfloor\tilde{a}\rfloor \rightarrow P\{\tilde{a}/\tilde{x}\}, \;\; \textit{if} \;\; K(\tilde{x}) := P$$

$$\frac{P \rightarrow P'}{P \mid Q \rightarrow P' \mid Q} \qquad \frac{P \rightarrow P'}{\nu a.P \rightarrow \nu a.P'} \qquad \frac{P \equiv Q \rightarrow Q' \equiv P'}{P \rightarrow P'}$$

We will use the notion of *derivatives* of a process introduced in [21]. For a process P with recursive definitions $K_i(\tilde{x}_i) := P_i$ for $i = 1,\ldots,n$, we define $\textit{derivatives}(P) = \textit{der}(P) \cup \bigcup_{i=1}^{n} \textit{der}(P_i)$, where $\textit{der}(\mathbf{0}) = \emptyset$, $\textit{der}(K\lfloor\tilde{a}\rfloor) =$

$\{K\lfloor\tilde{a}\rfloor\}$, $der(\sum_{i=1}^{n}\pi_i.P_i) = \{\sum_{i=1}^{n}\pi_i.P_i\} \cup \bigcup_{i=1}^{n} der(P_i)$ for $n > 0$, $der(P_1 \mid P_2) = der(P_1) \cup der(P_2)$, and $der(\nu a.P) = der(P)$.

The set of derivatives of a process is always finite, and it essentially corresponds to the set of its sequential subprocesses, but disregarding name restriction. As proved in [21], every reachable process can be built up by composing derivatives with its free names renamed.

Proposition 4. [21, Proposition 3] *Let P be a π-calculus process. Every Q reachable from P is structurally congruent with $\nu\tilde{a}.(Q_1\sigma_1 \mid \cdots \mid Q_n\sigma_n)$, where $Q_i \in derivatives(P)$ and $\sigma_i : fn(Q_i) \to fn(P) \cup \tilde{a}$.*

We will heavily rely on this result for our simulation of π-calculus processes by means of ν-MSRs. More precisely, we will consider the finite set of derivatives as predicates. If a derivative p has $x_1,\ldots,x_n$ as free names, then we will write $p(a_1,\ldots,a_n)$ to represent the derivative $p\{\tilde{x}_i/\tilde{a}_i\}$.

We assume that the sets of free names of derivatives are pairwise disjoint. Moreover, we remove repeated derivatives, in the sense that one can be obtained from another by renaming its free names.

Next we introduce some notations to deal with derivatives. In the first place, for a derivative $\tau.P + M$ with P equivalent to the process in standard form $\nu\tilde{a}.(D_1 \mid \cdots \mid D_k)$ we write $\tau.P + M \mapsto \nu\tilde{a}.(D_1 \mid \cdots \mid D_k)$. Let $K\lfloor\tilde{x}\rfloor$ be a derivative with $K(\tilde{x}) ::= P$, where P is equivalent to a process in standard form $\nu\tilde{a}.(D_1 \mid \cdots \mid D_k)$. Then we will write $K\lfloor\tilde{x}\rfloor \mapsto \nu\tilde{a}.(D_1 \mid \cdots \mid D_k)$.

Finally, if two derivatives D_1 and D_2 are equivalent to $M_1 + x_1\langle y_1\rangle.P_1$ and $M_2 + x_2(y_2).P_2$, respectively, and $P_i \equiv \nu\tilde{a}_i.(D_1^i \mid \cdots \mid D_{k_i}^i)$ for $i = 1,2$ then we will write

$$D_1 \mid D_2 \overset{x_1=x_2}{\longmapsto} \nu\tilde{a}_1,\tilde{a}_2.(D_1^1 \mid \cdots \mid D_{k_1}^1 \mid D_1^2\{y_2/y_1\} \mid \cdots \mid D_{k_2}^2\{y_2/y_1\})$$

Proposition 5. *For every π-calulus process P_0 there is a ν-MSR $H(P_0)$ (with H computable) such that P_0 and $H(P_0)$ are isomorphic (as transition systems).*

Proof. Let $\mathcal{P} = Derivatives(P_0)$ be the set of derivatives of P_0, which is finite. For any $P \in \mathcal{P}$, we use the atomic formula $P(x_1,\ldots,x_n)$ to represent P, provided $fn(P) = \{x_1,\ldots,x_n\}$. We will consider the following rules:

- For each $D \mapsto \nu\tilde{a}.(D_1 \mid \cdots \mid D_k)$ with $fn(D) = \tilde{x}$ and $fn(D_i) = \tilde{x}_i$, we consider the rule $D(\tilde{x}) \to \nu\tilde{a}.(D_1(\tilde{x}_1),\ldots,D_k(\tilde{x}_k))$.
- For each $D_1 \mid D_2 \overset{x=y}{\longmapsto} \nu\tilde{a}.(D_1' \mid \cdots \mid D_k')$ with $fn(D_i) = \tilde{x}_i$ and $fn(D_i') = \tilde{y}_i$, we consider the rule $D_1(\tilde{x}_1), D_2(\tilde{x}_2)\{y/x\} \to \nu\tilde{a}.(D_1'\{y/x\},\ldots,D_k'\{y/x\})$.

By Prop. 4, any reachable process P is equivalent to $\nu\tilde{a}.(D_1\sigma_1 \mid \cdots \mid D_n\sigma_n)$, with $D_i \in \mathcal{P}$, $fn(D_i) = \tilde{x}_i$, and $\sigma_i : \tilde{x}_i \to fn(P_0) \cup \tilde{a}$. Then, for any reachable P we can define the ν-MSR term P^* over $\mathcal{P}$ by structural induction: $K\lfloor\tilde{a}\rfloor^* = D(\tilde{a})$ for some $D(\tilde{x}) \in \mathcal{P}$, $(\sum_{i=1}^{n}\pi_i.P_i)^* = D(\tilde{a})$ for some $D(\tilde{x}) \in \mathcal{P}$, $(P_1 \mid P_2)^* = P_1^* \mid P_2^*$, and $(\nu a.P) = \nu a.P^*$. The mapping $(\cdot)^*$ is an isomorphism (between the quotients modulo $\equiv$) and $P_1 \to P_2$ if and only if $P_1^* \to P_2^*$.

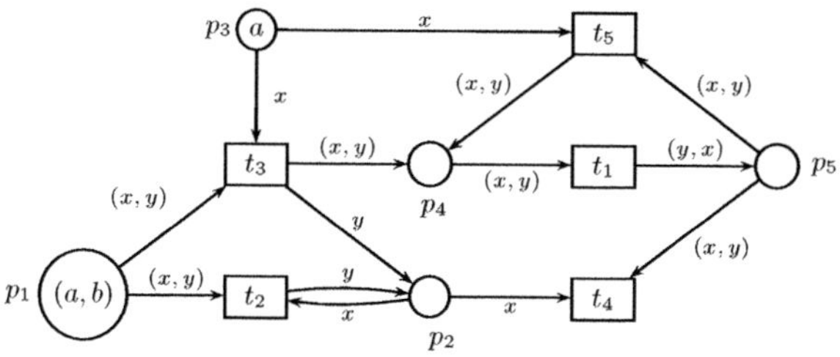

Fig. 2. $p\nu$-APN simulating the process in Example 1

Let us see that if P_1 is a reachable process and $P_1 \to P_2$ then $P_1^* \to P_2^*$. We see it by induction on the rules used to derive $P_1 \to P_2$. If $P_1 = K\lfloor\tilde{a}\rfloor$ and $P_2 = P\{\tilde{x}/\tilde{a}\}$ with $K(\tilde{x}) ::= P$ then there are derivatives $D_1, D_1', \ldots, D_k'$ such that $P_1^* = D_1(\tilde{a})$ and $P_2^* = \nu\tilde{b}.(D_1'(\tilde{a}_1, \tilde{b}_1), \ldots, D_k'(\tilde{a}_k, \tilde{b}_k)$ with $\bigcup \tilde{a}_i \subseteq \tilde{a}$ and $\bigcup \tilde{b}_i \subseteq \tilde{b}$ (assuming that $\tilde{a} \cap \tilde{b} = \emptyset$; otherwise, we just need to rename the names in $\tilde{b}$). By construction, we have a rule $D_1(\tilde{x}) \to \nu\tilde{b}.(D_1'(\tilde{x}_1, \tilde{b}_1), \ldots, D_k'(\tilde{x}_k, \tilde{b}_k))$, that can be applied for $P_1^* = D_1(\tilde{a})$, producing P_2^*. Let us now consider the case in which $P_1 = x(y).P_1' + M \mid x\langle z\rangle.P_2' + N$ and $P_2 = P_1'\{y/z\} \mid P_2'$. By Prop. 4 there are derivatives such that $P_1 = D_1\{\tilde{x}_i/\tilde{a}_i\} \mid D_2\{\tilde{x}_2/\tilde{a}_2\}$, $P_i' = \Pi_{j=1}^{k_i} D_j^i\{\tilde{y}_i/\tilde{a}_i'\}$, for $i = 1, 2$, and $D_1 \mid D_2 \overset{x_1=x_2}{\longmapsto} D_1' \mid D_2'$. By construction, there is a rule $D_1(\tilde{x}_1), D_2(\tilde{x}_2)\{x_2/x_1\} \to \nu\tilde{a}.(D_1'\{x_2/x_1\}, \ldots, D_k'\{x_2/x_1\})$, that can be instantiated for $P_1^* = D_1(\tilde{a}_1), D_2(\tilde{a}_2)$, yielding the term $\nu\tilde{a}.(D_1'(\tilde{a}_1'), D_2'(\tilde{a}_2'))$, which is P_2^*.

The rules for parallel composition and restriction are easy to check (they correspond to rules $(+)$ and (ν), respectively). The rule for $\equiv$ is trivial, because P^* is defined in the same way for all the processes belonging to the same equivalence class than P.

For the converse implication, it is enough to consider that any reachable term is of the form $M = \nu\tilde{a}.(D_1(\tilde{a}_1), \ldots, D_n(\tilde{a}_n))$ for some derivatives D_i. Then, $M = P^*$, with $P = \nu\tilde{a}.(D_1\{\tilde{x}_1/\tilde{a}_1\} \mid \ldots \mid D_n\{\tilde{x}_n/\tilde{a}_n\})$. Similarly as before, we can prove that $M_1 \to M_2$ implies $M_i = P_i^*$ with $P_1 \to P_2$.

Example 1. Let us consider $P = \nu b.a\langle b\rangle.b(x) \mid a(y).K\lfloor a, y\rfloor$, where $K(x, y) := y\langle x\rangle$. The set of derivatives of P is $\{p_1, p_2, p_3, p_4, p_5\}$, where $p_1 = x_1\langle y_1\rangle.y_1(z)$, $p_2 = x_2(z)$, $p_3 = x_3(z).K\lfloor x_3, z\rfloor$, $p_4 = K\lfloor x_4, y_4\rfloor$, and $p_5 = x_5\langle y_5\rangle$. The ν-MSR term corresponding to P is $P^* = \nu b.p_1(a, b), p_3(a)$. The derivatives can react in the following way:

1. $K\lfloor x_4, y_4\rfloor \mapsto y_4\langle x_4\rangle$
2. $x_1\langle y_1\rangle.y_1(z) \mid x_2(z) \overset{x_1=x_2}{\longmapsto} y_1(z)$
3. $x_1\langle y_1\rangle.y_1(z) \mid x_3(z).K\lfloor x_3, z\rfloor \overset{x_1=x_3}{\longmapsto} y_1(z) \mid K\lfloor x_3, y_1\rfloor$
4. $x_2(z) \mid x_5\langle y_5\rangle \overset{x_2=x_5}{\longmapsto} \mathbf{0}$
5. $x_3(z).K\lfloor x_3, y_3\rfloor \mid x_5\langle y_5\rangle \overset{x_3=x_5}{\longmapsto} K\lfloor x_3, y_5\rfloor$

Fig. 3. $p\nu$-APNs simulating the processes in Example 2

These reactions give rise to the following rules:

$$\begin{aligned}
&t_1 : p_4(x, y) \to p_5(y, x)\\
&t_2 : p_1(x, y), p_2(x) \to p_2(y)\\
&t_3 : p_1(x, y), p_3(x) \to p_2(y), p_4(x, y)\\
&t_4 : p_2(x), p_5(x, y) \to 0\\
&t_5 : p_3(x), p_5(x, y) \to p_4(x, y)
\end{aligned}$$

In turn, according to Prop. 3, we can write these rules as a $p\nu$-APN, which is depicted in Fig. 2. Its initial marking corresponds to the term P^*, with a token (a, b) in p_1 and a token a in p_3. Actually, the rules (and the net) obtained are the same for any process with derivatives in $p_1, \ldots, p_5$. Indeed, starting from the process P, one can check that the derivatives p_1 and p_2, or p_3 and p_5, will never be in parallel. Our construction is safe, so that it does consider the reaction rules t_2 and t_5, though they will never be enabled. Thus, any process whose set of derivatives coincides with that of P, is simulated by the same net, though with a different initial marking. Finally, notice that the resulting net does not have any arc labelled with any special variable, so that the names appearing in any reachable markings are taken from the finite set of names in the initial markings. In this situation, the net can be flattened to an equivalent P/T net.

Example 2. Let us consider the processes $P_1 = \nu a.L\lfloor a \rfloor$ and $P_2 = \nu a.K\lfloor a \rfloor$, with $L(x) := \nu b.(x\langle b \rangle \mid L\lfloor x \rfloor)$ and $K(x) := \nu a.(a\langle x \rangle \mid K\lfloor a \rfloor)$. *derivatives*$(P_1) = \{L\lfloor x \rfloor, y\langle z \rangle\}$ and *derivatives*$(P_2) = \{K\lfloor x \rfloor, y\langle z \rangle\}$. The only reaction of derivatives of P_1 is $L\lfloor x \rfloor \mapsto \nu b.(x\langle b \rangle \mid L\lfloor x \rfloor)$, where the call to procedure L is done. Analogously, the only reaction of derivatives of P_2 is $K\lfloor x \rfloor \mapsto \nu a.(a\langle x \rangle \mid K\lfloor a \rfloor)$, where procedure K is called. These reactions are simulated by the rules $p_1(x) \to \nu b.(p_1(x), p_2(x, b))$ and $p_1(x) \to \nu a.(p_1(a), p_2(a, x))$. The corresponding ν-MSRs give rise to two $p\nu$-APNs, which are shown in Fig. 3. Since each process has only two derivatives, the corresponding nets have two places, and since they can only react in one way, only one transition is produced for each.

The study of subclasses of the π-calculus where some properties become decidable is an active field of research. For instance, [5] considers *restriction-bounded* processes, which generate a finite number of restricted names. In the paper [22] *mix-bounded* processes are defined, which are the most expressive subclass of the π-calculus that can be mapped to finite P/T nets. Depth-bounded processes [20] are processes where the interdependence of names is bounded. We can consider this restriction in ν-MSR terms, thus obtaining depth-bounded terms. For instance, in the net on the right of Fig. 3, sequences of tokens or arbitrary length of the form $(a_0, a_1), (a_1, a_2), (a_2, a_3), \ldots$ can appear, so that the interdependence of names is not bounded (all the a_is are interdependent). Therefore, that net (or the corresponding ν-MSR) is not depth-bounded, but the one in the left of the same figure is depth-bounded.

In [20], it is proved that depth-bounded processes in the π-calculus produce WSTS. Actually, the same steps can be followed to prove that depth-bounded ν-MSR terms are WSTS, though we do not show the details in this paper due to lack of space. Thus, coverability, boundedness and termination are decidable for them. Therefore, this result carries over to any model that can be specified within ν-MSRs. Thus, depth-bounded $p\nu$-APNs, that is, $p\nu$-APNs in which the interdependence of names is bounded, are also WSTS, so that coverability, termination and boundedness are decidable for them.

5 ν-MSRs in Maude

Since the behavior of ν-MSR systems is specified in terms of a congruent rewriting relation (with respect to all the constructors), modulo the equational theory defined by $\equiv$, the translation from a ν-MSR specification to an equivalent rewrite specification is straightforward. Let us see some of the details of their representation in Maude. Moreover, the representation of ν-MSRs within Maude will allow us to use all the analysis machinery available for it. The syntax of terms is simply defined as follows:

```
sorts Predicate Term .
subsorts Predicate < Term .

op nil : -> Term .
op _(_) : Qid Tuple -> Predicate .
op __ : Term Term -> Term [comm assoc id: nil] .
op nu(_)_ : SeqQid Term -> Term .
```

As we proved in Prop. 1, we can use the standard normal form of a ν-MSR term to specify its full behavior, that is, an equational theory to obtain such a standard normal from is coherent with any set of rewriting rules. Therefore, our equational theory will reduce any term to its standard normal form, which amounts to pushing restrictions to the outermost position.

In order to deal with bound names, we can use de Bruijns indexes [11] in order to distinguish occurrences of the same name that are bound by different binding operators. Consider for instance a ν-MSR system composed of a single rule $p(x) \rightarrow \nu a.q(a)$. In that system, the term $p(a)p(b)$ rewrites in two steps to $\nu a.\nu b.q(a)q(b)$, which, using de Bruijn's indexes can be written as $\nu a.\nu a.q(a_0)q(a_1)$. Intuitively, for an indexed name a_i, i represents the number of intermediate a-bindings between the free occurrence and its binding occurrence. Therefore, names are defined as follows:

```
sort Name .
op _‘{_‘} : Qid Nat -> Name [prec 1] .
```

The side condition in the extrusion rule does not act like a restriction, since we are working modulo α-conversion. Instead, for the extrusion to happen, it enforces a renaming of a in case it occurs free in M_2, replacing in M_1 the name a by some fresh b (not free in M_2). In order to manage substitutions and indexed names, we use the explicit calculus for substitutions of [28], as done in other specifications of process algebra with name binding, as the π-calculus [30] or the Ambient Calculus [26].

```
eq  [ shiftup a ] a{n} = a{s(n)} .
ceq [ shiftup a ] b{n} = b{n}  if a =/= b .
eq  ( nu(x) NSP ) NSQ = nu(x)(NSP ([shiftup x] NSQ)) .
```

The standard normal form is unique not only up to commutativity, associativity and identity (which are the equational attributes of the multiset addition operator), but also up to rearrangement of the bounded names. In other words, the rule $\nu a.\nu b.M \equiv \nu b.\nu a.M$ cannot be directly specified in Maude, or the corresponding equational theory would be non-terminating. Instead of considering an artificial order (like lexicographic order over quoted identifiers) to obtain a unique normal form up to the equational attributes, we have chosen to allow sequences of quoted identifiers in restrictions, analogously to the $\nu\tilde{a}.M$ notation. Therefore, `SeqQid` is a commutative domain with a constant `eps` as empty sequence, and the standard normal form is unique up to commutativity of restricted names.

```
eq nu(at) nu(bt) P = nu(at . bt) P .
eq nu(eps) P = P .
```

Notice that our equational theory is an order-sorted equational specification (without membership equations). Each rule in a ν-MSR gives rise to an unconditional rewrite rule. For instance, the two ν-MSR systems obtained from the two π-calculus processes P_1 and P_2 in Example 2 (the corresponding equivalent $p\nu$-APNs are depicted in Figure 3), can be represented by the two following rules, respectively.

```
var x : Name .

rl [t] : 'p1(x) => nu('b)( 'p1( [shiftup 'b] x)
                           'p2( [shiftup 'b] x , 'b{0}) ) .

rl [t] : 'p1(x) => nu('a)( 'p1( 'a{0})
                           'p2( 'a{0} , [shiftup 'a] x ) ) .
```

Notice the application of the `shiftup` substitution in the right hand side of the rules. It ensures that the rules are applied without capturing free names. This is justified by the definition of the application of a rule in ν-MSR, where it may be necessary to rename some of the new bound names. Moreover, the variables in *Var* used in ν-MSR rules are just variables in the system module (that is, they are in the meta-level).

When we rewrite starting from the initial marking of the net in the left of Fig. 3 we obtain:

```
Maude> rew [3] 'p1( 'a{0} ) .
rewrite [3] in EX1 : 'p1('a{0}) .
rewrites: 114 in 0ms cpu (0ms real) (~ rewrites/second)
result NSTerm: nu('b . 'b . 'b)('p1('a{0})
               'p2('a{0},'b{0}) 'p2('a{0},'b{1}) 'p2('a{0},'b{2}))
```

In the case of the net in the right handside of Fig. 3, we get the following:

```
Maude> rew [3] 'p1( 'a{0} ) .
rewrite [3] in EX1 : 'p1('a{0}) .
rewrites: 108 in 0ms cpu (0ms real) (~ rewrites/second)
result NSTerm: nu('a . 'a . 'a)('p1('a{0})
               'p2('a{0},'a{1}) 'p2('a{1},'a{2}) 'p2('a{2},'a{3}))
```

Finally, let us comment that, in the case in which all predicates are 0-ary, that is, if only predicates are used in terms (instead of variables and names), then we simply have the classical interpretation of P/T nets as rewritings of multisets of elements taken from a finite set, and the implementation in Maude is equivalent to the one shown for instance in [18].

6 Conclusions and Future Work

In this paper we have defined ν-MSRs, where MSR stands for Multiset Rewriting. ν-MSRs encompass the multiset rewriting approach for concurrency, followed in [13], and the multiset rewriting approach for security, or name binding in general, followed in [9,8].

We have proved that ν-MSRs simulate, in a very natural way, two models of concurrency with name binding, as is the case of $p\nu$-APNs and π-calculus processes. The previous simulations establish that any result obtained for ν-MSRs can be translated both to the π-calculus and $p\nu$-APNs. For instance, we show that depth-boundedness, that was studied in [20] for the π-calculus, can be studied in the more general setting of ν-MSRs, thus obtaining the analogous results for ν-MSR in general, and $p\nu$-APNs in particular.

ν-MSRs establish a clean bridge between Petri nets and process algebra, that could be interesting in order to compare the natural concurrent (process) semantics of Petri nets to π-calculus processes.

As future work, we plan on coding the spi calculus and mobile ambients into ν-MSRs. Regarding the spi calculus, we believe that an analogous translation to the one carried out for π-calculus processes can be achieved for spi-calculus processes, thus bringing together two different approaches for the specification and analysis of security protocols, namely the spi-calculus and multiset rewriting.

We also plan to study how Mobile Ambient [7] processes can be encoded within ν-MSR. As a first approach for the translation of mobile ambients, it seems that a transfer (or broadcast) operation is needed to encode mobile ambients (since all the sequential processes within an ambient are affected by some operations on that ambient). However, under some conditions, transfers do not add any expressive power [2]. In the case of mobile ambients, it would certainly be interesting to compare in the common framework the decidability results obtained for restricted classes of mobile ambients, for instance in [6], to those obtained for the π-calculus.

Regarding the Maude implementation, we could use the π-calculus specification of [30] to automatically generate the system module corresponding to the ν-MSR that encode a process.

Finally, let us remark that it would be interesting to have an alternative presentation of ν-MSR in terms of monoidal constructions in named sets, or any equivalent theoretical model for name binding, like nominal sets or presheaves [10,15], which would allow us to obtain the analogous constructions to those studied in [19]. Moreover, this construction could allow us to formalize when name interdependence is preserved by an encoding.

Moreover, we argue that the results in [20] can also be obtained in the more general setting of ν-MSRs. Thus, we define depth-bounded ν-MSRs, that, intuitively, have a bounded interdependence of bounded names. Depth-bounded ν-MSRs can be proven to produce WSTS by following the same steps as in [20]. Moreover, this result can be translated to $p\nu$-APNs, so that depth-bounded $p\nu$-APNs are also WSTS, and coverability, boundedness and termination are also decidable for them.

References

1. Abadi, M., Gordon, A.D.: A Calculus for Cryptographic Protocols: The spi Calculus. Inf. Comput. 148(1), 1–70 (1999)
2. Abdulla, P.A., Delzanno, G., Begin, L.V.: Comparing the expressive power of well-structured transition systems. In: Duparc, J., Henzinger, T.A. (eds.) CSL 2007. LNCS, vol. 4646, pp. 99–114. Springer, Heidelberg (2007)
3. Baldan, P., Bonchi, F., Gadducci, F.: Encoding asynchronous interactions using open Petri nets. In: Bravetti, M., Zavattaro, G. (eds.) CONCUR 2009. LNCS, vol. 5710, pp. 99–114. Springer, Heidelberg (2009)
4. Boudol, G.: Some chemical abstract machines. In: de Bakker, J.W., de Roever, W.-P., Rozenberg, G. (eds.) REX 1993. LNCS, vol. 803, pp. 92–123. Springer, Heidelberg (1994)
5. Busi, N., Gorrieri, R.: Distributed semantics for the pi-calculus based on Petri nets with inhibitor arcs. J. Log. Algebr. Program. 78(3), 138–162 (2009)
6. Busi, N., Zavattaro, G.: Deciding reachability problems in turing-complete fragments of mobile ambients. Mathematical Structures in Computer Science 19(6), 1223–1263 (2009)
7. Cardelli, L., Gordon, A.D.: Mobile ambients. Theor. Comput. Sci. 240(1), 177–213 (2000)
8. Cervesato, I.: Typed MSR: Syntax and Examples. In: Gorodetski, V.I., Skormin, V.A., Popyack, L.J. (eds.) MMM-ACNS 2001. LNCS, vol. 2052, pp. 159–177. Springer, Heidelberg (2001)
9. Cervesato, I., Durgin, N.A., Lincoln, P., Mitchell, J.C., Scedrov, A.: A meta-notation for protocol analysis. In: CSFW, pp. 55–69 (1999)
10. Ciancia, V., Montanari, U.: A name abstraction functor for named sets. Electr. Notes Theor. Comput. Sci. 203(5), 49–70 (2008)

11. de Bruijn, N.: Lambda calculus with nameless dummies, a tool for automatic formula manipulation, with application to the church-rosser theorem. In: Proceedings Kninkl. Nederl. Akademie van Wetenschappen, vol. 75, pp. 381–392 (1972)
12. Delzanno, G.: An overview of MSR(C): A CLP-based framework for the symbolic verification of parameterized concurrent systems. Electr. Notes Theor. Comput. Sci. 76 (2002)
13. Delzanno, G.: Constraint multiset rewriting. Technical Report DISI-TR-05-08, University of Genova (2005)
14. Finkel, A., Schnoebelen, P.: Well-structured transition systems everywhere! Theor. Comput. Sci. 256(1-2), 63–92 (2001)
15. Gadducci, F., Miculan, M., Montanari, U.: About permutation algebras, (pre) sheaves and named sets. Higher-Order and Symbolic Computation 19(2-3), 283–304 (2006)
16. Gordon, A.D.: Notes on nominal calculi for security and mobility. In: Focardi, R., Gorrieri, R. (eds.) FOSAD 2000. LNCS, vol. 2171, pp. 262–330. Springer, Heidelberg (2001)
17. Lazic, R., Newcomb, T., Ouaknine, J., Roscoe, A.W., Worrell, J.: Nets with tokens which carry data. Fundam. Inform. 88(3), 251–274 (2008)
18. Meseguer, J.: Rewriting logic as a semantic framework for concurrency: a progress report. In: Sassone, V., Montanari, U. (eds.) CONCUR 1996. LNCS, vol. 1119, pp. 331–372. Springer, Heidelberg (1996)
19. Meseguer, J., Montanari, U.: Petri nets are monoids. Inf. Comput. 88(2), 105–155 (1990)
20. Meyer, R.: On boundedness in Depth in the pi-calculus. In: Ausiello, G., Karhumäki, J., Mauri, G., Ong, C.H.L. (eds.) IFIP TCS. LNCS, vol. 273, pp. 477–489. Springer, Heidelberg (1987)
21. Meyer, R.: A theory of structural stationarity in the *pi*-calculus. Acta Inf 46(2), 87–137 (2009)
22. Meyer, R., Gorrieri, R.: On the relationship between pi-calculus and finite place/transition Petri nets. In: Bravetti, M., Zavattaro, G. (eds.) CONCUR 2009. LNCS, vol. 5710, pp. 463–480. Springer, Heidelberg (2009)
23. Rosa-Velardo, F., de Frutos-Escrig, D.: Name creation vs. replication in Petri net systems. Fundam. Inform. 88(3), 329–356 (2008)
24. Rosa-Velardo, F., de Frutos-Escrig, D.: Decidability problems in Petri nets with name creation and replication (submitted)
25. Rosa-Velardo, F., de Frutos-Escrig, D., Alonso, O.M.: On the expressiveness of Mobile Synchronizing Petri Nets. Electr. Notes Theor. Comput. Sci. 180(1), 77–94 (2007)
26. Rosa-Velardo, F., Segura, C., Verdejo, A.: Typed mobile ambients in Maude. Electr. Notes Theor. Comput. Sci. 147(1), 135–161 (2006)
27. Sangiorgi, D., Walker, D.: The *pi*-calculus: a Theory of Mobile Processes. Cambridge University Press, Cambridge (2001)
28. Stehr, M.O.: CINNI - a generic calculus of explicit substitutions and its application to lambda-, varsigma- and pi-calculi. Electr. Notes Theor. Comput. Sci. 36 (2000)
29. Stehr, M.O., Meseguer, J., Ölveczky, P.C.: Rewriting logic as a unifying framework for Petri nets. In: Ehrig, H., Juhás, G., Padberg, J., Rozenberg, G. (eds.) APN 2001. LNCS, vol. 2128, pp. 250–303. Springer, Heidelberg (2001)
30. Thati, P., Sen, K., Martí-Oliet, N.: An executable specification of asynchronous pi-calculus semantics and may testing in Maude 2.0. Electr. Notes Theor. Comput. Sci. 71 (2002)

The Linear Temporal Logic of Rewriting Maude Model Checker

Kyungmin Bae and José Meseguer

Department of Computer Science,
University of Illinois at Urbana-Champaign, Urbana IL 61801
{kbae4,meseguer}@cs.uiuc.edu

Abstract. This paper presents the foundation, design, and implementation of the Linear Temporal Logic of Rewriting model checker as an extension of the Maude system. The Linear Temporal Logic of Rewriting (LTLR) extends linear temporal logic with spatial action patterns which represent rewriting events. LTLR generalizes and extends various state-based and event-based logics and aims to avoid certain types of mismatches between a system and its temporal logic properties. We have implemented the LTLR model checker at the C++ level within the Maude system by extending the existing Maude LTL model checker. Our LTLR model checker provides very expressive methods to define event-related properties as well as state-related properties, or, more generally, properties involving both events and state predicates. This greater expressiveness is gained without compromising performance, because the LTLR implementation minimizes the extra costs involved in handling the events of systems.

Keywords: Model checking, Rewriting Logic, Maude, Automata.

1 Introduction

The main motivation for the temporal logic of rewriting (TLR) [27] is to have a simple, yet expressive, temporal logic that can: (i) support *state-based* properties with all the good advantages of logics such as LTL, CTL, and CTL*; (ii) support just as easily *event-based* properties and, more generally, *mixed* properties involving states and events; and (iii) lift to the temporal logic level the extra expressiveness of rewriting logic specifications to describe events by *spatial action patterns*, that indicate not just that a rewrite event labeled l has happened, but *where* in the terms's geometry, and *how* (with what kind of instantiation). As explained in [27], meeting goals (i)–(iii) is a good way to obtain a suitable *tandem* between two logics: a *system specification logic*, where concurrent system models are specified —in this case rewriting logic— and a *property specification logic*, in which properties about such a model are specified —in this case TLR.

The problem with *unsuitable* tandems, is that there is a mismatch between the two logics, and this forces the specifier to "cook" the system specification in unnatural ways just to be able to encode in the model features not directly

P.C. Ölveczky (Ed.): WRLA 2010, LNCS 6381, pp. 208–225, 2010.

expressible in the property logic. In this sense, purely state-based logics, resp. purely event-based logics, are good as far as they go, but they become unsuitable tandem partners when event-based properties, resp. state-based properties, need to be dealt with. Section 5 describes in detail an example illustrating this kind of tandem mismatch, namely, the well-known bounded retransmission protocol (BRP). The point about BRP is that many of its relevant properties are related to various signaling messages exchanged between a sender and a receiver to inform each other about the status of the communication at each end. Since state-based temporal logics like LTL and CTL* have no way to directly express the sending of such signaling messages, a system specification of BRP has to *encode* such events in the system state in a way that makes the specification unnecessarily complex and obscures its meaning. With LTLR, as we explain, the problem completely disappears: all the BRP properties can be directly expressed without any need for complicating the protocol specification.

The *linear temporal logic of rewriting* (LTLR) is an attractive subset of TLR, because it extends naturally the quite easy-to-understand and widely used linear temporal logic (LTL). It is particularly attractive for the Maude system [10], because Maude already supports efficient explicit-state on-the-fly model checking of LTL formulas. Endowing Maude with LTLR model checking capabilities means keeping all the good state-based features of LTL *and* gaining the substantial expressiveness of spatial action patterns for rewrite events. In [2], a first implementation of an LTLR model checker for Maude was presented. However, the implementation in [2] reused the Maude LTL model checker *as given*, without any changes in its algorithm. It relied instead on a *theory and formula transformation* (see [2,27]), so that the LTLR model checking problem was transformed into an equivalent LTL model checking problem for a transformed Maude specification using Maude's metalevel features. While quite useful for experimental purposes, and still keeping the exact same performance as Maude's LTL model checker for its LTL fragment, the solution was less than optimal for event-based properties, because rewrite events had to be encoded in the state of the transformed Maude specification, leading to a considerable increase of the state space. In this work we present a completely new implementation and system that is based on a new LTLR model checking algorithm that does *not* change the given Maude specification, and therefore does not cause any increase in the state space. With this new algorithm we can have the best of both worlds: for state-based properties, we keep the exact same good performance as Maude's original LTL model checker; but, we gain also very good performance for event-based properties and for mixed state-and-event ones for two reasons: (i) the state space is exactly that of the given system; and (ii) the new LTLR model checking algorithm is implemented at the C++ level as an extension of Maude's C++ implementation of its LTL model checker.

This paper presents not only the implementation of the new LTLR model checker, but also the *automata-theoretic foundations* for its algorithm and its associated computational complexity. Specifically, these foundations show that model checking an LTLR formula φ for a given rewrite theory $\mathcal{R}$ is equivalent to deciding a language emptiness problem for a Büchi automaton obtained as a special synchronous product of the Büchi automaton $\mathscr{B}_{\neg\varphi}$ for the negation of φ and a labeled Kripke structure naturally associated to $\mathcal{R}$. Our approach is closely related to the automata-theoretic solution for model checking properties that are both state- and event-based proposed in [6]. Another topic studied in detail is the tool's support for both automated and user-defined *language extensions* to express LTLR properties. This is analogous to the need to extend the language of a given Maude specification with the desired state predicates for LTL model checking. The point is that something akin to a language of *proof terms* for the given rewrite theory $\mathcal{R}$ must be made available to the user in order to include spatial action patterns for $\mathcal{R}$ in LTLR formulas. The use of the LTLR model checker is then illustrated with the already-mentioned BRP protocol, and we give some the experimental results. We end the paper with a discussion of related work and some concluding remarks. The LTLR model checker itself, as well as some examples and some preliminary documentation is available at `http://www.cs.uiuc.edu/homes/kbae4/tlr`.

2 The Linear Temporal Logic of Rewriting

The linear temporal logic of rewriting (LTLR) extends LTL with *spatial action patterns* that describe the event of a rewrite step under certain spatial constraints. LTLR has the same syntax as LTL, except that an LTLR atom can be either a spatial action pattern or a state predicate. The semantics of LTLR is defined on a model of a rewrite theory described by Rewriting Logic, constituting the *Rewriting Logic/LTLR* tandem [29], similar to the usual tandem *Kripke/LTL*.

2.1 Rewrite Theories

A *rewrite theory* is a formal specification of a concurrent system [28], defined by a triple $\mathcal{R} = (\Sigma, E, R)$ such that:

- (Σ, E) is an underlying theory in *membership equational logic* [10] with Σ a signature and E a set of *conditional* equations and memberships. The initial algebra $T_{\Sigma/E}$ of (Σ, E) specifies the system's state space.
- R is a set of (possibly conditional) *rewrite rules* specifying the system's concurrent transitions between states, written $l : q \longrightarrow r$, where l is a *label*, and q and r are Σ-terms.

Each concurrent state is modeled as an E-equivalence class $[t]_E$ of ground terms, and rewriting happens *modulo* E. A *one-step rewrite* $[t]_E \rightarrow_{\mathcal{R}} [t']_E$ exists in $\mathcal{R}$ iff there exists $u \in [t]_E$ that can be rewritten to $v \in [t']_E$ using some rule in R. More precisely, if the one-step rewrite exists with a rule $l : q \longrightarrow r$, there is a subterm u' of $u \in [t]_E$ at a position p which is an instance of q with a substitution ϕ, and $u[\phi(r)]_p \in [t']_E$. Then, the one-step rewrite $[t]_E \rightarrow_{\mathcal{R}} [t']_E$ has the corresponding *one-step proof term* $\lambda = [u[l(\phi)]_p]_E$, and is denoted by $[t]_E \xrightarrow{\lambda} [t']_E$. Since $[t]_E \rightarrow^{*}_{\mathcal{R}} [t']_E$ is *undecidable* in general, we need to consider additional computability assumptions for a rewrite theory $\mathcal{R}$.

A rewrite theory $\mathcal{R} = (\Sigma, E \cup A, R)$ is *computable*, iff: (i) $(\Sigma, E \cup A)$ is ground terminating and confluent modulo the equational axioms A [11],[1] and (ii) R is ground coherent relative to E modulo A [35]. If $\mathcal{R}$ is computable, each $[t]_{E\cup A}$ has a unique E-canonical form $[can_{E/A}(t)]_A \in T_{\Sigma/A}$ which cannot be further rewritten with E modulo A. Moreover, each rewrite $[t]_{E\cup A} \xrightarrow{\lambda} [t']_{E\cup A}$ has an equivalent transition $[can_{E/A}(t)]_A \xrightarrow{\lambda'} [can_{E/A}(t')]_A$ in the canonical initial model associated to $\mathcal{R}$ with the corresponding canonical proof term λ' [29].

A *computation* (π, γ) in $\mathcal{R}$ is then a path $\pi(0) \xrightarrow{\gamma(0)} \pi(1) \xrightarrow{\gamma(1)} \pi(2) \xrightarrow{\gamma(1)} \cdots$ where $\pi(i) = [can_{E/A}(t_i)]_A$, $\gamma(i) = \lambda'_i$, $\pi(i) \xrightarrow{\gamma(i)} \pi(i+1)$ for each $i \in \mathbb{N}$, and all the t_i belong to the chosen kind k of states. Any computation in $\mathcal{R}$ expands infinitely if $\mathcal{R}$ is *deadlock-free*, i.e., there is no state that cannot be further rewritten by a rule in R. This is not a strong restriction since any rewrite theory whose rules do not have rewrites in their conditions can be transformed into a semantically equivalent deadlock-free theory [10,30]. Actually, our model checker does a similar transformation automatically (see Section 3.1).

2.2 Syntax and Semantics of LTLR Formulas

The only syntactic difference between LTLR and LTL is that an LTLR formula may include some spatial action patterns $\delta_1, \ldots, \delta_n$ as well as state propositions $p_1, \ldots, p_m$, and therefore may describe properties involving both states and events. The syntax of LTLR formulas is the following, where p ranges over atomic state propositions Π and δ ranges over spatial action patterns W:

$$\varphi ::= p \mid \delta \mid \neg\varphi \mid \varphi \wedge \varphi \mid \bigcirc\varphi \mid \Diamond\varphi \mid \Box\varphi \mid \varphi\mathcal{U}\varphi$$

Spatial action patterns describe properties of one-step rewrites (equivalently, one-step proof terms) in $\mathcal{R}$. For example, spatial action patterns to describe one-step rewrites with partial information are defined as follows [2], where $u_1, \ldots, u_m$ are ground terms,[2] and where if $l : t \to t' \in R$, then $\{x_1, \ldots, x_n\} \subseteq vars(t)$:

- l : one-step proof terms involving a rule in R with label l.
- $l(x_1\backslash u_1; \cdots; x_m\backslash u_m)$: one-step proof terms with a rule label l whose matching substitution ϕ satisfies $[\phi(x_i)]_{E\cup A} = [u_i]_{E\cup A}$.
- $t[l(x_1\backslash u_1; \cdots; x_m\backslash u_m)]_p$: one-step proof terms that are instances of the pattern $l(x_1\backslash u_1; \cdots; x_m\backslash u_m)$ where the corresponding rewrites happen at position p of $[t]_{E\cup A}$.

Many other examples of spatial action patterns are given in Section 5.

[1] In Maude, the axioms A are any combination of *associativity*, *commutativity*, and *identity* axioms for different binary operators. We assume that there exists a matching algorithm modulo A as part of the computability assumption for $\mathcal{R}$.

[2] The original definition in [27] allows spatial action patterns to have variables, and its semantics was specified by the matching relation with one-step proof terms. In this paper, we assume that spatial action patterns are ground terms. However, we do *not* lose any expressive power thanks to using equational semantics, since a matching relation can be defined by equations (see Definition 1).

To integrate rewriting logic and LTLR, a signature for actions and for state predicates, and their satisfaction definition are required. A rewrite theory $\mathcal{R} = (\Sigma \cup \Sigma', E \cup A \cup D, R)$ has a *support signature* Σ' with respect to equations D iff (i) $(\Sigma \cup \Sigma', E \cup A \cup D)$ protects $(\Sigma, E \cup A)$,[3] (ii) Σ' contains the following:

- a sort `State` for states of the system.
- a sort `ProofTerm` for one-step proof terms, and corresponding operators for rule labels and substitutions in proof terms.
- sorts `Prop` and `Action`, respectively, for state propositions Π and spatial action patterns W, with corresponding operators for predicates.
- a sort `Bool` with two distinct constants `true` and `false`.
- operators `_|=_ : State Prop -> Bool` and `_|=_ : ProofTerm Action -> Bool` to define the satisfaction relations for atoms.

and, (iii) the equations in D define the truth of each state predicate (resp., spatial action pattern) on each state (resp., one-step proof term) by means of the `_|=_` operators (see Section 4.1 for more details on the definition of satisfaction of action patterns). We assume that $\mathcal{R}$ has a support signature from now on.

On a computable deadlock-free rewrite theory $\mathcal{R}$, the semantics of LTLR formulas φ is defined by the satisfaction relation $\mathcal{R}, [t]_{E \cup A} \models \varphi$. By definition, $\mathcal{R}, [t]_{E \cup A} \models \varphi$ holds if and only if for each infinite computation (π, γ) starting at $[t]_{E \cup A}$ in $\mathcal{R}$, the path satisfaction relation $\mathcal{R}, (\pi, \gamma) \models \varphi$ holds. The path satisfaction relation for LTLR is quite similar to that of LTL. The key difference between the LTLR and the LTL semantics is the semantics of *spatial action patterns.* Specifically, the relation $\mathcal{R}, (\pi, \gamma) \models \delta$ holds iff the first proof term $\gamma(0)$ of the current computation satisfies the spatial action pattern δ.

Definition 1. *The path satisfaction relation $\mathcal{R}, (\pi, \gamma) \models \varphi$ for an LTLR formula φ is defined inductively as follows:*

- $\mathcal{R}, (\pi, \gamma) \models p$ *iff* $[\pi(0) \; \texttt{|=} \; p]_{E \cup A} = [\texttt{true}]_{E \cup A}$
- $\mathcal{R}, (\pi, \gamma) \models \delta$ *iff* $[\gamma(0) \; \texttt{|=} \; \delta]_{E \cup A} = [\texttt{true}]_{E \cup A}$
- $\mathcal{R}, (\pi, \gamma) \models \neg\varphi$ *iff* $\mathcal{R}, (\pi, \gamma) \not\models \varphi$
- $\mathcal{R}, (\pi, \gamma) \models \varphi \wedge \varphi'$ *iff* $\mathcal{R}, (\pi, \gamma) \models \varphi$ *and* $\mathcal{R}, (\pi, \gamma) \models \varphi'$
- $\mathcal{R}, (\pi, \gamma) \models \bigcirc\varphi$ *iff* $\mathcal{R}, (\pi, \gamma)^1 \models \varphi$ [4]
- $\mathcal{R}, (\pi, \gamma) \models \Diamond\varphi$ *iff for some* $k \geq 0$, $\mathcal{R}, (\pi, \gamma)^k \models \varphi$
- $\mathcal{R}, (\pi, \gamma) \models \Box\varphi$ *iff for all* $k \geq 0$, $\mathcal{R}, (\pi, \gamma)^k \models \varphi$
- $\mathcal{R}, (\pi, \gamma) \models \varphi \mathcal{U} \varphi'$ *iff there is some* $k \geq 0$ *such that* $\mathcal{R}, (\pi, \gamma)^k \models \varphi'$ *and for all* $0 \leq i < k$, $\mathcal{R}, (\pi, \gamma)^i \models \varphi$

Note that the above semantic definition of LTLR extends the semantics of LTL. Indeed, the semantics of an LTLR formula with no spatial action patterns is exactly the same as that of one of LTL on the underlying Kripke structure associated to the given rewrite theory [10,29].

---

[3] The unique Σ-homomorphism $T_{\Sigma/E} \to T_{\Sigma \cup \Sigma'/E \cup D}|_{\Sigma}$ induced by the theory inclusion $(\Sigma, E) \subseteq (\Sigma \cup \Sigma', E \cup D)$ should be bijective at each sort $s \in \Sigma$.

[4] $(\pi, \gamma)^i$ denotes the suffix of (π, γ) beginning at position $i \in \mathbb{N}$, i.e., $(\pi, \gamma)^i = (\pi \circ s^i, \gamma \circ s^i)$ with s the successor function.

3 Model Checking LTLR Formulas

In our previous work, we converted the LTLR model checking problem into an LTL model checking one by an automatic theory transformation to construct an associated Kripke structure [2,29]. In spite of the simplicity of the previous LTLR model checker, there is the problem of a blowup in the number of states,[5] and also some increase in the formula itself by addition of $\bigcirc$ operators. In the approach presented below, we avoid such a state-space blowup by directly constructing the *labeled Kripke structure* corresponding to a rewrite theory $\mathcal{R}$.

3.1 The Labeled Kripke Structure of a Rewrite Theory

A rewrite theory $\mathcal{R}$ in which atomic predicates have been defined has an underlying Kripke structure given by the total binary relation extending its one-step sequential rewrites [10,13]. Similarly, we can associate a labeled Kripke structure to a rewrite theory $\mathcal{R}$. A *labeled Kripke structure* (LKS) is a natural extension of a Kripke structure with transition labels [6], defined as follows:

Definition 2. *A* labeled Kripke structure *is a 6-tuple* $(S, S_0, AP, \mathcal{L}, ACT, T)$ *with* S *a set of* states, $S_0 \subseteq S$ *a set of* initial states, AP *a set of* atomic state propositions, $\mathcal{L} : S \to \mathcal{P}(AP)$ *a* state-labeling function, ACT *a set of* atomic events, *and* $T \subseteq S \times \mathcal{P}(ACT) \times S$ *a* labeled transition relation.

Note that each transition of an LKS is labeled by a *set* A of atomic events, which could be empty. We assume that the transition relation of a labeled Kripke structure is *total* so that every state has a next state, i.e., no deadlocks exist. A labeled transition $(s, A, s') \in T$ is often denoted by $s \xrightarrow{A} s'$.

A *path* (π, α) of an LKS is an infinite sequence $\langle \pi(0), \alpha(0), \pi(1), \alpha(1), \ldots \rangle$ such that $\pi(i) \in S$, $\alpha(i) \subseteq ACT$, and $\pi(i) \xrightarrow{\alpha(i)} \pi(i+1)$ for each $i \geq 0$. A *trace* $(\mathcal{L}(\pi), \alpha)$ of a path (π, α) is an infinite sequence $\langle \mathcal{L}(\pi(0)), \alpha(0), \mathcal{L}(\pi(1)), \alpha(1), \ldots \rangle$.

Given a computable rewrite theory $\mathcal{R} = (\Sigma, E, R)$, an initial state $[t]_E$ of sort `State`, a finite set $P \subseteq T_{\Sigma/E,Prop}$ of atomic propositions, and a finite set $W \subseteq T_{\Sigma/E,Action}$ of spatial action patterns, we can associate to $\mathcal{R}$ the LKS $\mathcal{K}_{P,W}(\mathcal{R})_{[t]_E} = (S, \{[t]_E\}, P, \mathcal{L}, W \uplus \{\star\}, T)$ such that:

- $S = T_{\Sigma/E,[State]}$, represented by the canonical terms of kind `[State]`.
- $\mathcal{L}(s)$ is the set of atomic state propositions in P that hold in the state s, that is, $\mathcal{L}([t]_E) = \{p \in P : [t \mid\!= p]_E = [\mathtt{true}]_E\}$.
- The labeled transition relation T specifies the corresponding transitions, labeled with events from W which are satisfied by a given one-step rewrite proof λ in $\mathcal{R}$. In addition, T has a self-loop for each deadlock state in order to be a total relation. That is, $\star$ is an event denoting a *deadlock*,

$$[t]_E \xrightarrow{A} [t']_E \in T \text{ iff } [t]_E \xrightarrow{\lambda}_{\mathcal{R}} [t']_E \text{ and } A = \{\delta \in W : [\lambda \mid\!= \delta]_E = [\mathtt{true}]_E\}.$$
$$[t]_E \xrightarrow{\{\star\}} [t]_E \in T \text{ iff } [t]_E \text{ cannot be further rewritten.}$$

[5] If $\mathcal{R}$ has n states and m transitions, the associated Kripke structure has $O(nm)$ states [29].

The semantics of an LTLR formula φ whose set of spatial action patterns is W and whose set of atomic propositions is P can be also defined over an LKS $\mathcal{K}_{P,W}(\mathcal{R})_{[t]_E}$. By construction, for each computation (π, γ) of $\mathcal{R}$, there is a path (π, α) of $\mathcal{K}_{P,W}(\mathcal{R})_{\pi(0)}$ with $\alpha(i) = \{\delta \in W : [\gamma(i) \models \delta]_E = [\mathtt{true}]_E\}$ for each $i \in \mathbb{N}$. Then, the LTLR semantics can be defined directly on $\mathcal{K}_{P,W}(\mathcal{R})_{\pi(0)}$ in a way entirely similar to Definition 1. For example, $\mathcal{K}_{P,W}(\mathcal{R})_{\pi(0)}, (\pi, \alpha) \models p$ iff $p \in \mathcal{L}(\pi(0))$, and $\mathcal{K}_{P,W}(\mathcal{R})_{\pi(0)}, (\pi, \alpha) \models \delta$ iff $\delta \in \alpha(0)$. This definition clearly satisfies the following equivalence lemma:

Lemma 1. *Given a computable deadlock-free rewrite theory $\mathcal{R}$, and an LTLR formula φ with set of atomic propositions P and set of action patterns W, $\mathcal{R}, (\pi, \gamma) \models \varphi$ iff $\mathcal{K}_{P,W}(\mathcal{R})_{\pi(0)}, (\pi, \alpha) \models \varphi$.*

If $\mathcal{R}$ has deadlock states, we can first transform $\mathcal{R}$ to the equivalent deadlock-free rewrite theory for the same result. In fact, the deadlock-free translation is similar to the deadlock completion in the above $\mathcal{K}_{P,W}(\mathcal{R})_{[t]_E}$ construction [30].

3.2 Automata-Based Verification of LTLR Formulas

The model checking problem of an LTLR formula φ on a rewrite theory $\mathcal{R}$ is now reduced to the satisfiability of φ on its associated LKS $\mathcal{K}_{P,W}(\mathcal{R})_{[t]_E}$. Given an LKS $M = (S, S_0, AP, \mathcal{L}, ACT, T)$ and an LTLR formula φ, we need to determine whether $M \models \varphi$, that is, to check the satisfaction relation $M, (\pi, \alpha) \models \varphi$ for all paths (π, α) of M starting at some $s \in S_0$.

The automata-based verification of an LTL formula φ on a given Kripke structure $\mathcal{K} = (S, S_0, AP, \mathcal{L}, T)$ uses the Büchi automaton $\mathscr{B}_{\neg\varphi}$ associated to the negated formula $\neg\varphi$, and checks the emptiness of the synchronous product $\mathcal{K} \times \mathscr{B}_{\neg\varphi}$ to determine whether $\mathscr{B}_{\neg\varphi}$ accepts any trace of $\mathcal{K}$ [3].

Definition 3. *A Büchi automaton is a 5-tuple $(S_B, S_{B_0}, P, T_B, F)$ such that S_B is a finite set of states, $S_{B_0} \subseteq S_B$ is a set of initial states, P is an alphabet of transition labels, $T_B \subseteq S_B \times P \times S_B$ is a transition relation, and $F \subseteq S_B$ is a set of accept states.*

The *synchronous product* of a Kripke structure $\mathcal{K}$ and a Büchi automaton $\mathscr{B}$ with an alphabet $\mathcal{P}(AP)$ is a Büchi automaton $(S \times S_B, S_0 \times S_{B_0}, \mathcal{P}(AP), T', S \times F)$ such that $(s, b) \xrightarrow{\mathcal{L}(s)} (s', b') \in T'$ iff $s \rightarrow s' \in T \ \wedge \ b \xrightarrow{\mathcal{L}(s)} b' \in T_B$. The essence of LTL model checking is expressed by the following theorem [9].

Theorem 1. *Given a Kripke structure M and an LTL formula φ, there is a Büchi automaton $\mathscr{B}_{\neg\varphi}$ such that $M \models \varphi$ iff $L(M \times \mathscr{B}_{\neg\varphi}) = \varnothing$.*

We generalize this automaton-based approach to characterize the LTLR model checking problem, following an approach similar to that of SE-LTL model checking [6]. If AP and ACT are disjoint, then there is a one-to-one correspondence between a trace $(\mathcal{L}(\pi), \alpha)$ and a union trace $\mathcal{L}(\pi) \cup \alpha$, where $(\mathcal{L}(\pi) \cup \alpha)(i) = \mathcal{L}(\pi)(i) \cup \alpha(i)$ for each $i \geq 0$. In such union traces, there is no difference between events and state propositions. Hence, we can check whether a union trace

$\mathcal{L}(\pi) \cup \alpha$ is accepted by a Büchi automaton for a formula $\neg\varphi$, using the same Büchi automata construction as in the LTL case. In fact, union traces of an LKS M induce an equivalent Kripke structure $\mathcal{D}(M)$, whose states are pairs consisting of a state and a transition label of M.

Definition 4. *Given an LKS $M = (S, S_0, AP, \mathcal{L}, ACT, T)$ with $AP \cap ACT = \varnothing$, its associated Kripke structure $\mathcal{D}(M)$ is $(S', S'_0, AP \cup ACT, \mathcal{L}', T')$ where*

- $S' = \{\langle s, A\rangle \in S \times \mathcal{P}(ACT) : \exists s' \in S.\ s \xrightarrow{A} s'\}$
- $S'_0 = (S_0 \times \mathcal{P}(ACT)) \cap S'$
- $\mathcal{L}'(\langle s, A\rangle) = \mathcal{L}(s) \cup A$
- $\langle s, A\rangle \to \langle s', A'\rangle \in T'$ *iff* $s \xrightarrow{A} s' \in T$ *and* $\langle s', A'\rangle \in S'$

It is clear that each trace of $\mathcal{D}(M)$ is a union trace of M. Furthermore, M and $\mathcal{D}(M)$ are equivalent in the sense of the satisfiability of a formula.

Lemma 2. *Given an LKS M and an LTLR formula φ, $M \models \varphi$ iff $\mathcal{D}(M) \models \varphi$, where any event in φ is regarded as a state proposition of $\mathcal{D}(M)$.*

Proof. It suffices to show that $M, (\pi, \alpha) \models \varphi$ iff $\mathcal{D}(M), \pi \cup \alpha \models \varphi$ for each path (π, α) of M. We can prove this by structural induction on φ. If φ is a spatial action pattern δ, then $M, (\pi, \alpha) \models \delta$ iff $\delta \in \alpha(0)$. Equivalently, $\delta \in \mathcal{L}(\pi(0)) \cup \alpha(0) = \mathcal{L}'(\langle \pi(0), \alpha(0)\rangle)$. Therefore, $\mathcal{D}(M), \pi \cup \alpha \models \delta$. The other cases are similar or follow easily from the induction hypothesis. □

In order to determine whether a Büchi automaton $\mathscr{B}_{\neg\varphi}$ accepts a union trace of an LKS M, we avoid the use of $\mathcal{D}(M) \times \mathscr{B}$ directly since it produces a state-space blowup. Instead, we define a special synchronous product $M \otimes \mathscr{B}_{\neg\varphi}$, which advances to the next state only if both state labels and event labels are accepted by the current transition of $\mathscr{B}_{\neg\varphi}$.

Definition 5. *Given an LKS $M = (S, S_0, AP, \mathcal{L}, ACT, T)$ and a Büchi automaton $\mathscr{B} = (S_B, S_{B_0}, P, T_B, F)$ with $P = \mathcal{P}(AP \cup ACT)$, the* state/event product *is a Büchi automaton $M \otimes \mathcal{B} = (S \times S_B, S_0 \times S_{B_0}, P, T', S \times F)$ such that $(s, b) \xrightarrow{\mathcal{L}(s) \cup A} (s', b') \in T'$ iff $s \xrightarrow{A} s' \in T$ and $b \xrightarrow{\mathcal{L}(s) \cup A} b' \in T_B$.*

The following lemma shows that the language emptiness problem for $M \otimes \mathscr{B}$ is equivalent to that of $\mathcal{D}(M) \times \mathscr{B}$.

Lemma 3. *Given an LKS M and a Büchi automaton $\mathscr{B}$, $L(M \otimes \mathscr{B}) = \varnothing$ iff $L(\mathcal{D}(M) \times \mathscr{B}) = \varnothing$.*

Proof. Let $\rho \in L(M \otimes \mathscr{B})$. By definition, $\rho = (\mathcal{L}(s_1), b_1)(\mathcal{L}(s_2), b_2)\ldots$ such that $\mathcal{L}(s_1)\mathcal{L}(s_2)\mathcal{L}(s_3)\ldots$ is a trace of M and $b_1 b_2 \ldots \in L(\mathscr{B})$, where $s_i \xrightarrow{A_i} s_{i+1}$ and $b_i \xrightarrow{\mathcal{L}(s_i) \cup A_i} b_{i+1}$ for each $i \geq 0$. Clearly, $(\mathcal{L}(s_1) \cup A_1)\ (\mathcal{L}(s_2) \cup A_2)\ldots$ is a trace of $\mathcal{D}(M)$. Therefore, the union trace $\rho' = (\mathcal{L}(s_1) \cup A_1, b_1)(\mathcal{L}(s_2) \cup A_2, b_2)\ldots$ is in $L(\mathcal{D}(M) \times \mathscr{B})$. The proof for the implication in the other direction is similar. □

As a result, for an LTLR formula φ and an LKS M, we can conclude that $M \models \varphi$ iff $L(M \otimes \mathscr{B}_{\neg\varphi}) = \varnothing$, where $\mathscr{B}_{\neg\varphi}$ is a Büchi automaton for $\neg\varphi$ constructed in exactly the same way as in the LTL case. Consequently, by the LKS construction associated to a rewrite theory $\mathcal{R}$, we have:

Theorem 2. *Given an LTLR formula φ with set of atomic propositions P and set of spatial action patterns W, a computable deadlock-free rewrite theory $\mathcal{R}$, and an initial state $[t]_E$,*

$$\mathcal{R}, [t]_E \models \varphi \quad \Leftrightarrow \quad L(\mathcal{K}_{P,W}(\mathcal{R})_{[t]_E} \otimes \mathscr{B}_{\neg\varphi}) = \varnothing$$

Note that the cost for LTLR model checking using an LKS is $O((n+m)\cdot 2^f)$ with n states, m transitions, and f the size of the formula. The m factor here is added since each transition needs to test each spatial action pattern in a formula, where a labeled Kripke structure may have several transitions with different labels between two states. If there are no spatial action patterns in a formula, then the cost is $O(n \cdot 2^f)$, exactly the same as for LTL model checking. Recall that our previous implementation leads to a model checking complexity of $O(n \cdot m \cdot 2^f)$.

4 The Maude LTLR Model Checker

The Maude LTLR model checker extends the existing LTL model checker in Maude, which contains support signatures for LTL model checking in the predefined module `MODEL-CHECKER`. The Maude LTL model checker supports on-the-fly LTL model checking for an initial state $[t]_E$ with sort `State` of a computable rewrite theory $\mathcal{R} = (\Sigma, E, R)$ such that the set of all states *reachable* from $[t]_E$ is *finite* [13]. If such a rewrite theory $\mathcal{R}$ is specified in Maude by a system module `M` with an initial state `init` of sort $\texttt{State}_M$, the following procedure is used for model checking of LTL properties beginning at the initial state `init`:

- Define a new module, say `CHECK-M`, which includes both the module `M` and the module `MODEL-CHECKER`.
- Give a *subsort* declaration, `subsort` $\texttt{State}_M$ `< State`, where the sort `State` is in the support signature in `MODEL-CHECKER`.
- Define the *syntax* of (parameterized) *state predicates* of sort `Prop`, which is a subsort of the sort `Formula` for LTL formulas.
- Define the *semantics* of the state predicates using the operator `_|=_ : State Prop -> Bool`. The semantics of each (parameterized) state predicate `p` is then given by a set of (conditional) equations of the form:

  `ceq` State_i `|= p(`arg_{i1}`, ..., `arg_{im}`) = true if` Cond_i `.`

  where State_i are *patterns* of sort $\texttt{State}_M$, `p(`$\mathit{arg}_{i1}, \ldots, \mathit{arg}_{im}$`)` are patterns of sort `Prop`, and Cond_i are conjunctions of equalities and memberships.
- The model checking command `reduce modelCheck(init,`*formula*`)` returns `true` or a counterexample.

Our LTLR model checker involves a similar process and interface as that summarized above for an LTL formula. However, for LTLR formulas we need to extend the module's signature to support spatial action patterns.

4.1 Support Signature for LTLR Formulas

The support signature for the LTLR model checker is defined in the system module `LTLR-MODEL-CHECKER` which includes the definitions of one-step proof terms and spatial action patterns. A one-step proof term $u[l(\phi)]_p$ is represented as a triple consisting of a rule label l, a substitution ϕ as a set of single assignments, and a *context term* of u that has a hole `[]` at position p. A substitution has the form $\texttt{var}_1\backslash\texttt{value}_1$; ... ; $\texttt{var}_k\backslash\texttt{value}_k$. In our LTLR model checker, one-step proof terms have sort `ProofTerm` and are defined by the following operator:

```
op {_|_:_}  : StateContext RuleName Substitution -> ProofTerm [ctor...] .
```

To denote a deadlock event $\star$ in the LKS construction (see Section 3.1), the `deadlock` constant with sort `ProofTerm` is defined as well.

Spatial action patterns have sort `Action`, and both `Prop` and `Action` are subsorts of `Formula`. The syntax and the semantics of spatial action patterns is defined in a way similar to that of state propositions. By default, we define useful spatial action patterns related to partial information of one-step proof terms.

```
subsorts ProofTerm < Action .                       --- {u[]_p | l : x1\t1;...}
op {_}    : RuleName -> Action .                    --- {l}
op {_:_}  : RuleName Substitution -> Action .       --- {l : x1\t1;...}
op {_|_}  : StateContext RuleName -> Action .       --- {u[]_p | l}
op top    : RuleName -> Action .                    --- top(l)
op top    : RuleName Substitution -> Action .       --- top(l, x1\t1;...)
```

Terms of sort `ProofTerm` are also viewed as spatial action patterns that describe one-step proof terms containing the given partial substitution in the spatial action pattern. Furthermore, other action patterns are described by rule labels, or by rule labels with an associated substitution. Sometimes we also want to consider patterns corresponding to rewrites at the top level. The satisfaction relation between a proof term and a spatial action pattern is then defined by equations involving the operator `_|=_ : ProofTerm Action -> Bool`.

```
var C : StateContext . var R : RuleName . var S S' : Substitution .

eq {C  | R : S}      |= {R}         = true .
eq {C  | R : S ; S'} |= {R : S}     = true .
eq {C  | R : S ; S'} |= {C | R : S} = true .
eq {C  | R : S}      |= {C | R}     = true .
eq {[] | R : S}      |= top(R)      = true .
eq {[] | R : S ; S'} |= top(R, S)   = true .
```

In addition, users can define their own (parameterized) spatial action pattern `sp` by giving a set of (conditional) equations of the form:

`ceq` $\textit{Proofterm}_i$ `|= sp(`$arg_{i,1}$, ..., $arg_{i,m}$`) = true if` $\textit{Cond}_i$.

where $\textit{Proofterm}_i$ are patterns for one-step proof terms.

4.2 Theory Extension for One-Step Proof Terms

The signature of context terms and substitutions for one-step proof terms depends on the given rewrite theory $\mathcal{R}$. A rewrite can happen at any position of a state term by a rule in R, which implies that a hole symbol `[]` in a context term can have any sort of the left-sides of the rules in R. Moreover, variables in rules can have any sort in Σ, and this is needed for assignments in substitutions. Therefore, additional operators regarding context holes and assignment symbols are required to generate one-step proof terms for model checking purposes.

Given a computable rewrite theory $\mathcal{R} = (\Sigma, E, R)$, the maximal signature $P_\Omega(R)$ required for one-step proof terms can be generated from the rules R and the subsignature Ω of constructors[6]. For each rewrite rule[7] $l : q \to r$ with q, r of sort B, the signature $P_\Omega(R)$ extends Ω by adding incrementally:

– assignment operators for each sort B_i of variables x_i in q.

```
op _\_ : Qid Bi -> Assignment [ctor...] .
```

– A hole operator with a new sort Context$$B$ related to the sort B:

```
op [] : -> Context$B [ctor...] .
```

– For each operator o : A1 ...Am -> A in Ω, a set of operators:

```
op o : Context$A1 A2 A3 ... Am -> Context$A [ditto] .
op o : A1 Context$A2 A3 ... Am -> Context$A [ditto] .
...
op o : A1 A2 A3 ... Context$Am -> Context$A [ditto] .
```

where Context$$A_1$,..., Context$$A_m$ and Context$A are new sorts related to each sort in the operator declaration. These operator declarations guarantee that each context term should contain only one hole symbol.

The new signature $P_\Omega(R)$ defines sorts of context terms for all operators in Ω. Since one-step proof terms contain only context terms of states, the sort Context$$State_M$ corresponding to sort $State_M$ for states should be finally defined as a subsort of StateContext. Note that the full signature of $P_\Omega(R)$ is not needed in general; only those operators involved in the spatial action patterns of a given LTLR formula ϕ are needed.

The theory extension for one-step proof terms is either manually defined by the user, or automatically generated. A user can define the minimal required subset of $P_\Omega(R)$ for a given LTLR formulas at the Core Maude level. The automatic generation of $P_\Omega(R)$ is also provided at the Full Maude level. We define a theory transformation PROOF : $\mathcal{R} \to \mathcal{R} \cup P_\Omega(R)$, by extending Full Maude [12] using Maude's reflective capability [10]. In the Full Maude interface, a module PROOF[M] contains the signature $P_\Omega(R)$ where $\mathcal{R}$ is specified by a system module M in Maude. In addition, the Full Maude interface provides the function to selectively generate context terms for optimization purposes, since spatial action patterns do not use context information in many cases according to our experience, as illustrated by the example in Section 5.

[6] If $\Omega \subseteq \Sigma$ is the subsignature of *constructors* (specified with the `[ctor]` keyword), then every canonical ground term in $\mathcal{R}$ should be an Ω-term.

[7] We do not consider the condition part of a rewrite rule for a one-step proof term.

4.3 The LTLR Model Checker Implementation

Our LTLR model checker reuses the modules of the existing C++ LTL model checker implementation, which uses a very weak alternating automaton to generate the Büchi automaton [17] with the strongly connected component optimizations [34], and the nested depth first method [20] for the emptiness checking algorithm (see [13,14] for the details). For automata-based verification of LTLR formulas, since events do not need to be distinguished from state propositions as discussed in Section 3.2, an LTLR formula is transformed into an LTL formula by regarding events as state propositions, and the same algorithm as in the LTL case is then used to generate a Büchi automaton. The emptiness checking algorithm is also the same as the one for the LTL model checker, but the synchronous product is constructed from an LKS instead of from a Kripke structure.

A labeled Kripke structure is generated on-the-fly, so that only requested (or reachable) states or transitions are created. Each state or transition generated keeps two bit vectors to record:

1. which state propositions (or events) have been tested, and
2. which state propositions (or events) were satisfied in the state (or transition).

In addition, both the LTL and the LTLR implementations require three extra bit vectors for each state regarding the synchronous product search, which depends on the search algorithm. For space optimization, whenever the one-step proof term is generated, we test all possible spatial action patterns, and the full term graph representation of the one-step proof term is discarded, except that we need it later to generate a counterexample with one-step proof terms.

5 Example: The Bounded Retransmission Protocol

In this example we show how a complex system description under only state-based design can be greatly simplified using our LTLR model checker. We show the modeling of the bounded retransmission protocol (BRP), which is an extension of the alternating bit protocol where a limit is placed on the number of transmissions of the messages [1]. Descriptions of this protocol such as those given in [30] are quite complex due to the use of a state-based logic, which forces the specification of the protocol to *encode* a lot of action information in the state. With LTLR this encoding need completely disappears, leading to a much simpler protocol specification. For example, the previous rewriting logic specification had 36 rewrite rules,[8] but we show here the same model with only 14 rewrite rules and with a more readable state representation.

The BRP protocol description is as follows [1]. At the sender side the protocol requests a sequence of data elements $d_1, \dots, d_n$ (action REQ) and communicates a confirmation which can be either SOK, SNOK, or SDNK. SOK means that the file has been transferred successfully, SNOK means that the file has not

[8] The specification in [30] has 35 rules, however, by comparing with our new specification, we found that one rule was missing there.

been transferred completely, and SDNK means that the file may not have been transferred completely. At the receiver side the protocol delivers each correctly received datum with an indication which can be either RFST, RINC, ROK, or RNOK. RFST means that the delivered datum is the first one and more data will follow, RINC means that the datum is an intermediate one, and ROK means that this was the last datum and the file is completed. However, when the connection with the sender is broken, an indication RNOK is delivered. The Maude specification of BRP in [30] is adapted from the *untimed* model which appeared in [1], and our following specification simplifies it.

States of the system are represented by terms of sort `Conf` with a 6-tuple operator `<_,_,_,_,_,_> : Sender Bool MsgL MsgL Bool Receiver -> Conf`. The first and the sixth components describe the current status of the sender and the receiver, respectively. The second and fifth components are boolean values used by the sender and the receiver for synchronization purposes. The third and fourth components correspond to the two (ordered) lossy channels through which the sender and the receiver communicate.

Each message is one of 0, 1, *fst*, *last*, where *fst* denotes the first datum and *last* the last datum. The sender's status is one of `idle`, `snd`(α) and `acc`(α), where `snd`(α) means that the sender is sending a message α, and `acc`(α) indicates that the sender gets an acknowledgement of the acceptance of a message α. The receiver can have status `wait` or `rec`(α), where `rec`(α) denotes that the receiver gets a message α. The initial state is a 6-tuple `< idle, false, nil, nil, false, wait >`.

The behavior of the protocol is described by rewrite rules. The client side behavior of the protocol is defined by the following rewrite rules. The auxiliary status `set`(α) denotes that the sender is about to send a message α, and it is equationally reduced to the state with status `snd`(α) with one α sent.

```
var S : Sender . var R : Receiver .
vars M M' : Msg . vars K L : MsgL . vars A T : Bool .

rl  [req]  : < idle,      A,     nil,    nil,     false, R >
          => < set(fst), false, nil,    nil,     false, R > .
rl  [snd]  : < snd(M),   A,     K,      L,       T,     R >
          => < snd(M),   A,     K ; M, L,       T,     R > .
rl  [acc]  : < snd(M),   A,     K,      M' ; L, T,     R >
          => < M # M',   A,     K,      L,       T,     R > .
crl [loss] : < snd(M),   A,     K,      nil,     T,     R >
          => < idle,     true,  K,      nil,     T,     R > if M =/= fst .
eq M # M' = if M == M' then acc(M) else snd(M) fi .
eq < set(M), A, K, L, T, R > = < snd(M), A, K ; M, L, T, R > .
```

The following rewrite rules describe the choice of the next message.

```
rl [sel] : acc(fst)  => set(0) .  rl [sel] : acc(fst) => set(last) .
rl [sel] : acc(0)    => set(1) .  rl [sel] : acc(0)   => set(last) .
rl [sel] : acc(1)    => set(0) .  rl [sel] : acc(1)   => set(last) .
rl [sel] : acc(last) => idle .
```

The server-side behavior of the protocol is defined as follows. In the rule `rec`, when a received datum is `fst`, the server flag is set to true.

```
crl [rec]  : < S, false, M ; K, L,      T,      R      >
          => < S, false, K,      L ; M, M ? T, rec(M) > if R =/= rec(M) .
rl  [ign]  : < S, A,     M ; K, L,      T,      rec(M) >
          => < S, A,     K,      L ; M, T,      rec(M) > .
crl [nil]  : < S, A,     nil,   L,      T,      rec(M) >
          => < S, A,     nil,   L,      false, wait   > if M == last or
                                                             A == true .
eq M ? T = if M == fst then true else T fi .
```

The BRP protocol should satisfy the following properties:

1. A request REQ must be followed by a confirmation (SOK, SNOK, or SDNK) before the next request.
2. An RFST indication must be followed by one of the two indications ROK or RNOK before the beginning of a new transmission (new request of a sender).
3. An SOK confirmation must be preceded by an ROK indication.
4. An RNOK indication must be preceded by an SNOK or SDNK confirmation (abortion).

Events occurring in the above properties can be defined by equations as follows:

```
(mod BRP-CHECK is
  protecting PROOF[BRP] .
  including LTLR-MODEL-CHECKER .

  subsort Conf < State .
  subsorts Context$Conf < StateContext .
  ops req sok snok sdnk rfst rinc rok rnok : -> Action .
  var M : Msg . var C : StateContext . var SS : Substitution .

  eq  {C | 'req  : SS}                |= req  = true .
  eq  {C | 'acc  : 'M  \ last ;
                   'M' \ last ; SS}   |= sok  = true .
  ceq {C | 'loss : 'M  \ M ; SS}      |= snok = true if M =/= last .
  eq  {C | 'loss : 'M  \ last ; SS}   |= sdnk = true .
  eq  {C | 'rec  : 'M  \ fst ; SS}    |= rfst = true .
  ceq {C | 'rec  : 'M  \ M ; SS}      |= rinc = true if M == 0 or M == 1 .
  eq  {C | 'rec  : 'M  \ last ; SS}   |= rok  = true .
  ceq {C | 'nil  : 'M  \ M ; SS}      |= rnok = true if M =/= last .
endm)
```

The system has an infinite number of states, but by equational abstraction, we can collapse the set of states to a finite number (see [30]). Then, the properties (1)–(4) can be model-checked as follows:

```
Maude> (red modelCheck(init,[](req -> O(~ req W(sok \/ snok \/ sdnk)))).)
result Bool : true

Maude> (red modelCheck(init,[](rfst -> (~ req W(rok \/ rnok)))) .)
result Bool : true
```

```
Maude> (red modelCheck(init,[](req -> (~ sok W rok))) .)
result Bool : true

Maude> (red modelCheck(init,[](req -> (~ rnok W(snok \/ sdnk)))) .)
result Bool : true
```

6 Experimental Results

We experiment with three LTLR model checking cases to compare the performance of the new algorithm in contrast to our old implementation. As both LTL and LTLR model checking use the same algorithms for the Büchi automata generation and the emptiness checking, only the size of the model is important to evaluate the performance. To compute the size of models for the old implementation, we count the number of states and transitions in the transition graph induced by a rewrite theory, for the reason that a Kripke structure associated to the rewrite theory is created from the transition graph. The size of models in the new implementation is measured by the number of states and transitions in the labeled Kripke structure.

The models and properties that we experimented with are: (i) the BRP protocol and the first property in Section 5, (ii) a simple client-server model and the liveness property described in [29], and (iii) Dekker's mutual exclusion algorithm and the weak fairness property from [2]. Table 1 summarizes the results for each model. We can see that both the number of states and transitions for the new algorithm are considerably smaller than those for the old algorithm.

Table 1. The number of states and transitions

Model	Old #state/#trans	New #state/#trans
BRP protocol	283/ 1034	122/ 372
Client-Server	141/ 1140	48/ 272
Dekker's algorithm	263/ 586	152/ 336

Besides, the previous specification of BRP in [30] has 122 states and 539 transitions for the same property. In this case, the number of states is not increased since the specification itself is optimized, but the number of transition is still increased due to the large number of rules.

7 Related Work and Conclusions

The family of TLR logics incorporating spatial action patterns is introduced in [27,29]. Besides *LTLR*, the most general one of these logics is TLR^*, which generalizes the state-based logic CTL^*. Many well-known state-based logics such as *LTL*, *CTL*, and CTL^* [9,25], and event-based logics such as Hennessy-Milner's logic [19], or De Nicola and Vaandrager's $A\text{-}CTL^*$ [32], can be viewed as spatial cases of TLR^*, either in the literal sense or in the sense of existing faithful mappings of tandems [27]. The mixed state/event logic *SE-LTL* in [6,7] can be also considered as a special case of TLR^*, in particular, *LTLR*, and our model checking algorithm are clearly related to the one in [6].

There are many approaches to combine state-based and event-based formulas. In [4,15,18,33], several extensions of either $A\text{-}CTL^*$ or $A\text{-}CTL$ are discussed. Three other approaches proposing mixed logics with both state-predicates and actions are: (i) the extension of the *SE-LTL* in [6,7] to a universally path quantified logic involving ω-regular expressions [5]; (ii) the *ESTL* logic of events and states for Petri nets of [22]; and (iii) the Kripke modal transition systems of [21], and their use in the verification of safety and liveness properties in the context of the modal μ-calculus [23] (μL).[9] Two other logics that combine actions and state-based formulas are the UNITY logic of Chandy and Misra [8], and Misra's logic for Seuss [31]; however, actions as such do not appear in temporal logic formulas, which remain state-based. The work most closely related to TLR^* is that on *VLRL* [16,26], but the *VLRL* solution was less general and did not consider model checking aspects. Lamport's Temporal Logic of Actions (*TLA*) [24] is also able to specify properties related to both states and events, however, there is no division of labor between system and property specification logics. Additionally, actions in *TLA* are interpreted as binary relations between states, so that one cannot distinguish between two actions having the same outcomes from a given state.

After reviewing the syntax and semantics of LTLR, we have presented the automata-theoretic foundations and the implementation of the new Maude LTLR model checker, explained its support of language extensions for spatial action patterns, and illustrated its use with the BRP example. The tool's implementation is already quite mature; after a more detailed documentation is completed, we plan to make it available to Maude users before WRLA 2010. This will make possible a much wider range of experiments by different users, which will provide very valuable experience for further improving both its implementation and its user interface.

Acknowledgments. A considerable part of this work was carried out during a Summer internship of Kyungmin Bae at SRI's Computer Science Laboratory. We thank Steven Eker, Mark-Oliver Stehr, and Carolyn Talcott for the very fruitful discussions, particularly on issues pertaining to the Maude's LTL model checker, that greatly facilitated the task of developing the LTLR implementation. This work has been partially supported by NSF Grants CNS 07-16638 and CCF 09-05584.

References

1. Abdulla, P., Annichini, A., Bouajjani, A.: Symbolic verification of lossy channel systems: Application to the bounded retransmission protocol. In: Cleaveland, W.R. (ed.) TACAS 1999. LNCS, vol. 1579, pp. 208–222. Springer, Heidelberg (1999)

[9] μL is in some ways more powerful than TLR^*, but lacks spatial action patterns. In fact, it is possible to generalize μL to a modal μ-calculus of rewriting μLR [27].

2. Bae, K., Meseguer, J.: A rewriting-based model checker for the temporal logic of rewriting. In: Proc. 9th Inte. Workshop on Rule-Based Programming. ENTCS, Elsevier, Amsterdam (2008)
3. Baier, C., Katoen, J.P.: Principles of Model Checking. The MIT Press, Cambridge (2007)
4. ter Beek, M.H., Fantechi, A., Gnesi, S., Mazzanti, F.: An action/state-based model-checking approach for the analysis of communication protocols for service-oriented applications. In: Leue, S., Merino, P. (eds.) FMICS 2007. LNCS, vol. 4916, pp. 133–148. Springer, Heidelberg (2008)
5. Chaki, S., Clarke, E., Grumberg, O., Ouaknine, J., Sharygina, N., Touili, T., Veith, H.: State/event software verification for branching-time specifications. In: Romijn, J.M.T., Smith, G.P., van de Pol, J. (eds.) IFM 2005. LNCS, vol. 3771, pp. 53–69. Springer, Heidelberg (2005)
6. Chaki, S., Clarke, E., Ouaknine, J., Sharygina, N., Sinha, N.: State/event-based software model checking. In: Boiten, E.A., Derrick, J., Smith, G.P. (eds.) IFM 2004. LNCS, vol. 2999, pp. 128–147. Springer, Heidelberg (2004)
7. Chaki, S., Clarke, E., Ouaknine, J., Sharygina, N., Sinha, N.: Concurrent software verification with states, events, and deadlocks. Formal Aspects of Computing 17, 461–483 (2005)
8. Chandy, K.M., Misra, J.: Parallel Program Design: a Foundation. Addison-Wesley, Reading (1988)
9. Clarke, E.M., Grumberg, O., Peled, D.A.: Model Checking. The MIT Press, Cambridge (2001)
10. Clavel, M., Durán, F., Eker, S., Meseguer, J., Lincoln, P., Martí-Oliet, N., Talcott, C.: All About Maude - A High-Performance Logical Framework. LNCS, vol. 4350. Springer, Heidelberg (2007)
11. Dershowitz, N., Jouannaud, J.P.: Rewrite systems. In: van Leeuwen, J. (ed.) Handbook of Theoretical Computer Science, vol. B, pp. 243–320. North-Holland, Amsterdam (1990)
12. Durán, F., Meseguer, J.: Maude's module algebra. Science of Computer Programming 66, 125–153 (2007)
13. Eker, S., Meseguer, J., Sridharanarayanan, A.: The Maude LTL model checker. In: Gadducci, F., Montanari, U. (eds.) Proc. 4th. Intl. Workshop on Rewriting Logic and its Applications. ENTCS. Elsevier, Amsterdam (2002)
14. Eker, S., Meseguer, J., Sridharanarayanan, A.: The Maude LTL model checker and its implementation. In: Ball, T., Rajamani, S.K. (eds.) SPIN 2003. LNCS, vol. 2648, pp. 230–234. Springer, Heidelberg (2003)
15. Fantechi, A., Gnesi, S., Lapadula, A., Mazzanti, F., Pugliese, R., Tiezzi, F.: A model checking approach for verifying cows specifications. In: Fiadeiro, J.L., Inverardi, P. (eds.) FASE 2008. LNCS, vol. 4961, pp. 230–245. Springer, Heidelberg (2008)
16. Fiadeiro, J., Martí-Oliet, N., Maibaum, T., Meseguer, J., Pita, I.: Towards a verification logic for rewriting logic. In: Bert, D., Choppy, C., Mosses, P.D. (eds.) WADT 1999. LNCS, vol. 1827, pp. 438–458. Springer, Heidelberg (2000)
17. Gastin, P., Oddoux, D.: Fast ltl to büchi automata translation. In: Berry, G., Comon, H., Finkel, A. (eds.) CAV 2001. LNCS, vol. 2102, pp. 53–65. Springer, Heidelberg (2001)
18. Gnesi, S., Mazzanti, F.: A model checking verification environment for uml statecharts. In: Proceedings XLIII AICA Annual Conference, University of Udine - AICA (2005), http://fmt.isti.cnr.it/WEBPAPER/gmaica2005.pdf

19. Hennessy, M., Milner, R.: Algebraic laws for nondeterminism and concurrency. Journal of the Association for Computing Machinery 32(1), 137–172 (1985)
20. Holzmann, G., Peled, D., Yannakakis, M.: On nested depth first search (extended abstract). In: The Spin Verification System, pp. 23–32. American Mathematical Society, Providence (1996)
21. Huth, M., Jagadeesan, R., Schmidt, D.: Modal transition systems: A foundation for three-valued program analysis. In: Sands, D. (ed.) ESOP 2001. LNCS, vol. 2028, pp. 155–169. Springer, Heidelberg (2001)
22. Kindler, E., Vesper, T.: ESTL: A temporal logic for events and states. In: Desel, J., Silva, M. (eds.) ICATPN 1998. LNCS, vol. 1420, pp. 365–384. Springer, Heidelberg (1998)
23. Kozen, D.: Results on the propositional mu-calculus. Theoretical Computer Science 27, 333–354 (1983)
24. Lamport, L.: A temporal logic of actions. ACM Trans. on Prog. Lang. and Systems 16(3), 872–923 (1994)
25. Manna, Z., Pnueli, A.: The Temporal Logic of Reactive and Concurrent Systems – Specification. Springer, Heidelberg (1992)
26. Martí-Oliet, N., Pita, I., Fiadeiro, J.L., Meseguer, J., Maibaum, T.S.E.: A verification logic for rewriting logic. J. Log. Comput. 15(3), 317–352 (2005)
27. Meseguer, J.: The temporal logic of rewriting. Tech. Rep. UIUCDCS-R-2007-2815, CS Dept., University of Illinois at Urbana-Champaign (February 2007) (revised) (November 2007)
28. Meseguer, J.: Conditional rewriting logic as a unified model of concurrency. Theoretical Computer Science 96(1), 73–155 (1992)
29. Meseguer, J.: The temporal logic of rewriting: A gentle introduction. In: Degano, P., De Nicola, R., Meseguer, J. (eds.) Concurrency, Graphs and Models. LNCS, vol. 5065, pp. 354–382. Springer, Heidelberg (2008)
30. Meseguer, J., Palomino, M., Martí-Oliet, N.: Equational abstractions. In: Baader, F. (ed.) CADE 2003. LNCS (LNAI), vol. 2741, pp. 2–16. Springer, Heidelberg (2003)
31. Misra, J.: A Discipline of Multiprogramming. Springer, Heidelberg (2001)
32. Nicola, R.D., Vaandrager, F.W.: Action versus state based logics for transition systems. In: Guessarian, I. (ed.) LITP 1990. LNCS, vol. 469, pp. 407–419. Springer, Heidelberg (1990)
33. Pecheur, C., Raimondi, F.: Symbolic model checking of logics with actions. In: Edelkamp, S., Lomuscio, A. (eds.) MoChArt IV. LNCS (LNAI), vol. 4428, pp. 113–128. Springer, Heidelberg (2007)
34. Somenzi, F., Bloem, R.: Efficient büchi automata from ltl formulae. In: Emerson, E.A., Sistla, A.P. (eds.) CAV 2000. LNCS, vol. 1855, pp. 248–263. Springer, Heidelberg (2000)
35. Viry, P.: Equational rules for rewriting logic. Theoretical Computer Science 285, 487–517 (2002)

Enhancing the Debugging of Maude Specifications*

Adrian Riesco, Alberto Verdejo, and Narciso Martí-Oliet

Facultad de Informática, Universidad Complutense de Madrid, Spain
ariesco@fdi.ucm.es, {alberto,narciso}@sip.ucm.es

Abstract. Declarative debugging is a semi-automatic technique that locates a program fragment responsible for the error by building a tree representing the computation and guiding the user through it to find the error. Two different kinds of errors are considered for debugging: *wrong answers*—a wrong result obtained from an initial value—and *missing answers*—a term that should be reachable but cannot be obtained from an initial value—, where the latter has only been considered in nondeterministic systems. However, we consider that missing answers can also appear in deterministic systems, when we obtain correct results that do not provide all the expected information, which corresponds, in the context of Maude modules, to terms whose normal form is not reached and to terms whose computed least sort is, although correct, bigger than the expected one. We present in this paper a calculus to deduce normal forms and least sorts, and a proper abbreviation of the trees obtained with it. These trees increase both the causes (missing equations and memberships) and the errors (erroneous normal forms and least sorts) detected in our debugging framework.

Keywords: declarative debugging, Maude, rewriting logic, membership equational logic, wrong answers, missing answers.

1 Introduction

Declarative debugging (also known as declarative diagnosis or algorithmic debugging) [17] is a debugging technique that abstracts the computation details to focus on results. It starts from an incorrect computation, the error symptom, and locates a program fragment responsible for the error. To find this error the debugger represents the computation as a *debugging tree* [10], where each node stands for a computation step and must follow from the results of its child nodes by some logical inference. This tree is traversed by asking questions to an external oracle (generally the user) until a *buggy node*—a node containing an erroneous result, but whose children are all correct—is found. Traditional debugging techniques are devoted to fixing errors in specifications when an erroneous result, called a wrong answer, is found. Declarative debugging of this

* Research supported by MICINN Spanish project *DESAFIOS10* (TIN2009-14599-C03-01) and Comunidad de Madrid program *PROMETIDOS* (S2009/TIC-1465).

P.C. Ölveczky (Ed.): WRLA 2010, LNCS 6381, pp. 226–242, 2010.

kind of errors has been widely studied in the logic [9,19], functional [11,12], and multi-paradigm [3,7] programming languages. Another kind of errors, called *missing answers* [4,1], appears in nondeterministic systems when a term that should be reachable cannot be obtained from an initial one. This kind of errors has been less studied because it can only be applied to nondeterministic systems and because the associated calculus may be much more complicated than the one associated to wrong answers, making the debugging process unbearable.

Maude [5] is a high-level language and high-performance system supporting both equational and rewriting logic computation. Maude modules correspond to specifications in *rewriting logic* [8], a logic that allows the representation of many models of concurrent and distributed systems. This logic is an extension of *membership equational logic* [2], an equational logic that, in addition to equations, allows to state *membership axioms* characterizing the elements of a sort. Rewriting logic extends membership equational logic by adding rewrite rules, that represent transitions in a concurrent system. The Maude system supports several approaches for debugging: tracing, term coloring, and using an internal debugger [5, Chap. 22]. As part of an ongoing project to develop a declarative debugger for Maude specifications, we have already studied wrong answers in both functional and system modules [14] and missing answers in rewrites [15]. We now extend our framework by developing a calculus to deduce normal forms and least sorts seeing that the errors associated to these deductions correspond to missing answers in a deterministic framework. With this calculus we can detect errors due not only to wrong statements in a given specification but also to statements that the user *forgot* to specify,[1] indicating in this last case the operator at the top that the statement needs. These features improve our debugger in two ways: allowing to debug missing answers in the equational part of Maude modules and increasing the range of errors detected by the tool. For example, we can now debug missing answers when a rule cannot be applied because the term does not reach its normal form due to a missing equation or because the lefthand side does not match the term because it has a wrong least sort. We illustrate this improvement in Section 3 with a system module that, if debugged with the previous version of our tool, would print `Error: With the given information (labeling, correct module, and answers) it is impossible to debug.`, while in the current version the error is located.

The rest of the paper is organized as follows: after briefly introducing Maude modules with an example, Section 2 presents the calculus for missing answers and how the proof trees built with it are pruned in order to obtain appropriate debugging trees. Section 3 presents our tool by debugging some examples, while Section 4 concludes and outlines some future work.

The Maude source of the debugger, a user guide [13], additional examples, and other papers on this subject, including detailed proofs of the results [16], are all available from the webpage `http://maude.sip.ucm.es/debugging`.

[1] Note that the treatment of these missing statements is more powerful than the one currently applied in the Maude sufficient completeness checker [6], because it can be used with conditional and non left-linear statements.

1.1 An Example: Heaps

We show in this section how to specify in Maude binary heaps, that is, binary trees fulfilling that (1) all levels of the tree, except possibly the last one, are complete and, if the last level of the tree is not complete, the nodes of that level are filled from left to right; and (2) the value in each node is greater than the value in each of its children. The module HEAP defines binary trees (BTree) and Heaps and its nonempty variants (NeBTree and NeHeap), using a theory TH (not shown here) that defines the functions min, max, and a total order _<_ over the elements of the sort Elt:

```
(fmod HEAP{X :: TH} is
  pr NAT .

  sorts BTree Heap NeBTree NeHeap .
  subsort NeHeap < NeBTree Heap < BTree .

  op mt : -> Heap [ctor] .
  op ___ : BTree X$Elt BTree -> NeBTree [ctor] .
```

We state by means of memberships when a binary tree is a heap:

```
  vars E E' : X$Elt .                     vars BT BT' : BTree .
  vars L L' R R' : Heap .                 vars NL NR : NeHeap .

  cmb [h1] : NL E mt : NeHeap
   if max(NL) < E /\ depth(NL) == 1 .
  cmb [h2] : NL E NR : NeHeap
   if max(NL) < E /\ max(NR) < E /\
      (depth(NL) == depth(NR) and complete(NL)) or
      (depth(NL) == s(depth(NR)) and complete(NR)) .
```

where the auxiliary function depth computes the depth of a binary tree; max returns the value at the root of a nonempty heap (i.e., its maximum); and complete checks whether a binary tree is complete:

```
  op depth : BTree -> Nat .
  eq [dp1] : depth(mt) = 0 .
  eq [dp2] : depth(BT N BT') = max(depth(BT), depth(BT')) + 1 .

  op max : NeHeap -> X$Elt .
  ceq [max] : max(L E R) = E if L E R : NeHeap .

  op complete : BTree -> Bool .
  eq [cmp1] : complete(mt) = true .
  eq [cmp2] : complete(BT E BT') = complete(BT) and complete(BT') and
                                   depth(BT) == depth(BT') .
```

The function insert introduces a new element in a heap by sinking it to the appropriate position:

```
  op insert : X$Elt Heap ~> NeHeap .
  eq [ins1] : insert(E, mt) = mt E mt .
  ceq [ins2] : insert(E, L E' R) = L' max(E, E') R
   if L E' R : NeHeap /\
      not complete(L) or ((depth(L) > depth(R)) and complete(R)) /\
      L' := insert(min(E, E'), L) .
  ceq [ins3] : insert(E, L E' R) = L max(E, E') R'
   if L E' R : NeHeap /\
      not complete(R) or (depth(L) > depth(R)) and complete(L) /\
      R' := insert(min(E, E'), R) .
endfm)
```

We use a view HN (not shown here) to instantiate the values of the heap as natural numbers and we define a constant `heap` for testing:

```
(fmod NAT-HEAP is
  pr HEAP{HN} .
  op heap : -> NeHeap .
  eq heap = (mt 4 mt) 5 (mt 3 mt) .
endfm)
```

If we check in our specification the type of the constant `heap`:

```
Maude> (red heap .)
result NeBTree : (mt 4 mt) 5 (mt 3 mt)
```

we realize that although it has a correct sort (it is a `NeBTree`) its expected least sort, `NeHeap`, has not been obtained. We will show in Section 3 how to debug it.

2 Debugging Trees for Normal Forms and Least Sorts

We present in this section a calculus to compute the normal form and the least sort of a given term. The proof trees computed with this calculus contain the information proving why the term has been reduced to this normal form or this sort has been inferred (positive information) and also why the term has not been further reduced or a lesser sort has not been computed (negative information). The calculus is introduced as an extension of the calculus in [14] that allowed to deduce judgments corresponding to oriented equations $t \to t'$ and memberships $t : s$, and improves the calculus of missing answers of [15] by adding new causes to the errors debugged thus far. Once this extended calculus is presented, we show how to use it to define appropriate debugging trees.

2.1 A Calculus for Normal Forms and Least Sorts

From now on, we assume a rewrite theory $\mathcal{R} = (\Sigma, E, R)$ satisfying the Maude executability requirements, i.e., E is confluent and terminating, maybe modulo some equational attributes such as associativity and commutativity, while R is coherent with respect to E. Equations corresponding to the equational attributes form the set A and the equations in $E - A$ can be oriented from left to right.

Throughout this paper we only consider a special kind of conditions and substitutions that operate over them, called *admissible*. They correspond to the ones used in Maude modules and are defined as follows:

Definition 1. *A condition* $C_1 \wedge \cdots \wedge C_n$ *is* admissible *if, for* $1 \leq i \leq n$, C_i *is*

- *an equation* $u_i = u'_i$ *or a membership* $u_i : s$ *and* $vars(C_i) \subseteq \bigcup_{j=1}^{i-1} vars(C_j)$, *or*
- *a matching condition* $u_i := u'_i$, u_i *is a pattern and* $vars(u'_i) \subseteq \bigcup_{j=1}^{i-1} vars(C_j)$, *or*
- *a rewrite condition* $u_i \Rightarrow u'_i$, u'_i *is a pattern and* $vars(u_i) \subseteq \bigcup_{j=1}^{i-1} vars(C_j)$.

Note that the lefthand side of matching conditions and the righthand side of rewrite conditions can contain extra variables that will be instantiated once the condition is solved.

Definition 2. *A* kind-substitution, *denoted by* κ, *is a mapping from variables to terms of the form* $v_1 \mapsto t_1; \ldots; v_n \mapsto t_n$ *such that* $\forall_{1 \leq i \leq n} . kind(v_i) = kind(t_i)$, *that is, each variable has the same kind as the term it binds.*

Definition 3. *A* substitution, *denoted by* θ, *is a mapping from variables to terms of the form* $v_1 \mapsto t_1; \ldots; v_n \mapsto t_n$ *such that* $\forall_{1 \leq i \leq n} . sort(v_i) \geq ls(t_i)$, *that is, the sort of each variable is greater than or equal to the least sort of the term it binds. Note that a substitution is a special type of kind-substitution where each term has the sort appropriate to its variable.*

Definition 4. *Given an atomic condition* C, *we say that a substitution* θ *is* admissible for C *if*

- C *is an equation* $u = u'$ *or a membership* $u : s$ *and* $vars(C) \subseteq dom(\theta)$, *or*
- C *is a matching condition* $u := u'$ *and* $vars(u') \subseteq dom(\theta)$, *or*
- C *is a rewrite condition* $u \Rightarrow u'$ *and* $vars(u) \subseteq dom(\theta)$.

The calculus presented in this section (Figures 1 and 2) will be used to deduce the following judgments, that we introduce together with their meaning for a Σ-term model [8,16] $\mathcal{T}' = \mathcal{T}_{\Sigma/E',R'}$ defined by equations and memberships E' and by rules R':

- Given a term t and a kind-substitution κ, $\mathcal{T}' \models adequateSorts(\kappa) \rightsquigarrow \Theta$ when either $\Theta = \{\kappa\} \wedge \forall v \in dom(\kappa).\mathcal{T}' \models \kappa[v] : sort(v)$ or $\Theta = \emptyset \wedge \exists v \in dom(\kappa).\mathcal{T}' \not\models \kappa[v] : sort(v)$, where $\kappa[v]$ denotes the term bound by v in κ. That is, when all the terms bound in the kind-substitution κ have the appropriate sort, then κ is a substitution and it is returned; otherwise (at least one of the terms has an incorrect sort), the kind-substitution is not a substitution and the empty set is returned.
- Given an admissible substitution θ for an atomic condition C, $\mathcal{T}' \models [C, \theta] \rightsquigarrow \Theta$ when $\Theta = \{\theta' \mid \mathcal{T}', \theta' \models C \text{ and } \theta' \upharpoonright_{dom(\theta)} = \theta\}$, that is, Θ is the set of substitutions that fulfill the atomic condition C and extend θ.

$$\frac{\theta(t_2) \rightarrow_{norm} t' \quad adequateSorts(\kappa_1) \rightsquigarrow \Theta_1 \quad \ldots \quad adequateSorts(\kappa_n) \rightsquigarrow \Theta_n}{[t_1 := t_2, \theta] \rightsquigarrow \bigcup_{i=1}^{n} \Theta_i} \; \mathsf{PatC}$$
$$\text{if } \{\kappa_1, \ldots, \kappa_n\} = \{\kappa\theta \mid \kappa(\theta(t_1)) \equiv_A t'\}$$

$$\frac{t_1 : sort(v_1) \quad \ldots \quad t_n : sort(v_n)}{adequateSorts(v_1 \mapsto t_1; \ldots; v_n \mapsto t_n) \rightsquigarrow \{v_1 \mapsto t_1; \ldots; v_n \mapsto t_n\}} \; \mathsf{AS}_1$$

$$\frac{t_i :_{ls} s_i}{adequateSorts(v_1 \mapsto t_1; \ldots; v_n \mapsto t_n) \rightsquigarrow \emptyset} \; \mathsf{AS}_2 \; \text{ if } s_i \not\leq sort(v_i)$$

$$\frac{\theta(t) : s}{[t : s, \theta] \rightsquigarrow \{\theta\}} \; \mathsf{MbC}_1 \qquad \frac{\theta(t) :_{ls} s'}{[t : s, \theta] \rightsquigarrow \emptyset} \; \mathsf{MbC}_2 \; \text{ if } s' \not\leq s$$

$$\frac{\theta(t_1) \downarrow \theta(t_2)}{[t_1 = t_2, \theta] \rightsquigarrow \{\theta\}} \; \mathsf{EqC}_1 \qquad \frac{\theta(t_1) \rightarrow_{norm} t'_1 \quad \theta(t_2) \rightarrow_{norm} t'_2}{[t_1 = t_2, \theta] \rightsquigarrow \emptyset} \; \mathsf{EqC}_2 \; \text{ if } t'_1 \not\equiv_A t'_2$$

$$\frac{\theta(t_1) \rightsquigarrow_{n+1}^{t_2 := \circledast} S}{[t_1 \Rightarrow t_2, \theta] \rightsquigarrow \{\theta'\theta \mid \theta'(\theta(t_2)) \in S\}} \; \mathsf{RIC}$$
$$\text{if } n = min(x \in \mathbb{N} : \forall i \geq 0 \; (\theta(t_1) \rightsquigarrow_{x+i}^{t_2 := \circledast} S))$$

$$\frac{[C, \theta_1] \rightsquigarrow \Theta_1 \quad \cdots \quad [C, \theta_m] \rightsquigarrow \Theta_m}{\langle C, \{\theta_1, \ldots, \theta_m\}\rangle \rightsquigarrow \bigcup_{i=1}^{m} \Theta_i} \; \mathsf{SubsCond}$$

Fig. 1. Calculus for substitutions

- Given a set of admissible substitutions Θ for an atomic condition C, $\mathcal{T}' \models \langle C, \Theta \rangle \rightsquigarrow \Theta'$ when $\Theta' = \{\theta' \mid \mathcal{T}', \theta' \models C \text{ and } \theta' \upharpoonright_{dom(\theta)} = \theta \text{ for some } \theta \in \Theta\}$, that is, Θ' is the set of substitutions that fulfill the condition C and extend any of the admissible substitutions in Θ.
- Given an equation or membership a and a term t, $\mathcal{T}' \models disabled(a, t)$ when a cannot be applied to t at the top.
- Given two terms t and t', $\mathcal{T}' \models t \rightarrow_{red} t'$ when $\mathcal{T}' \models t \rightarrow^{1}_{E'} t'$ or $\mathcal{T}' \models t_i \rightarrow^{!}_{E'} t'_i$, with $t_i \neq t'_i$, for some subterm t_i of t such that $t' = t[t_i \mapsto t'_i]$, that is, the term t is either reduced one step at the top or reduced by substituting a subterm by its normal form.
- Given two terms t and t', $\mathcal{T}' \models t \rightarrow_{norm} t'$ when $\mathcal{T}' \models t \rightarrow^{!}_{E'} t'$, that is, t' is in normal form with respect to the equations E'.
- Given a term t and a sort s, $\mathcal{T}' \models t :_{ls} s$ when $\mathcal{T}' \models t : s$ and moreover s is the least sort with this property (with respect to the ordering on sorts obtained from the signature Σ and the equations and memberships E' defining the Σ-term model $\mathcal{T}'$).

We introduce in Figure 1 the inference rules defining the relations $[C, \theta] \rightsquigarrow \Theta$, $\langle C, \Theta \rangle \rightsquigarrow \Theta'$, and $adequateSorts(\kappa) \rightsquigarrow \Theta$. Intuitively, these judgments will provide positive information when they lead to nonempty sets (indicating that the condition holds in the first two judgments or that the kind-substitution is a substitution in the third one) and negative information when they lead to the empty set (indicating respectively that the condition fails or the kind-substitution is not a substitution):

- Rule PatC computes all the possible substitutions that extend θ and satisfy the matching of the term t_2 with the pattern t_1 by first computing the normal form t' of t_2, obtaining then all the possible kind-substitutions κ that make t' and $\theta(t_1)$ equal modulo axioms (indicated by $\equiv_A$), and finally checking that the terms assigned to each variable in the kind-substitutions have the appropriate sort with $adequateSorts(\kappa)$. The union of the set of substitutions thus obtained constitutes the set of substitutions that satisfy the matching.
- Rule AS_1 checks whether the terms of the kind-substitution have the appropriate sort to match the variables. In this case the kind-substitution is a substitution and it is returned.
- Rule AS_2 indicates that, if the least sort of any of the terms in the kind-substitution is bigger than the required one, then it is not a substitution and thus the empty set of substitutions is returned.
- Rule MbC_1 returns the current substitution if a membership condition holds.
- Rule MbC_2 is used when the membership condition is not satisfied. It checks that the least sort of the term is not less than or equal to the required one, and thus the substitution does not satisfy the condition and the empty set is returned.
- Rule EqC_1 returns the current substitution when an equality condition holds, that is, when the two terms can be joined with equations, abbreviated as $t_1 \downarrow t_2$.
- Rule EqC_2 checks that an equality condition fails by obtaining the normal forms of both terms and then examining that they are different.
- Rewrite conditions are handled by rule RlC. This rule extends the set of substitutions by computing all the reachable terms that satisfy the pattern (using the relation $t \rightsquigarrow_n^C S$ explained in [16]) and then using these terms to obtain the new substitutions.
- Finally, rule SubsCond computes the extensions of a set of admissible substitutions $\{\theta_1, \ldots, \theta_n\}$ by using the rules above with each of them.

We use these judgments to define the inference rules of Figure 2, that describe how the normal form and the least sort of a term are computed:

- Rule Dsb indicates when an equation or membership a cannot be applied to a term t. It checks that there are no substitutions that satisfy the matching of the term with the lefthand side of the statement and that fulfill its condition. Note that we check the conditions from left to right, following the same order as Maude and making all the substitutions admissible.
- Rule Rdc_1 reduces a term by applying one equation when it checks that the conditions can be satisfied, where the matching conditions are included in the equality conditions. While in the previous rule we made explicit the evaluation from left to right of the condition to show that finally the set of substitutions fulfilling it was empty, in this case we only need one substitution to fulfill the condition and the order is unimportant.
- Rule Rdc_2 reduces a term by reducing a subterm to normal form (checking in the side condition that it is not already in normal form).

$$\frac{[l := t, \emptyset] \rightsquigarrow \Theta_0 \quad \langle C_1, \Theta_0\rangle \rightsquigarrow \Theta_1 \quad \ldots \quad \langle C_n, \Theta_{n-1}\rangle \rightsquigarrow \emptyset}{disabled(a, t)} \ \mathsf{Dsb}$$
$$\text{if } a \equiv l \to r \Leftarrow C_1 \wedge \ldots \wedge C_n \in E \text{ or}$$
$$a \equiv l : s \Leftarrow C_1 \wedge \ldots \wedge C_n \in E$$

$$\frac{\{\theta(u_i) \downarrow \theta(u'_i)\}_{i=1}^{n} \quad \{\theta(v_j) : s_j\}_{j=1}^{m}}{\theta(l) \to_{red} \theta(r)} \ \mathsf{Rdc}_1 \ \text{ if } l \to r \Leftarrow \bigwedge_{i=1}^{n} u_i = u'_i \wedge \bigwedge_{j=1}^{m} v_j : s_j \in E$$

$$\frac{t \to_{norm} t'}{f(t_1, \ldots, t, \ldots, t_n) \to_{red} f(t_1, \ldots, t', \ldots, t_n)} \ \mathsf{Rdc}_2 \ \text{ if } t \not\equiv_A t'$$

$$\frac{disabled(e_1, f(t_1, \ldots, t_n)) \quad \ldots \quad disabled(e_l, f(t_1, \ldots, t_n)) \quad t_1 \to_{norm} t_1 \quad \ldots \quad t_n \to_{norm} t_n}{f(t_1, \ldots, t_n) \to_{norm} f(t_1, \ldots, t_n)} \ \mathsf{Norm}$$
$$\text{if } \{e_1, \ldots, e_l\} = \{e \in E \mid e \ll_K^{top} f(t_1, \ldots, t_n)\}$$

$$\frac{t \to_{red} t_1 \quad t_1 \to_{norm} t'}{t \to_{norm} t'} \ \mathsf{NTr}$$

$$\frac{t \to_{norm} t' \quad t' : s \quad disabled(m_1, t') \quad \ldots \quad disabled(m_l, t')}{t :_{ls} s} \ \mathsf{Ls}$$
$$\text{if } \{m_1, \ldots, m_l\} = \{m \in E \mid m \ll_K^{top} t' \ \wedge \ sort(m) < s\}$$

Fig. 2. Calculus for normal forms and least sorts

- Rule Norm states that the term is in normal form by checking that no equations can be applied at the top considering the variables at the kind level (which is indicated by $\ll_K^{top}$) and that all its subterms are already in normal form.
- Rule NTr describes the transitivity for the reduction to normal form. It reduces the term with the relation $\to_{red}$ and the term thus obtained then is reduced to normal form by using again $\to_{norm}$.
- Rule Ls computes the least sort of the term t. It computes a sort for its normal form (that has the least sort of the terms in the equivalence class) and then checks that memberships deducing lesser sorts, applicable at the top with the variables considered at the kind level, cannot be applied.

In these rules Dsb provides the negative information, proving why the statements (either equations or membership axioms) cannot be applied, while the remaining rules provide the positive information indicating why the normal form and the least sort are obtained.

Theorem 1. *The calculus of Figures 1 and 2 is correct w.r.t.* $\mathcal{R} = (\Sigma, E, R)$ *in the sense that for any judgment* φ, φ *is derivable in the calculus if and only if* $\mathcal{T}_{\Sigma/E,R} \models \varphi$, *with* $\mathcal{T}_{\Sigma/E,R}$ *being the corresponding initial model.*

Once these rules have been presented, we can compute the proof tree associated to the erroneous computation shown in Section 1.1 for the heaps example. Remember that the least sort of the term `heap`, that should be `NeHeap`, was instead `NeBTree`. Figures 3 and 4 show the associated proof tree, where h stands for the term `(mt 4 mt) 5 (mt 3 mt)`, l for the lefthand side of the membership

$$\dfrac{\dfrac{\dfrac{}{\texttt{heap} \rightarrow_{red} h}\,\mathsf{Rdc}_1 \quad \dfrac{}{h \rightarrow_{norm} h}\,\mathsf{Norm}}{\texttt{heap} \rightarrow_{norm} h}\,\mathsf{NTr} \qquad \dfrac{}{h : \texttt{NeBTree}}\,\mathsf{Mb} \qquad T_1}{\texttt{heap} :_{ls} \texttt{NeBTree}}\,\mathsf{Ls}$$

Fig. 3. Proof tree for the heap example

$$\dfrac{\dfrac{\dfrac{}{h \rightarrow_{norm} h}\,\mathsf{Norm} \quad \dfrac{\dfrac{\nabla}{\texttt{mt 4 mt} :_{ls} \texttt{NeBTree}}\,\mathsf{Ls}}{adequateSorts(l,\theta)}\,\mathsf{AS}_2}{[l := h] \rightsquigarrow \emptyset}\,\mathsf{PatC} \quad \dfrac{}{\langle C_1, \emptyset\rangle \rightsquigarrow \emptyset}\,\mathsf{SubsCond} \quad \ldots \quad \dfrac{}{\langle C_n, \emptyset\rangle \rightsquigarrow \emptyset}\,\mathsf{SubsCond}}{disabled(\texttt{h2}, h)}\,\mathsf{Dsb}$$

Fig. 4. Proof tree T_1, proving the matching with `h2`

`h2`, namely `NL E NR` with `NL` and `NR` variables of sort `NeHeap` and `E` a natural number, C_1 and C_n are respectively the first condition and last condition of `h2`, θ is $\texttt{NL} \mapsto \texttt{mt 4 mt}; \texttt{E} \mapsto 5; \texttt{NR} \mapsto \texttt{mt 3 mt}$, and ∇ represents a tree similar to the one depicted in Figure 3.

The tree shown in Figure 3 illustrates that to compute the least sort of `heap` first it obtains its normal form and then it checks that no memberships can be applied to this term (and thus the sort is inferred by using the operator declarations). To check that no memberships are applied it only checks whether `h2` is used, because the other membership does not match the term with the variables at the kind level. The tree T_1, depicted in Figure 4, is in charge of this proof, that is, it provides the negative information proving that the membership cannot be applied. First, it checks that the lefthand side of the membership does not match the term because `mt 4 mt` has as least sort `NeBTree` and hence it does not match the variable `NL`, that has sort `NeHeap`. Since the empty set of substitutions is computed for this matching, the rest of conditions of the membership cannot be fulfilled, which is proved by the nodes associated with the rule SubsCond.

Following the approach shown in [14], we assume the existence of an *intended interpretation* $\mathcal{I}$ of the given rewrite theory $\mathcal{R} = (\Sigma, E, R)$. This intended interpretation is a Σ-term model corresponding to the model that the user had in mind while writing the specification $\mathcal{R}$. We say that a judgment is *valid* when it holds in $\mathcal{I}$, and *invalid* otherwise. The basis of declarative debugging consists in searching *buggy nodes* (invalid nodes with all its children valid) [10] in a debugging tree standing for a problematic computation. In our debugging framework, we are able to locate wrong equations, wrong memberships, missing equations, and missing memberships,[2] which are defined as follows:

[2] It is important not to confuse wrong and missing answers with wrong and missing statements. The former are the initial symptoms that indicate the specifications fails, while the latter are the errors that generated this misbehavior.

- Given a statement $A \Leftarrow C_1 \wedge \cdots \wedge C_n$ (where A is either an equation $l = r$ or a membership $l : s$) and a substitution θ, the *statement instance* $\theta(A) \Leftarrow \theta(C_1) \wedge \cdots \wedge \theta(C_n)$ is *wrong* when all the atomic conditions $\theta(C_i)$ are valid in $\mathcal{I}$ but $\theta(A)$ is not.
- Given a term t, there is a *missing equation for* t if the computed normal form of t does not correspond with the one expected in $\mathcal{I}$.
- A specification has a *missing equation* if there exists a term t such that there is a missing equation for t.
- Given a term t, there is a *missing membership for* t if the computed least sort for t does not correspond with the one expected in $\mathcal{I}$.
- A specification has a *missing membership* if there exists a term t such that there is a missing membership for t.

Regarding missing statements, what the debugger reports is that a statement is missing *or* the conditions in the remaining statements are not the intended ones (thus they are not applied when expected and another one would be needed), but the error *is not located* in the statements used in the conditions, since they are also checked during the debugging process.

Proposition 1. *Let N be a buggy node in some proof tree in the calculus of Figures 1 and 2 w.r.t. an intended interpretation $\mathcal{I}$. Then the error associated to N is a wrong equation, a missing equation, or a missing membership.*

Although these are the errors detected by the calculus presented in this paper, since it is integrated with both the calculus of wrong answers [14] and the calculus for missing answers [15], the debugger as a whole can also detect wrong memberships and wrong and missing rules.

2.2 Abbreviated Proof Trees

We describe in this section how the proof trees shown in the previous section can be abbreviated in order to ease the questions posed to the user while keeping the completeness and correctness of the technique. To achieve this aim we extend the notion of $APT(T)$ introduced in [14]; $APT(T)$ (from *Abbreviated Proof Tree*) is obtained by a transformation based on deleting nodes whose correctness only depends on the correctness of their children. For example, nodes related to judgments about sets of substitutions, that can be complicated due to matching modulo, are removed.

The rules to compute the abbreviated proof tree, which are assumed to be applied in order (i.e., a rule cannot be applied if there is another one with a lower index that can be used), are described in Figure 5:

- Rule ($\mathbf{APT}_1$) keeps the root of the tree and applies the general function APT', that returns a set of trees, to the tree.
- Rule ($\mathbf{APT}_2$) improves the questions presented to the user when the inference rule NTr is used. This abbreviation associates the equation applied in the left branch (in the inference rule Rdc_1) to the judgment rooting the tree. In this way we ask about reductions to normal form instead of reductions in one step.

$$(\mathbf{APT}_1)\ APT\left(\frac{T_1 \dots T_n}{aj} R_1\right) = \frac{APT'\left(\frac{T_1 \dots T_n}{aj} R_1\right)}{aj} R_1$$

$$(\mathbf{APT}_2)\ APT'\left(\frac{\frac{T_1\ \dots\ T_n}{t \to t''}\mathsf{Rdc}_1 \quad T'}{t \to t'}\mathsf{NTr}\right) = \left\{\frac{APT'(T_1)\ \dots\ APT'(T_n)\ APT'(T')}{t \to t'}\mathsf{Rdc}_1\right\}$$

$$(\mathbf{APT}_3)\ APT'\left(\frac{T_{t \to_{norm} t'}\ T_1 \dots T_n}{t :_{ls} s}\mathsf{Ls}\right) = \left\{\frac{APT'(T_{t \to_{norm} t'})\ APT'(T_1)\ \dots\ APT'(T_n)}{t' :_{ls} s}\mathsf{Ls}\right\}$$

$$(\mathbf{APT}_4)\ APT'\left(\frac{T_1 \dots T_n}{aj} R_2\right) = \left\{\frac{APT'(T_1)\ \dots\ APT'(T_n)}{aj} R_2\right\}$$

$$(\mathbf{APT}_5)\ APT'\left(\frac{T_1 \dots T_n}{aj} R_1\right) = APT'(T_1) \cup \dots \cup APT'(T_n)$$

R_1 any inference rule $\quad$ R_2 Rdc_1, or Norm $\quad$ aj any judgment

Fig. 5. APT rules

- Rule ($\mathbf{APT}_3$) improves the questions about least sorts by asking about the normal form of the term and thus the user is not in charge of computing it.
- Rule ($\mathbf{APT}_4$) keeps the conclusion of the inference rules that contain debugging information.
- Rule ($\mathbf{APT}_5$) discards the conclusion of the rules which do not contain debugging information.

Theorem 2. *Let T be a finite proof tree representing an inference in the calculus of Figures 1 and 2 w.r.t. some rewrite theory $\mathcal{R}$. Let $\mathcal{I}$ be an intended interpretation of $\mathcal{R}$ such that the root of T is invalid in $\mathcal{I}$. Then:*

- *$APT(T)$ contains at least one buggy node (completeness).*
- *Any buggy node in $APT(T)$ has an associated wrong equation, missing equation, or missing membership axiom in $\mathcal{R}$ (correctness).*

The abbreviated proof tree obtained by applying these rules to the proof tree depicted in Figures 3 and 4 is shown in Figure 6. This proof tree has been obtained by combining different features available in our tool:

- Judgments of the form $t \to_{norm} t$, that indicate that t is in normal form, are dropped from the proof tree if they are built only with constructors. In our example, the nodes corresponding to $h \to_{norm} h$ have been removed.
- Only labeled statements generate nodes in the abbreviated proof tree. For example, the equation to reduce the constant `heap` is not labeled and thus the node `heap` $\to_{red} h$ (or its corresponding abbreviation) does not appear in the abbreviated tree. Moreover, the debugger provides some other trusting mechanisms: statements and imported modules can be trusted before starting the debugging process; statements can also be trusted on the fly; and a correct module, introduced before starting the debugging process, can be used as oracle before asking the user.

$$\cfrac{\cfrac{\cfrac{}{(\ddagger)\ \texttt{mt} :_{ls} \texttt{Heap}}\ \mathsf{Ls}}{(\dagger)\ h :_{ls} \texttt{NeBTree}}\ \mathsf{Ls}}{\texttt{heap} :_{ls} \texttt{NeBTree}}\ \mathsf{Ls}$$

Fig. 6. Abbreviated proof tree for the heap example

- The signature is always considered correct, and hence judgments inferred by using it do not appear in the abbreviated tree. For example, the membership inference h : `BTree` only uses operator declarations and thus it does not appear in the final tree.
- The rest of nodes have been pruned by the *APT* rules. For example, they prevent all the judgments using substitutions from being asked.

Furthermore, the user can also follow some strategies to reduce the size of the debugging tree:

- If an error is found using a complex initial term, this error can probably be reproduced with a simpler one. Using this simpler term leads to easier debugging sessions.
- When facing a problem with both wrong and missing answers, it is usually better to debug first the wrong answers, because questions related to them are easier to answer and fixing them can also solve the missing answers problem.
- The Maude profiler [5, Chap. 22] indicates the most frequently used statements for a given computation. Trusting these statements will greatly reduce the size of the tree, although it requires the user to make sure that these statements are indeed correct.

Once the tree has been abbreviated we only have a subset of the original nodes and hence only the correctness of the judgments in these nodes concerns the debugging process. We present here the questions derived only from the calculus presented here, while the rest of the questions asked by the debugger can be found in [13]:

- When a term cannot be further reduced and it is not built only by constructors the debugger asks "Is t in normal form?," which is correct if the user expected t to be a normal form.
- When a term t has been reduced by using equations to another term t', the debugger asks questions of the form "Is this reduction correct? $t \to t'$." These judgments are correct if the user expected t to be reduced to t'.
- When a sort s is inferred for a term t, the debugger prompts questions of the form "Is this membership correct? $t : s$." This judgment is correct if t has sort s.
- When the judgment refers to the least sort ls of a term t, the tool makes questions of the form "Did you expect t to have least sort ls?." In this case, the judgment is correct if the intended least sort of t is exactly ls.

3 A Debugging Session

We describe in this section how to debug the specification shown in Section 1.1. To debug the error discovered in this specification (the least sort of the term heap is NeBTree) we use the command:

```
Maude> (missing heap : NeBTree .)
```

This command builds the tree depicted in Figure 6 and asks the following question, associated with the node marked with (†) in the figure:[3]

```
Is NeBTree the least sort of mt 4 mt ?
Maude> (no .)
```

Since we expected the term to have sort NeHeap the judgment is erroneous and the next question, that is associated to the node (‡) in Figure 6, is:

```
Is Heap the least sort of mt ?
Maude> (yes .)
```

With this answer the node (‡) disappears from the tree and the node (†) becomes buggy, because it is associated to an incorrect judgment and it has no children. The debugger presents the following message:

```
The buggy node is:
The least sort of mt 4 mt is NeBTree
Either the operator ___ needs more membership axioms or the conditions of
the current axioms are not written in the intended way.
```

Actually, if we check the specification we notice that the membership corresponding to the case when both heaps are empty was not stated. We should add to the specification the membership axiom:

```
  mb [h3] : mt E mt : NeHeap .
```

We can use now these heaps to implement another application. We present here a very simple specification of an auction. The module AUCTION defines the sort People as a multiset of Person (a pair of names and bids) and an Auction as some people and a heap, defined in NS-HEAP, containing elements of the form [N,S], where N is a natural number standing for the bid and S a String with the name of the bidder. The winner of the auction will be the person on the top of the heap:

```
(mod AUCTION is
  pr NS-HEAP .

  sorts Person People Auction .
```

[3] Although the debugger provides two different navigation strategies, in this simple tree both of them choose the same node.

```
  subsort Person < People .

  op <_`,_> : String Nat -> Person [ctor] .
  op nobody : -> People [ctor] .
  op __ : People People -> People [ctor comm assoc id: nobody] .
  op _`[_`] : People Heap -> Auction [ctor] .
```

The rule bid inserts a bid into the heap:

```
  var N : Nat .                         var H : Heap .
  var P : People .                      var S : String .

  rl [bid] : (P < S, N >) [H] => P [insert([N,S], H)] .
endm)
```

If we search now for the possible winners of an auction, where `initial` stands for `< "aida", 5 > < "nacho", 4 > < "charlie", 3 > [mt]`:

```
Maude> (search in AUCTION : initial =>!
                            nobody [L:Heap [N:Nat, S:String] R:Heap] .)
No solution.
```

no solutions are found. Since one solution is expected, we debug the specification with the command:

```
Maude> (missing initial =>! nobody [ L:Heap [N:Nat, S:String] R:Heap ] .)
```

This command builds the corresponding debugging tree and traverses it with the default divide and query strategy, that each time selects the node whose subtree's size is the closest one to half the size of the whole tree, keeping only this subtree if its root is incorrect, and deleting the whole subtree otherwise. The first question is:

```
Are the following terms all the reachable terms from
(< "aida", 5 > < "charlie", 3 > < "nacho", 4 >)[mt] in one step?
1 (< "aida", 5 > < "nacho", 4 >)[mt [3, "charlie"] mt]
2 (< "aida", 5 > < "charlie", 3 >)[mt [4, "nacho"] mt]
3 (< "charlie", 3 > < "nacho", 4 >)[mt [5, "aida"] mt]
Maude> (yes .)
```

The rule has inserted each person into the heap and thus the transition is correct. After some other questions related to rewrites in the style of [15], the debugger asks:

```
Is insert([4,"nacho"],mt[3,"charlie"]mt) in normal form?
Maude> (no .)
```

This term is not in normal form because we expected `insert` to be reduced. The next questions are also related to normal forms:[4]

[4] Note that, in these cases, the `String` values are not built with constructors and thus this question is not automatically removed by the debugger. If we defined our own constants for the names with the `ctor` attribute, these questions would not appear.

```
Is mt [3, "charlie"] mt in normal form?
Maude> (yes .)

Is [4,"nacho"] in normal form?
Maude> (yes .)
```

In these cases the judgment is correct because no equations should be applied to them. The next questions refer to reductions:

```
Is this reduction (associated with the equation dp1) correct?
depth(mt) -> 0
Maude> (trust .)

Is this reduction (associated with the equation cmp1) correct?
complete(mt) -> true
Maude> (trust .)
```

Since these reductions were associated to simple equations we have used the command `trust` to prevent the debugger from asking questions related to these equations again. The next question deals with memberships:

```
Is this membership (associated with the membership h3) correct?
mt [3, "charlie"] mt : NeHeap
Maude> (yes .)
```

The membership is correct because it only contains the value at the root. With this information the debugger finds the following bug:

```
The buggy node is:
insert([4,"nacho"], mt [3, "charlie"] mt) is in normal form.
Either the operator insert needs more equations or the conditions of
the current equations are not written in the intended way.
```

If we carefully inspect the equations for `insert` we notice that we have not treated the case where the tree is complete and a new level has to be started. We can add the appropriate equation or fix the equation ins2, that distinguishes a case that cannot occur in heaps. If we choose the latter, it should be fixed as follows:

```
ceq [ins2] : insert(E, L E' R) = L' max(E, E') R
 if L E' R : NeHeap /\
    not complete(L) or ((depth(L) == depth(R)) and complete(R)) /\
    L' := insert(min(E, E'), L) .
```

4 Future Work

In this paper we have presented a calculus to debug erroneous normal forms and least sorts by abbreviating the proof trees obtained with it. This calculus, besides allowing to debug these new errors, improves the former versions of our debugger by allowing the debugging of new causes of missing answers in rewrites: missing equations and memberships. These debugging features have also been integrated with the graphical user interface [13].

Although the current version of the tool allows the user to introduce a correct but maybe incomplete module in order to shorten the debugging session [14], we also want to add a new command to introduce *complete* modules, which would greatly reduce the number of questions asked to the user. We also intend to add new navigation strategies like the ones shown in [18] that take into account the number of different potential errors in the subtrees, instead of their size.

Finally, we plan to use the new narrowing features of Maude to implement a test generator for Maude specifications. This generator would allow to check Maude specifications and then to invoke the debugger when one of the test cases fails.

Acknowledgements. We cordially thank Martin Wirsing for encouraging us to investigate the causes for missing answers in a deterministic context, and the referees for their useful comments.

References

1. Alpuente, M., Comini, M., Escobar, S., Falaschi, M., Lucas, S.: Abstract diagnosis of functional programs. In: Leuschel, M. (ed.) LOPSTR 2002. LNCS, vol. 2664, pp. 1–16. Springer, Heidelberg (2003)
2. Bouhoula, A., Jouannaud, J.-P., Meseguer, J.: Specification and proof in membership equational logic. Theoretical Computer Science 236, 35–132 (2000)
3. Caballero, R.: A declarative debugger of incorrect answers for constraint functional-logic programs. In: Proceedings of the 2005 ACM SIGPLAN Workshop on Curry and Functional Logic Programming (WCFLP 2005), Tallinn, Estonia, pp. 8–13. ACM Press, New York (2005)
4. Caballero, R., Rodríguez-Artalejo, M., del Vado Vírseda, R.: Declarative diagnosis of missing answers in constraint functional-logic programming. In: Garrigue, J., Hermenegildo, M.V. (eds.) FLOPS 2008. LNCS, vol. 4989, pp. 305–321. Springer, Heidelberg (2008)
5. Clavel, M., Durán, F., Eker, S., Lincoln, P., Martí-Oliet, N., Meseguer, J., Talcott, C.: All About Maude - A High-Performance Logical Framework. LNCS, vol. 4350. Springer, Heidelberg (2007)
6. Hendrix, J., Meseguer, J., Ohsaki, H.: A sufficient completeness checker for linear order-sorted specifications modulo axioms. In: Furbach, U., Shankar, N. (eds.) IJCAR 2006. LNCS (LNAI), vol. 4130, pp. 151–155. Springer, Heidelberg (2006)
7. MacLarty, I.: Practical declarative debugging of Mercury programs. Master's thesis, University of Melbourne (2005)
8. Meseguer, J.: Conditional rewriting logic as a unified model of concurrency. Theoretical Computer Science 96(1), 73–155 (1992)
9. Naish, L.: Declarative diagnosis of missing answers. New Generation Computing 10(3), 255–286 (1992)
10. Naish, L.: A declarative debugging scheme. Journal of Functional and Logic Programming (3) (1997)

11. Nilsson, H.: How to look busy while being as lazy as ever: the implementation of a lazy functional debugger. Journal of Functional Programming 11(6), 629–671 (2001)
12. Pope, B.: A Declarative Debugger for Haskell. PhD thesis, The University of Melbourne, Australia (2006)
13. Riesco, A., Verdejo, A., Caballero, R., Martí-Oliet, N.: A declarative debugger for Maude specifications - User guide. Technical Report SIC-7-09, Dpto. Sistemas Informáticos y Computación, Universidad Complutense de Madrid (2009), `http://maude.sip.ucm.es/debugging`
14. Riesco, A., Verdejo, A., Caballero, R., Martí-Oliet, N.: Declarative debugging of rewriting logic specifications. In: Corradini, A., Montanari, U. (eds.) WADT 2008. LNCS, vol. 5486, pp. 308–325. Springer, Heidelberg (2009)
15. Riesco, A., Verdejo, A., Martí-Oliet, N.: Declarative debugging of missing answers in rewriting logic. Technical Report SIC-6-09, Dpto. Sistemas Informáticos y Computación, Universidad Complutense de Madrid (2009), `http://maude.sip.ucm.es/debugging`
16. Riesco, A., Verdejo, A., Martí-Oliet, N., Caballero, R.: Declarative debugging of rewriting logic specifications. Technical Report SIC-02-10, Dpto. Sistemas Informáticos y Computación, Universidad Complutense de Madrid (2010), `http://maude.sip.ucm.es/debugging`
17. Shapiro, E.Y.: Algorithmic Program Debugging. In: ACM Distinguished Dissertation. MIT Press, Cambridge (1983)
18. Silva, J.: A comparative study of algorithmic debugging strategies. In: Puebla, G. (ed.) LOPSTR 2006. LNCS, vol. 4407, pp. 143–159. Springer, Heidelberg (2007)
19. Tessier, A., Ferrand, G.: Declarative diagnosis in the CLP scheme. In: Deransart, P., Hermenegildo, M.V., Maluszynski, J. (eds.) DiSCiPl 1999. LNCS, vol. 1870, pp. 151–174. Springer, Heidelberg (2000)

The Third Rewrite Engines Competition

Francisco Durán[1], Manuel Roldán[1], Jean-Christophe Bach[2], Emilie Balland[2], Mark van den Brand[3], James R. Cordy[4], Steven Eker[5], Luc Engelen[3], Maartje de Jonge[7], Karl Trygve Kalleberg[6], Lennart C.L. Kats[7], Pierre-Etienne Moreau[2], and Eelco Visser[7]

[1] LCC, Universidad de Málaga, Málaga, Spain
[2] INRIA Nancy - Grand Est, Villers-lès-Nancy Cedex, France
[3] Dpt. of Mathematics and Comp. Sci., Eindhoven U. of Technology, The Netherlands
[4] School of Computing, Queen's University at Kingston, Canada
[5] Computer Science Laboratory, SRI International, Menlo Park, CA, USA
[6] University of Bergen, Bergen, Norway
[7] Dpt. of Software Technology, Delft University of Technology, The Netherlands

Abstract. This paper presents the main results and conclusions of the Third Rewrite Engines Competition (REC III). This edition of the competition took place as part of the 8th Workshop on Rewriting Logic and its Applications (WRLA 2010), and the systems ASF+SDF, Maude, Stratego/XT, Tom, and TXL participated in it.

1 Introduction

As in the 2006 and 2008 editions of the Workshop on Rewriting Logic and its Applications [9,13], in WRLA 2010 a rewrite engines competition was organized, with the aim of bringing to the community the different rewrite engines available, with the main purpose of showing the strengths of each of the participating systems. And as in WRLA 2006 and WRLA 2008, the 2010 edition of the workshop included a session on the competition, in which, in addition to a presentation on the organization, development, and results of the competition, the developers of each of the systems in it had the opportunity of presenting their systems. The discussion and questions from the audience where without any doubt the most interesting part of the session. The present paper tries to summarize such a competition and session, providing additional details on the way the competition was organized and conducted, and trying to complete on the discussion and comparison of the different systems and the results obtained.

The Third Rewrite Engines Competition counted with the particiption of five systems, namely ASF+SDF [20,19], represented by M. van den Brand and L. Engelen; Maude [4,5], represented by F. Durán and S. Eker; Stratego/XT [22,2], represented by M. de Jonge, K. T. Kalleberg, L. Kats, and E. Visser; Tom [1], represented by J.-C. Bach, E. Balland, and P.-E. Moreau; and TXL [7,6], represented by J. Cordy. The second edition gathered the same number of participants (ASF+SDF, Maude, Stratego, TermWare [14] and Tom) and two in the first one (ASF+SDF and Maude). We would have liked to gather more systems,

P.C. Ölveczky (Ed.): WRLA 2010, LNCS 6381, pp. 243–261, 2010.

although it is not easy, for different reasons. We would like to thank the developers of Kiama [15], Rascal [12] and TermWare, who showed their interest in being involved, but for one reason or another, were not able to get to the end. Developers of other systems were also invited, but kindly refused their participation. Our apologies to any other system that should have been invited but was not... perhaps in the next one!

This edition of the competition, as the previous ones, was very illustrative, since it showed that each of the engines focusses on very specific problems, and that they are very good at them. More than a competition, REC is an opportunity to show the different systems and their strengths, and why not, their weaknesses, to the rewriting community.

The first competition [10], which was organized by G. Roşu, focused on efficiency, specifically speed, memory management and built-ins use. There were only two participants, ASF+SDF, represented by M. van den Brand, and Maude, represented by S. Eker, but awoke interest on such a kind of event and opened the door to the subsequent competitions. For this first edition of the competition, a number of test examples were compiled, all of them using features supported by both systems. Most of the problems used came from the benchmarks of the two systems. The paper [10] includes very interesting discussions on the technical details why Maude and ASF+SDF behaved like they did on the different tests run in the 1st REC. Since many of the problems used in it are again in this 3rd REC, the discussions there are a very useful complement to the present paper.

For the second edition [11], the possibility of having some bigger problems to develop was considered. Several ideas were considered, as the development of a small theorem prover, the exploration of a search space, a transformation of XML or a tree... Among all these problems, first steps in the world of program transformations were taken. In the end, a common language to specify term rewrite systems, called REC, was developed, and the development of an interpreter for REC was proposed. Then, the set of rewrite problems proposed was expressed in this REC language.

The REC language and its use to run the problems in the competition was maintained on this 3rd competition. Some new problems were included in our benchmark, but basically the efficiency of the systems was compared running the different problems in this REC syntax on the interpreters developed for REC in the participating systems. To be able to reuse the interpreters developed for the 2008 competition, the syntax of the REC language, which is described in Section 3.1, was not modified.

After the experience with the REC language, and since some of the systems in the competition specialize on program transformation, we decided to include in this edition some additional problems related to the definition of programming languages, and to the generation, analysis and transformation of programs, which is one of the key application areas of term rewriting. Following a suggestion by J. Cordy, we decided to include some problems from the TIL Chairmarks, developed by J. Cordy and E. Visser. The TIL language and the TIL Chairmarks problems used in the competition are described in Sections 3.2 and 5.

2 The Systems in the Competition

The systems in the 3rd REC are of a very different nature. We have compilers and interpreters, we have specific-purpose and general-purpose systems, we have embedded rewriting systems and stand-alone systems, ... The results here should not be taken as a final comparison of the systems, but just as a starting point on some very specific issues. In fact, there are many strong points in each of the systems that are not considered in the competition. For example, SDF+SDF, Stratego, Tom, and TXL have very sophisticated facilities for program manipulation, with, e.g., very powerful parsers and pretty-printing tools; Tom is embedded into different generalist programming languages (e.g. C, Java, Python, C++, C#); Maude supports matching modulo any combination of associativity, commutativity, and identity, and unification modulo commutativity and associativity-commutativity, and provides a suite of formal tools. In this section we introduce the main features of each of the systems.

2.1 ASF+SDF

ASF+SDF is a general-purpose, executable, algebraic specification formalism based on (conditional) term rewriting. Its main application areas are the definition of the syntax and the static semantics of (programming) languages, program transformations and analysis, and for defining translations between languages.

The ASF+SDF formalism [21] is a combination of two formalisms: ASF (the Algebraic Specification Formalism) and SDF (the Syntax Definition Formalism). SDF is used to define the concrete syntax of a language, whereas ASF is used to define conditional rewrite rules; the combination ASF+SDF allows the syntax defined in the SDF part of a specification to be used in the ASF part, thus supporting the use of user-defined syntax when writing ASF equations. ASF+SDF also supports modular structuring of specifications using names modules, and thus enabling reuse.

The ASF+SDF and the ASF+SDF Meta-Environment have been applied in a broad range of applications. The application areas can be characterized as: prototyping of domain specific languages, software renovation, and code generation. An overview of some of the applications is given in [17]. The ASF+SDF system, its documentation, and related papers are available at `http://www.meta-environment.org/`. ASF+SDF is no longer maintained and is replaced by Rascal, see `http://www.rascal-mpl.org/`.

2.2 Maude

Maude is a language and a system based on rewriting logic [4,5,3]. Maude modules are rewrite theories, while computation with such modules corresponds to efficient deduction by rewriting. Since rewriting logic contains equational logic, Maude also supports equational specification and programming in its sublanguage of functional modules and theories. The underlying equational logic of

Maude is membership equational logic, that has sorts, subsorts, operator overloading, and partiality definable by membership and equality conditions. Because of its logical basis and its initial model semantics, a Maude module defines a precise mathematical model. This means that Maude and its formal tool environment can be used in three, mutually reinforcing ways: as a declarative programming language, as an executable formal specification language, and as a formal verification system. The Maude system, its documentation, and related papers and applications are available from the Maude website `http://maude.cs.uiuc.edu`.

Maude provides very efficient support for rewriting modulo any combination of associativity, commutativity, and identity axioms, and provides two built-in rewrite strategies: top-down rule fair and position fair. Maude's rewrite engine makes extensive use of advanced semi-compilation techniques and sophisticated data structures supporting rewriting modulo. Besides supporting efficient execution, Maude also provides a range of formal tools and algorithms to analyze rewrite theories and verify their properties including a search facility for doing breadth first search with cycle detection, and a linear time temporal logic model checker.

2.3 Stratego/XT

Stratego/XT is a language and toolset for program transformation. The Stratego language provides rewrite rules for expressing basic transformations, programmable rewriting strategies for controlling the application of rules, concrete syntax for expressing the patterns of rules in the syntax of the object language, and dynamic rewrite rules for expressing context-sensitive transformations, thus supporting the development of transformation components at a high level of abstraction.

The XT toolset offers a collection of extensible, reusable transformation tools, such as powerful parser and pretty-printer generators and grammar engineering tools. Stratego/XT supports the development of program transformation infrastructure, domain-specific languages, compilers, program generators, and a wide range of meta-programming tasks.

Stratego has two backends: one for generating C code (StrC), and another for generating Java code (StrJ). The Stratego/XT system, its documentation, and related papers are available at `http://strategoxt.org/`.

2.4 Tom

Tom [1] is an extension of Java which adds support for algebraic data-types and pattern matching. Contrary to other languages, Tom does not enforce any particular tree representation for the objects being matched. To make this possible, Tom provides a mapping definition formalism to describe the relationship between the concrete Java implementation and the algebraic view, which allows to define transformations directly on existing Java data-structures. The other features of the Tom language are mainly a powerful pattern-matching construct (matching modulo theory, list-matching, anti-patterns, XML notation,...); support for private types in Java; an efficient implementation of typed and maximally shared terms, an extension for term-graph rewriting and a strategy language inspired by Elan and Stratego.

To conclude, the main originality of Tom is that it is piggybacked on top of Java, which allows to integrate smoothly declarative transformation code in existing Java programs. It has been used to implement many large and complex applications, among them the compiler itself. Tom is used in academic projects to prototype models based on rewriting but it is also successfully integrated in industrial products (for example, database request translation in SAP's software). The Tom systems is available at `http://tom.loria.fr/`.

2.5 TXL

TXL [6,7] is a special-purpose programming language designed for creating, manipulating and rapidly prototyping language descriptions, tools and applications using source transformation. TXL is designed to allow explicit programmer control over the interpretation, application, order and backtracking of both parsing and rewriting rules. Using first order functional programming at the higher level and term rewriting at the lower level, TXL provides for flexible programming of traversals, guards, scope of application and parameterized context. This flexibility has allowed TXL users to express and experiment with both new ideas in parsing, such as robust, island and agile parsing, and new paradigms in rewriting, such as XML markup, rewriting strategies and contextualized rules, without any change to TXL itself. TXL's website is `http://txl.ca`.

3 The REC and TIL Languages

With different goals in mind, two different languages, REC and TIL, have been used in the competition. We present these simple languages in the following sections. Section 3.3 discusses the lexical analysis and parsing tools developed for these languages as part of the competition.

3.1 The REC Language

REC is a term rewriting language, that was defined for the second rewrite engines competition as a common language in which to write the rewrite tasks to pose to the participant systems. The REC language is many-sorted, does not have any built-ins, uses prefix syntax, does not support overloading, allows conditional rules, and includes syntax for `assoc`, `comm`, `id`, and `strat` attributes *à la* OBJ. A BNF description of the syntax of the language is given in Figure 1. Figure 2 shows the REC specification of the factorial function, with the natural numbers, with plus and times operations, represented using Peano notation.

Each of the participants was asked to build a program transforming the problems in this REC syntax to the language of their corresponding tools. Those

```
⟨spec⟩          ::= REC-SPEC ⟨id⟩
                      [ SORTS ⟨idlist⟩ ]
                      [ VARS ⟨vardecllist⟩ ]
                      [ OPS ⟨opdecllist⟩ ]
                      [ RULES ⟨rulelist⟩ ]
                    END-SPEC
⟨idlist⟩        ::= ⟨id⟩ ⟨idlist⟩ | ϵ
⟨vardecllist⟩   ::= ⟨idlist⟩ : ⟨id⟩ ⟨vardecllist⟩ | ϵ
⟨opdecllist⟩    ::= ⟨opdecl⟩ ⟨opdecllist⟩ | ϵ
⟨opdecl⟩        ::= op ⟨id⟩ : ⟨idlist⟩ -> ⟨id⟩
                  | op ⟨id⟩ : ⟨idlist⟩ -> ⟨id⟩  ⟨opattrlist⟩
⟨opattrlist⟩    ::= ⟨opattr⟩ ⟨opattrlist⟩ | ϵ
⟨opattr⟩        ::= assoc | comm | id( ⟨term⟩ ) | strat( ⟨intlist⟩ )
⟨rulelist⟩      ::= ⟨rule⟩ ⟨ruleslist⟩ | ϵ
⟨rule⟩          ::= ⟨term⟩ -> ⟨term⟩ | ⟨term⟩ -> ⟨term⟩ if ⟨condlist⟩
⟨condlist⟩      ::= ⟨cond⟩ | ⟨cond⟩ , ⟨condlist⟩
⟨cond⟩          ::= ⟨term⟩ -><- ⟨term⟩                        % ==
                  | ⟨term⟩ ->/<- ⟨term⟩                       % =/=
⟨term⟩          ::= ⟨id⟩ | ⟨id⟩ ( ) | ⟨id⟩ ( ⟨termlist⟩ )
⟨termlist⟩      ::= ⟨term⟩ | ⟨term⟩ , ⟨termlist⟩
⟨intlist⟩       ::= ⟨int⟩ ⟨intlist⟩ | ϵ
⟨command⟩       ::= get normal form for: ⟨term⟩
                  | check the confluence of: ⟨term⟩ -><- ⟨term⟩
```

⟨*id*⟩ are non-empty sequences of any characters except ' ', '(', ')', '{', '}', '"' and ','; and excluding ':', '->', '-><-', '->/<-', 'if', and keywords REC-SPEC, SORTS, VARS, OPS, RULES, and END-SPEC.
⟨*int*⟩ are non-empty sequences of digits.
Comments are given using '%'. Text written in the line after a '%' is discarded.

Fig. 1. BNF description of the syntax of the REC language

that already developed this program transformer for REC II were able to use the same tool, since the syntax of the language did not change. This was one of the reasons for developing such a language in 2008. ASF+SDF and TXL had to build it from scratch for REC III.

3.2 TIL

The Tiny Imperative Language (TIL) is a very small imperative language with assignments, conditionals, and loops, designed by J. Cordy and E. Visser, as a basis for small illustrative example transformations. These example transformations define the benchmark transformation tasks they propose as the TIL Chairmarks. As we will explain in Section 5, a selection of the TIL Chairmarks has been used in this 3rd REC. The syntax of TIL is given in Figure 3. A some more detailed description of the language is available at `http://www.program-transformation.org/Sts/TinyImperativeLanguage`.

```
REC-SPEC Factorial
SORTS Nat
OPS
  0 : -> Nat                   % zero
  s : Nat -> Nat               % succesor
  plus : Nat Nat -> Nat        % addition
  times : Nat Nat -> Nat       % product
  fact : Nat -> Nat            % factorial
VARS N M : Nat
RULES
  plus(0, N) -> N
  plus(s(N), M) -> s(plus(N, M))
  times(0, N) -> 0
  times(s(N), M) -> plus(M, times(N, M))
  fact(0) -> s(0)
  fact(s(N)) -> times(s(N), fact(N))
END-SPEC
```

Fig. 2. REC specification of the factorial function

3.3 Lexical Analysis and Parsing

We had two different approaches in the competition for the implementation of the translators requested for REC and TIL. While ASF+SDF, Stratego/XT, Tom, and TXL representatives built programs that transformed the original programs and commands, and were later loaded and executed, in Maude a programming environment was built, able to read REC programs and commands and give outputs. Maude does not have facilities to handle files, what complicates the reading of input files and the generation of output files with the resulting programs. However, Maude has some facilities for building execution environments, that was the approach followed in that case.

Maude has some limitations at the lexical level, what forced the Maude representatives to alter the input files (enclosing the input programs in parentheses and removing comments). Maude and ASF+SDF does not offer constructs to read input from the command line while rewriting, which makes it impossible to implement the interpreter for TIL as it is implemented in Tom or TXL. Alternatively, in the Maude and ASF+SDF cases, interpreters that take a program and a list of values as input, and provide the output for that program given the input as its result, were implemented.

No lexical or parsing problems were encountered in the cases of ASF+SDF, Stratego/XT, Tom, and TXL. ASF+SDF and Stratego/XT are based on SDF and SGLR,[1] and support the full class of context-free grammars. Tom uses the ANTLR parser generator;[2] the abstract syntax tree (AST) produced by ANTLR

[1] SGLR (Scannerless Generalized LR Parser) is an implementation of the Generalized LR algorithm [16] with extensions for scannerless parsing.

[2] The web site of ANTLR (ANother Tool for Language Recognition) is at http://www.antlr.org/

```
⟨program⟩ ::= ⟨statement_list⟩
⟨statement_list⟩ ::= ⟨statement⟩ ⟨statement_list⟩ | ϵ
⟨statement⟩ ::= ⟨declaration⟩ | ⟨assignment_statement⟩ | ⟨if_statement⟩
              | ⟨while_statement⟩ | ⟨for_statement⟩ | ⟨read_statement⟩
              | ⟨write_statement⟩
⟨declaration⟩ ::= var ⟨identifier⟩ ;                    % Untyped variables
⟨assignment_statement⟩ ::= ⟨identifier⟩ := ⟨expression⟩ ;
⟨if_statement⟩ ::= if ⟨expression⟩ then ⟨statement_list⟩ end
                 | if ⟨expression⟩ then ⟨statement_list⟩
                   else ⟨statement_list⟩ end
⟨while_statement⟩ ::= while ⟨expression⟩ do ⟨statement_list⟩ end
⟨for_statement⟩ ::= for ⟨identifier⟩ := ⟨expression⟩ to ⟨expression⟩ do
                      ⟨statement_list⟩
                    end
⟨read_statement⟩ ::= read ⟨identifier⟩ ;
⟨write_statement⟩ ::= write ⟨expression⟩ ;
⟨expression⟩ ::= ⟨primary⟩ | ⟨expression⟩ ⟨op⟩ ⟨expression⟩
⟨primary⟩ ::= ⟨identifier⟩ |  ⟨integer⟩ | ⟨string⟩ | ( ⟨expression⟩ )
⟨op⟩ ::=   = | != | + | - | * | /   % from lowest to highest priority
```

Fig. 3. Grammar for Tiny Imperative Language (TIL)

can be directly reused in the Tom system. TXL has its own top-down programmable parser that the user can control directly [8] as part of the TXL program.

4 The REC Problems

The REC language presented in Section 3.1 has been used in two different ways in this 3rd rewrite engines competition. First, the participants were asked to write interpreters for it, so that REC can be used as a common language in which to write the problems used to compare their performance. Since all interpreters for all the systems were provided, there was no need for hand-made transformations. In the 2008 competition some of the systems did not develop such interpreters, and solutions were provided by hand; the rest of the systems were allowed to provide optimizations of the automatically generated rewrite systems. In the 2006 competition all the specifications were written by hand in each of the participating systems.

Translating the REC specifications to their counterparts in the different systems is an easy task, and the automatic translations take little time. The implementations of these translations are quite straightforward in all the systems, and optimization was not attempted in any case. In all the cases, all terms are represented by their concrete syntax all the time. E.g., natural numbers are represented using Peano notation. Manual optimizations using built-ins, memoization, etc. could have been considered for all the systems but were not.

4.1 Disclaimer

We must acknowledge that there was perhaps somewhat of a mismatch between the REC test cases and the normal applications of the Stratego and TXL systems. These systems are not traditional rewrite engines, and are typically not applied for traditional rewriting problems but for other applications such as program transformation and analysis. Our test set in this section focuses purely on raw rewriting power, and as such may be biased towards traditional rewriting systems.

For Stratego, an *innermost* strategy is used to emulate the behavior of a true term rewriting engine. Likewise, for TXL, term rewriting is implemented using a global transformation rule that globally applies the entire ruleset to a fixed point. REC rewrite rules are directly mapped to rules in the different systems, but in the case of Stratego and TXL, the individual rules are combined using functional composition. Although the order of application can affect performance, no attempt has been made to optimize this order in the mechanical translation from REC. Stratego and TXL do not apply memoization when evaluating the rules.

Maximal sharing of identical subterms ensures efficient memory usage and constant time comparison at the cost of slightly increased time spent when constructing new terms. Since the tests in our benchmark involve large terms with repeating subterms and do not use line numbers or other context information, systems that employ maximal sharing may be at the advantage. Stratego (when compiled to C) and ASF+SDF implement maximal sharing based on the ATerm library [18]. TXL and the Java version of Stratego do not employ maximal sharing. Tom provides an efficient implementation of typed and maximally shared terms in Java.

As in REC II, the rewriting problems are organized in four categories: unconditional rewriting (TRS), conditional rewriting (CTRS), rewriting modulo (Modulo), and context-sensitive rewriting/rewriting with local strategies (CS). Only Maude has support for the features needed to be in all these categories. ASF+SDF, Stratego and TXL only participate in the TRS and CTRS categories. Tom supports rewriting modulo associativity since its first version. In a recent release it also provides support for rewriting modulo associativity-commutativity. However, although the implementation is correct, it is not yet very efficient.

4.2 Results for the Rewriting Problems

We now present the results for each of the rewrite examples considered in the competition. Although we have five participants, namely ASF+SDF, Maude, Stratego, Tom and TXL, two different versions were considered for both Maude and Stratego. In the case of Maude we used 32-bits and 64-bits binaries, and for Stratego we tested a C implementation and a newly developed Java version.

The five systems were installed on a 64-bits Linux 2.40GHz/4GB Intel Core 2 Quad. The installation of the systems was done by M. Roldán, who also ran most of the tests.

For each case, after a brief description of the problem, a table with the times used in the computations is presented. In these tables, all times are given in milliseconds. Those test cases that either took long (more than one hour), ran out of memory, or produced an internal error show as '—'.

In most of these cases, a manual implementation in the system's language, rather than a (naive) automatic translation of the REC specification, would be more appropriate. In some cases we may get huge improvements by reordering the equations, saving partial computations, using memoization, etc. In the 2nd REC we consider both an automatic translation and a handwritten optimized version for each of the problems in the competition. In this edition we are only considering the automatic translation. See [10] for the results and comparison in the 2008 edition of the competition.

In most cases, the numbers are self explanatory. In some of them we give some explanations or provide some pointers for a discussion on them. We present a selection of the results in this paper, and refer to the web site of the competition, at `http://www.lcc.uma.es/rewriting_competition`, for further details. All the files and results of the competition are available in this web site, where one can find a table that includes, for each of the problems, the specification and the tests run on it in REC syntax, and the corresponding problems in the syntax of each the participant systems, together with the times consumed in their computation and the solutions given.

TRS: unconditional rewriting. In this category we have rewrite systems for the calculation of the factorial of a natural number, the n-th number in the Fibonacci sequence, a function reversing a list, an artificial rewrite system to test garbage collection algorithms, and an ASF+SDF benchmark for the study of resource usage in brute-force rewriting (no built-ins, no strategies).

Factorial. The specification of the factorial of a natural number was presented in Figure 2. The factorial function is calculated for values 6, 8, 10, and 12.

	ASF+SDF	Maude32	Maude64	StrC	StrJ	Tom	TXL
6	17	0	0	0	20	5	4,566
8	26	4	5	50	170	—	—
10	32,466	544	754	—	—	—	—
12	—	—	—	—	—	—	—

The reason why the Maude interpreter outperforms the ASF+SDF and Tom compilers is probably because of the term representation they used. See [10] for a more in depth discussion on this case for the ASF+SDF and Maude systems.

Fibonacci. The Fibonacci sequence is specified by the following three rules:

```
fibb(0) -> s(0)
fibb(s(0)) -> s(0)
fibb(s(s(N))) -> plus(fibb(s(N)), fibb(N))
```

The `fibb` function is calculated for values 10, 20, 30, 40, and 50.

	ASF+SDF	Maude32	Maude64	StrC	StrJ	Tom	TXL
10	10	0	0	0	20	2	7
20	86	10	7	20	90	—	108,196
30	10,788	2,273	2,505	—	—	—	—
40	—	—	—	—	—	—	—

Garbage collection. This rewrite system consists of the following rules:

```
c(0, Y) -> Y
c(s(X), Y) -> s(c(X,Y))
f(X, Y, Z, T, U) -> f(X, Y, Z, Y, Z, T, U)
f(X, Y, s(Z), N, P, T, U) -> f(X, Y, Z, N, P, c(T, T), U)
f(X, s(Y), 0, N, P, T, U) -> f(X, Y, P, N, P, T, T)
f(s(X), 0, 0, N, P, T, U) -> f(X, N, P, N, P, 1, 0)
f(0, 0, 0, N, P, T, U) -> T
```

The different tests run consist in the reduction of terms of the form `f(m,n,p,0,1)`, with different values for `m`, `n`, and `p`.

	ASF+SDF	Maude32	Maude64	StrC	StrJ	Tom	TXL
f(2,2,2,0,1)	14	0	0	0	10	17	7
f(2,2,4,0,1)	39	1	1	0	0	50	13,378
f(2,4,2,0,1)	20	0	1	—	—	26	661
f(2,4,4,0,1)	9,019	261	300	—	—	—	—
f(4,2,2,0,1)	15	0	0	—	—	18	8
f(4,2,4,0,1)	44	2	1	—	—	57	14,495
f(4,4,2,0,1)	16	1	0	—	—	26	727
f(4,4,4,0,1)	8,918	459	512	—	—	—	—

Notice that all the systems behave quite well for all the tests except for those with `n` = 4 and `p` = 4.

List reverse. Given lists represented with constructors `cons : Nat List -> List` and `nil : -> List`, the following `rev` function reverses the elements of a list of natural numbers.

```
conc(cons(E, L), L') -> cons(E, conc(L, L'))
conc(nil, L') -> L'
reverse(cons(E, L)) -> conc(reverse(L), cons(E, nil))
reverse(nil) -> nil
```

The tests are run on lists of 10^2, 10^3, and 10^4 elements.

	ASF+SDF	Maude32	Maude64	StrC	StrJ	Tom	TXL
10^2	23	0	0	0	10	39	1,495
10^3	780	46	31	—	—	622	—
10^4	69,403	4,520	3,714	—	—	105,930	—

ASF+SDF benchmark for brute force rewriting. In these tests we include three different functions: symbolic evaluation of 2^n modulo 17 (sym), for testing speed of rewriting with almost no memory usage; symbolic evaluation of 2^n modulo 17 after expanding the expression (eval), to test memory management; and computation on huge 2^n, not-alike trees (tree), also to test memory management. The specification of these problems can be found in [19]. An interesting discussion on the behavior of ASF+SDF and Maude on these tests can be found in [10].

	ASF+SDF	Maude32	Maude64	StrC	StrJ	Tom	TXL
sym(10)	22	3	3	0	10	90	4,714
sym(20)	337	3,162	2,506	1,830	4,480	5,877	—
eval(10)	16	—	—	0	80	84	—
eval(20)	346	—	—	2,190	10,210	4,488	—
tree(10)	56,734	5	5	—	—	113	—
tree(20)	—	9,674	12,480	—	—	5,818	—

ASF+SDF performs much better than the others for `sym` and `eval`. However, rewriting `tree(10)` and `tree(20)` take a lot of time in all the systems using the automatically generated specifications because many computations are repeated. It is remarkable that Tom and Maude perform better than ASF+SDF in these tests. Saving the computations to avoid the repetition of the evaluations would result in big improvements for all the systems. E.g., just by introducing variables that store the rewritten result of such subterms `tree(20)` takes 15 milliseconds in ASF+SDF.

CTRS: conditional term rewrite systems. In this category we find bubblesort, mergesort, quicksort, a bit matrix closure algorithm, an odd/even artificial problem, and a specification of the towers of Hanoi problem.

Bubblesort. Given lists of natural numbers defined by `cons` and `nil` as above, and given a less-than function `lt`, the bubblesort algorithm is specified by the single following rule:

```
cons(N, cons(M, L)) -> cons(M, cons(N, L)) if lt(M, N) -><- true
```

The following results are obtained for lists of 10, 100, and 1,000 elements in reverse order:

	ASF+SDF	Maude32	Maude64	StrC	StrJ	Tom	TXL
10	13	0	0	0	50	35	19
100	26	85	74	110	500	88	—
1,000	1,550	383,815	450,887	130	330	5,299	—

Maude performs so badly in this case because of the very ineffective way in which it treats conditional rules.

Mergesort. Given lists of natural numbers defined by `cons` and `nil` as above, and a less-than-or-equal predicate on natural numbers `lte`, the `mergesort` function is specified as follows:

```
merge(nil, L) -> L
merge(L, nil) -> L
merge(cons(X, L1), cons(Y, L2)) -> cons(X, merge(L1, cons(Y, L2)))
  if lte(X, Y) -><- true
merge(cons(X, L1), cons(Y, L2)) -> cons(Y, merge(cons(X, L1), L2))
  if lte(X, Y) -><- false
split(cons(X, cons(Y, L)))
  -> pair(cons(X, p1(split(L))), cons(Y, p2(split(L))))
split(nil) -> pair(nil, nil)
split(cons(X, nil)) -> pair(cons(X, nil), nil)
mergesort(nil) -> nil
mergesort(cons(X, nil)) -> cons(X, nil)
mergesort(cons(X, cons(Y, L)))
  -> merge(mergesort(cons(X, p1(split(L)))),
           mergesort(cons(Y, p2(split(L)))))
p1(pair(L1, L2)) -> L1
p2(pair(L1, L2)) -> L2
```

The following results are obtained for lists of 10, 100, and 1,000 elements in reverse order:

	ASF+SDF	Maude32	Maude64	StrC	StrJ	Tom	TXL
10	9	1	0	0	70	50	8
100	—	9	9	—	—	—	—
1,000	—	9,134	10,721	—	—	—	—

The reason why most of the systems perform so badly is because the equations for `split` and `merge` are not right-linear. The rewriting of the `split(L)` terms is repeated if the sharing is not detected as in Maude. Simple modifications in the specifications, using memoization or intermediate variables, would lead to big improvements. E.g., in ASF+SDF, the use of these variables takes the computation times to 11/7/20.

Quicksort. The following results are obtained for lists of 10, 100, and 1,000 elements in reverse order:

	ASF+SDF	Maude32	Maude64	StrC	StrJ	Tom	TXL
10	10	0	1	0	230	—	305
100	—	42	39	—	—	—	—
1,000	—	193,616	227,166	—	—	—	—

As for the mergesort function above, the reason for such results is that many computations are repeated many times. By introducing new variables to avoid re-computations in, e.g., ASF+SDF makes the times to go down to 15/19/32.

Bit matrix closure. This rewrite system calculates the reflective and transitive closure of a bits matrix. The results for sizes 10x10, 20x20 and 30x30 are:

	ASF+SDF	Maude32	Maude64	StrC	StrJ	Tom	TXL
10x10	19	2	1	—	—	28	1,504
20x20	32	12	10	—	—	84	56,907
30x30	8	59	43	—	—	103	809,494

Odd/even. This is an artificial example to test the exponential explosion that can result due to conditional rewriting.

```
odd(0) -> false
even(0) -> true
odd(s(N)) -> true if even(N) -><- true
even(s(N)) -> true if odd(N) -><- true
odd(s(N)) -> false if even(N) -><- false
even(s(N)) -> false if odd(N) -><- false
```

The results obtained are the following:

	ASF+SDF	Maude32	Maude64	StrC	StrJ	Tom	TXL
odd(15)	13	69	60	0	10	15	0
odd(20)	11	0	0	0	0	14	5,880
odd(25)	14	66,828	53,864	0	0	18	0

ASF+SDF and Tom do well in this example because they optimize the compiled code to avoid re-computation in conditions. Maude rewrites the computations as given. In the case of TXL, evaluating the odd function on an odd number results in an almost immediate success, while evaluating it on an even number results in an exponential search. In the case of Maude it is the other way around, because the rules are considered is a different order.[3] The use of memoization, or simply changing the order of the rules, significantly improves the efficiency of Maude in this case.

Hanoi towers. This rewrite system solves the traditional problem of the towers of Hanoi. The solutions were executed for 4 and 16 disks.

	ASF+SDF	Maude32	Maude64	StrC	StrJ	Tom	TXL
4	8	0	0	0	20	26	6
16	950	188	210	—	—	378	—

Modulo: rewriting modulo associativity and/or commutativity and/or identity. Maude and Tom are the only systems between the participants providing some form of rewriting modulo. Maude supports rewriting modulo any combination of associativity, commutativity and identity. Tom supports rewriting modulo associativity, and a first attempt for rewriting modulo associativity-commutativity in its latest release.

[3] The program transformation implemented for Maude uses a set of rules instead of a list, and in cases like this it may change the order in which the rules are considered.

Tautology-hard, darts, 3-value logic, and permutations. The tautology-hard rewrite system evaluates Boolean expressions with associative and commutative `and`, `xor`, `or`, and `iff` operations. Logic3 defines a 3-value logic, and darts operates on sets. Their specifications include several associative and commutative operators. The permutations specification defines a function that calculates all the permutations of a list. It uses two operators which are declared associative and with identity element. The tautology-hard rewrite system is evaluated on three expressions of different sizes. These are all the results obtained:

	Maude32	Maude64	Tom
tautology-hard 1	10	9	451
tautology-hard 2	200	173	—
tautology-hard 3	523	471	—
darts	2	2	56
logic 3	11	10	—
permutations	14	20	21

CS: context sensitive rewriting. Although other participants provide support for very sophisticated strategies, Maude is the only system among the participants supporting local strategies *à la* OBJ.

Sieve of Eratosthenes. The specification of the sieve of Erathostenes algorithm is used to compute the first 20, 100, and 1,000 prime numbers.

	Maude32	Maude64
20	2	2
100	152	125
1,000	165,039	135,639

5 The TIL Chairmarks

In addition to the problems used in the previous competition (see Section 4), we included a few transformation problems from the TIL Chairmarks, by J. Cordy and E. Visser. Detailed information on the TIL Chairmark is available in the web site at `http://www.program-transformation.org/Sts/TILChairmarks`. As Cordy and Visser explain in this web page, "They are called *chairmarks* because they are too small to be called *benchmarks*". From all the tests proposed there, we chose six of them, trying to cover different kinds of problems. Examples illustrating some of the transformations proposed are included here, see `http://www.pro\discretionary-gram-transformation.org/Sts/TILChairmarks` for examples and additional explanations on the rest, and also for the rest of the transformations proposed.

The problems chosen, with the numbers as in the TIL Chairmarks site, are:

2.2 For to whiles: This transformation restructures all for-loops in a TIL program to their while equivalents. Figure 4 shows an example of the application on this transformation to a TIL program.

```
for i := 1 to 9 do              var i;
   for j := 1 to 10 do         i := 1;
       write i * j;            while i != 9 + 1 do
   end                             var j;
end                                 var j;
                                   j := 1;
                                   while j != 10 + 1 do
                                       write i * j;
                                       j := j + 1;
                                   end
                                   i := i + 1;
                               end
```

Fig. 4. Program that outputs the first 10 multiples of numbers 1 through 9. The program in the right-hand side is the result of applying transformation 2.2 to the program on the left.

2.4 Declarations to local: Declaration are moved to its most local context.
3.2 Common subexpression elimination: Common subexpressions are recognized and factored out to new temporary variables.
4.1 Redundant declarations: Unused declarations are detected and removed.
4.2 Statistics: The number of statements of different kinds (declarations, assignments, ifs, whiles, fors, reads, and writes) in a program are counted.
5.1 Interpretation: TIL programs are executed by source transformation/rewriting.

Notice that, although clearly stated, the problems can be solved in different ways, and the outputs given in different forms. The outputs were not systematically checked. The outputs given and the program transformations proposed by the different systems are available at the competition's web site at `http://www.lcc.uma.es/rewriting_competition`. Given the interpreter provided as solution of the task 5.1, we can at least think of checking that both programs give the same result. But it was not done in this edition.

In ASF+SDF, implementing Tasks 2.2, 4.1 and 4.2 is straightforward. Task 2.4 requires a way of swapping statements; once it is clear how this should be done, the solution can be specified quite easily. The algorithmics needed to solve Task 3.2 are not trivial. Indeed, in the case of ASF+SDF most of the time was spent on implementing this 'chairmark'. Finally, the interpreter for Task 5.1 would take some time to implement without prior experience, but there exists an interpreter specified using ASF+SDF for a similar imperative language that can be used to understand the general idea behind such an interpreter.

For Maude the situation is very similar to the one for ASF+SDF. In this case, all the experience gathered along the years in giving semantics and defining execution environments for different languages is of great help.

Stratego appears to be a suitable language for the implementation of the TIL chairmarks. Simple transformations like 2.2 are defined with help of rewrite rules that are applied in a traversal strategy. This can be a general traversal strategy like topdown (2.2), or a custom traversal (for example 5.1). The separation of rules and strategies enables reuse. An example of reuse can be found in Task 4.2 where the *occurrences* strategy is used to collect statistic data. Sometimes the application of a rewrite rule depends on contextual information. Context information is handled with help of dynamic rules which are created during a traversal and can be scoped. Dynamic rules have been specifically designed for concisely handling problems as seen in the chairmarks, making Stratego highly effective at solving these problems. Dynamic rules are used in Tasks 2.4, 3.2, and 4.1 to implement lookup tables for variables and declarations.

The TIL chairmarks are typical applications for the Tom system. By using ANTLR it was straightforward to implement a parser. Then, given the produced AST, Tom appeared very appropriate to describe and implement the various transformations and optimizations: we have used the notion of rule (elementary strategy) to describe the transformations, and the user defined strategy language to describe how to apply the rules. E.g, Task 3.2 was solved using two strategies and a Java HashMap; Task 4.1 was solved, in less than 100 lines, using two strategies (one parameterized by a String) and a topdown Task 4.2 was solved, in around 70 lines, using a *count* strategy, integer counters and a topdown.

The TIL source transformation tasks are the kind of problems that TXL was designed for, and all of them are relatively straightforward for an experienced TXL programmer as self-contained TXL programs with no need for external tools or support routines. Task 4.1 is a single rewrite rule of 9 lines in TXL's vertical rule layout, using a scoped searching guard. Task 2.4 in TXL uses a sorting strategy in two parts, moving declarations to the first statement that uses them, and then moving them inside if it is a compound statement, using about 100 lines. Task 3.2 is a bit more challenging, using TXL rule parameters and scoped application to find and replace subexpressions with a searching guard to insure non-interference, for a total of 80 lines. Task 4.2 exploits the TXL built-in type extract and count rules to solve the problem in one rule of 37 lines. Finally, the full TIL interpreter in TXL (Task 5.1) uses a pure rewriting interpretation with global terms to store the state, taking 286 lines.

Given the facilities provided by the different systems and the simplicity of most of the tasks, the tasks were solved in a short time, being most of the time spent in designing the solutions and debugging them.

6 Conclusions

As in the previous Rewrite Engines Competitions, we believe that both rewrite engines users and developers have benefited from this third edition of the competition. Although in edition we took a great step forward, by having five systems, focusing on program transformations without forgetting performance, and on automation, there is still a lot to be done towards having a real competition and really showing the potential of all the participating systems. In any case, our main goals were satisfied: we got to know each of the systems better, some of the strengths and weaknesses of the engines were shown, and we got more motivation to go on working on our respective systems.

And one wish for the competition: More automatization is required! For entering the programs, time capturing, results table generation, etc.

Acknowledgements

We thank P. Ölveczky, as organizer of WRLA 2010, for inviting us to organize the competition, and to G. Roşu for getting the ball rolling in the 1st REC. And, of course, we have to thank all the people who have participated in the development of all the rewrite engines in the competition. F. Durán and M. Roldán have been supported by Research Projects TIN2008-03107 and P07-TIC-03184.

References

1. Balland, E., Brauner, P., Kopetz, R., Moreau, P.-E., Reilles, A.: Tom: Piggybacking rewriting on java. In: Baader, F. (ed.) RTA 2007. LNCS, vol. 4533, pp. 36–47. Springer, Heidelberg (2007)
2. Bravenboer, M., Kalleberg, K.T., Vermaas, R., Visser, E.: Stratego/XT 0.17. A language and toolset for program transformation. Science of Computer Programming 72(1-2), 52–70 (2008)
3. Clavel, M., Durán, F., Eker, S., Escobar, S., Lincoln, P., Martí-Oliet, N., Meseguer, J., Talcott, C.L.: Unification and narrowing in Maude 2.4. In: Treinen, R. (ed.) RTA 2009. LNCS, vol. 5595, pp. 380–390. Springer, Heidelberg (2009)
4. Clavel, M., Durán, F., Eker, S., Lincoln, P., Martí-Oliet, N., Meseguer, J., Quesada, J.: Maude: Specification and programming in rewriting logic. Theoretical Computer Science 285, 187–243 (2002)
5. Clavel, M., Durán, F., Eker, S., Lincoln, P., Martí-Oliet, N., Meseguer, J., Talcott, C.: All About Maude - A High-Performance Logical Framework. LNCS, vol. 4350. Springer, Heidelberg (2007)
6. Cordy, J.R.: The TXL source transformation language. Science of Computer Programming 61(3), 190–210 (2006)
7. Cordy, J.R., Halpern, C., Promislow, E.: TXL: A rapid prototyping system for programming language dialects. Computer Languages 16(1), 97–107 (1991)
8. Dean, T.R., Cordy, J.R., Malton, A.J., Schneider, K.A.: Agile parsing in TXL. Automated Software Engineering 10(4), 311–336 (2003)
9. Denker, G., Talcott, C. (eds.): 6th Intl. Workshop on Rewriting Logic and its Applications, vol. 176. Elsevier, Amsterdam (2007)
10. Denker, G., Talcott, C., Roşu, G., van den Brand, M., Eker, S., Şerbănuţă, T.F.: Rewriting logic systems. Electronic Notes in Theoretical Computer Science 176(4), 233–247 (2007)
11. Durán, F., Roldán, M., Balland, E., van den Brand, M., Eker, S., Kalleberg, K.T., Kats, L.C.L., Moreau, P.-E., Schevchenko, R., Visser, E.: The second rewrite engines competition. In: Roşu, G. (ed.) Procs. 7th Intl. Workshop on Rewriting Logic and its Applications (WRLA 2008). Electronic Notes in Theoretical Computer Science, vol. 238, pp. 281–291. Elsevier, Amsterdam (2008)
12. Klint, P., van der Storm, T., Vinju, J.: RASCAL: a domain specific language for source code analysis and manipulation. In: 9th IEEE Intl. Working Conf. on Source Code Analysis and Manipulation, pp. 168–177 (2009)

13. Roşu, G. (ed.): Procs. 7th Intl. Workshop on Rewriting Logic and its Applications (WRLA 2008). Electronic Notes in Theoretical Computer Science. Elsevier, Amsterdam (2008)
14. Shevchenko, R., Doroshenko, A.: A rewriting framework for rule-based programming dynamic applications. Fundamenta Informaticae 72(1-3), 95–108 (2006)
15. Sloane, A.: Experiences with domain-specific language embedding in Scala. In: Lawall, J., Reveillere, L. (eds.) Procs. of the 2nd Intl. Workshop on Domain-Specific Program Development (2008)
16. Tomita, M.: LR parsers for natural languages. In: Procs. of the 10th Intl. Conf. on Computational Linguistics and 22nd Annual Meeting of Assoc. for Computational Linguistics (ACL-22), pp. 354–357. Assoc. for Computational Linguistics (1984)
17. van den Brand, M.: Applications of the ASF+SDF meta-environment. In: Lämmel, R., Saraiva, J., Visser, J. (eds.) GTTSE 2005. LNCS, vol. 4143, pp. 278–296. Springer, Heidelberg (2006)
18. van den Brand, M.G.J., de Jong, H.A., Klint, P., Olivier, P.A.: Efficient annotated terms. Software: Practice and Experience 30(3), 259–291 (2000)
19. van den Brand, M.G.J., Heering, J., Klint, P., Olivier, P.A.: Compiling language definitions: the ASF+SDF compiler. ACM Transactions on Programming Languages and Systems 24(4), 334–368 (2002)
20. van den Brand, M.G.J., van Deursen, A., Heering, J., Jong, H., Jonge, M., Kuipers, T., Klint, P., Moonen, L., Olivier, P., Scheerder, J., Vinju, J., Visser, E., Visser, J.: The ASF+SDF Meta-Environment: a component-based language development environment. In: Wilhelm, R. (ed.) CC 2001. LNCS, vol. 2027, pp. 365–370. Springer, Heidelberg (2001)
21. van Deursen, A., Heering, J., Klint, P.: Language Prototyping: An Algebraic Specification Approach. World Scientific, Singapore (1996)
22. Visser, E.: Stratego: A language for program transformation based on rewriting strategies. In: Middeldorp, A. (ed.) RTA 2001. LNCS, vol. 2051, pp. 357–361. Springer, Heidelberg (2001)

Author Index

GPSR Compliance

The European Union's (EU) General Product Safety Regulation (GPSR) is a set of rules that requires consumer products to be safe and our obligations to ensure this.

If you have any concerns about our products, you can contact us on ProductSafety@springernature.com

In case Publisher is established outside the EU, the EU authorized representative is:

Springer Nature Customer Service Center GmbH
Europaplatz 3
69115 Heidelberg, Germany

Batch number: 09490872

Printed by Printforce, the Netherlands